RECENT DEVELOPMENTS IN THE STUDY OF RECRYSTALLIZATION

Edited by **Peter Wilson**

Recent Developments in the Study of Recrystallization
http://dx.doi.org/10.5772/56031
Edited by Peter Wilson

Contributors

Li, Tomi Laurila, Toni Mattila, Hongbo Xu, Mervi Paulasto-Kröckel, Quan Guo-Zheng, Cho, Taitel-Goldman, Robert N. Ben, Chantelle Capicciotti, Malay Doshi, Dong Nyung Lee, Heung Nam Han, Mohamed Gepreel

Published by InTech

Janeza Trdine 9, 51000 Rijeka, Croatia

Publishing Process Manager Ana Pantar
Technical Editor InTech DTP team
Cover InTech Design team

Additional hard copies can be obtained from orders@intechopen.com

Recent Developments in the Study of Recrystallization, Edited by Peter Wilson
p. cm.
ISBN 978-953-51-0962-4

For EU product safety concerns:
IN TECH d.o.o., Prolaz Marije Krucifikse Kozulić 3, 51000 Rijeka, Croatia,
info@intechopen.com or visit our website at intechopen.com.

Meet the editor

Professor Wilson gained his undergraduate degree in physics and then postgraduate degrees in electronics and in optics from Otago University, in New Zealand. His interests then changed to soft condensed matter and the antifreeze proteins polar fishes use to survive ice-laden waters. Postdoctoral work included measuring the thickness of the ice-water interface and developing techniques for measuring levels of recrystallization in ice. For much of his career Professor Wilson has been studying the nucleation of supercooled solutions and how best to measure subsequent growth and recrystallization of ice. In 2005 Roskilde University, in Denmark, awarded him the degree Doctor Scientiarum. Professor Wilson now works in the Faculty of Life and Environmental Science at the University of Tsukuba.

Contents

Preface

Recrystallization is a phenomenon that is moderately well documented in the geological and metallurgical literature. This book provides a timely overview of the latest research and methods in a variety of fields where recrystallization is studied and is an important factor. Perhaps the main advantage of a new look at these fields is the rapid increase in modern techniques, such as TEM, spectrometers and modeling capabilities which are providing us with far better images and analysis than ever previously possible.

Section 1 includes two chapters giving a general overview of state of the art in research and techniques involving recrystallization. In Chapter 1 Lee and Han discuss the process whereby recrystallization takes place through nucleation and growth. Nucleation during recrystallization can be defined as the formation of strain-free crystals, in a high energy matrix, that are able to grow under energy release by a movement of high-angle grain boundaries. They argue though that the definition is broad and that crystallization of amorphous materials is called recrystallization by some people and can be confused with the abnormal grain growth. They present a theory which is able to determine whether grains surviving deformation can act as nuclei.

In Chapter 2 Quan gives us an overview of microstructures of alloys and grain boundary migration. Metals and alloys have properties of importance including high strength, relatively good ductility and good corrosion resistance. The author describes how optimization of the thermo-mechanical process can be achieved through an understanding of the entire forming process and the metallurgical variables affecting the micro-structural features occurring during deformation.

He concludes that at a fixed temperature, as deformation strain rate increases, the microstructure of the billet becomes more and more refined due to increasing migration energy stored in grain boundaries and decreasing grain growth time.

Section 2 consists of a variety of chapters generally related to recrystallization in metals and alloys. In Chapter 3 Li et al. look at solder joints from the perspective of recrystallization. Solder alloys are widely used bonding materials in the electronics industry and issues with reliability for solder interconnections are rising, with the increasing use of highly integrated components in portable electronic products. They discuss how recrystallization is a source of deformation and thermomechanical stress in the solder interconnection.

In Chapter 4 Gepreel gives an in-depth look at β-type titanium alloys. Typically they have high strength, low density, good cold-workability, heat treatability and corrosion resistance. In this chapter, results from studies of different groups of β-type Ti-alloys with different level of β-phase stability containing different alloying elements are discussed. A strategy to design alloys and how to control the phase's stability are also discussed. Chapter 5, Cho and Kang provide an in-depth look at magnesium alloys. They present the evolution of texture and microstructure during deformation and recrystallization in various magnesium alloys. Both of these two chapters and work will be invaluable to workers trying to produce stronger and lighter alloys.

Section 3 deals with recrystallization in real environmental situations and in Chapter 6 Taitel-Goldman gives us wonderful insight into iron oxides found in sand dunes, soils, sediments and the like. This chapter introduces a fascinating look at recrystallization in places not commonly considered by workers in the field.

The final Section and Chapter 7 by Capicciotti and co-workers provides a thorough and very timely look at recrystallization in water-ice. Recrystallization in ice is often defined as the growth of large ice crystals, or grains, at the expense of small ones. The industrial significance and the benefits of preventing this process have been realized for hundreds of years, in areas such as glaciology, food preservation and cryo-medicine. The authors give us a very clear overview of the state of the art in what is known about inhibiting recrystallization in ice, including a very nice look at inhibition by biological antifreeze proteins, novel synthetic peptides, glycopeptides, polymers and small molecules. The chapter concludes with a summary of the role of ice recrystallization in cryo-injury.

Peter W. Wilson
Professor,
Faculty of Life and Environmental Sciences,
University of Tsukuba
Japan

Section 1

General Topics in Recrystallization

Chapter 1

Recrystallization Textures of Metals and Alloys

Dong Nyung Lee and Heung Nam Han

Additional information is available at the end of the chapter

http://dx.doi.org/10.5772/54123

1. Introduction

Recrystallization (Rex) takes place through nucleation and growth. Nucleation during Rex can be defined as the formation of strain-free crystals, in a high energy matrix, that are able to grow under energy release by a movement of high-angle grain boundaries. The nucleus is in a thermodynamic equilibrium between energy released by the growth of the nucleus (given by the energy difference between deformed and recrystallized volume) and energy consumed by the increase in high angle grain boundary area. This means that a critical nucleus size or a critical grain boundary curvature exists, from which the newly formed crystal grows under energy release. This definition is so broad and obscure that crystallization of amorphous materials is called Rex by some people, and Rex can be confused with the abnormal grain growth when grains with minor texture components can grow at the expense of neighboring grains with main texture components because the minor-component grains can be taken as nuclei. Here we will present a theory which can determine whether grains survived during deformation act as nuclei and which orientation the deformed matrix is destined to assume after Rex. A lot of Rex textures will be explained by the theory.

2. Theories for evolution of recrystallization textures

Rex occurs by nucleation and growth. Therefore, the evolution of the Rex texture must be controlled by nucleation and growth. In the oriented nucleation theory (ON), the preferred activation of a special nucleus determines the final Rex texture [1]. In the oriented growth theory (OG), the only grains having a special relationship to the deformed matrix can preferably grow [2]. Recent computer simulation studies tend to advocate ON theory [3]. This comes from the presumption that the growth of nuclei is predominated by a difference in energy between the nucleus and the matrix, or the driving force. In addition to this, the weakness of the conventional OG theory is in much reliance on the grain boundary mobility.

One of the present authors (Lee) advanced a theory for the evolution of Rex textures [4] and elaborated later [5,6]. In the theory, the Rex texture is determined such that the absolute maximum stress direction (AMSD) due to dislocation array formed during fabrication and subsequent recovery is parallel to the minimum Young's modulus direction (MYMD) in recrystallized (Rexed) grains and other conditions are met, whereby the strain energy release can be maximized. In the strain-energy-release-maximization theory (SERM), elastic anisotropy is importantly taken into account.

In what follows, SERM is briefly described. Rex occurs to reduce the energy stored during fabrication by a nucleation and growth process. The stored energy may include energies due to vacancies, dislocations, grain boundaries, surface, etc. The energy is not directional, but the texture is directional. No matter how high the energy may be, the defects cannot directly be related to the Rex texture, unless they give rise to some anisotropic characteristics. An effect of anisotropy of free surface energy due to differences in lattice surface energies can be neglected except in the case where the grain size is larger than the specimen thickness in vacuum or an inert atmosphere. Differences in the mobility and/or energy of grain boundaries must be important factors to consider in the texture change during grain growth. Vacancies do not seem to have an important effect on the Rex texture due to their relatively isotropic characteristics. The most important driving force for Rex (nucleation and growth) is known to be the stored energy due to dislocations. The dislocation density may be different from grain to grain. Even in a grain the dislocation density is not homogeneous. Grains with low dislocation densities can grow at the expanse of grains with high dislocation densities. This may be true for slightly deformed metals as in case of strain annealing. However, the differences in dislocation density and orientation between grains decrease with increasing deformation. Considering the fact that strong deformation textures give rise to strong Rex textures, the dislocation density difference cannot be a dominant factor for the evolution of Rex textures. Dislocations cannot be related to the Rex texture, unless they give rise to anisotropic characteristics.

The dislocation array in fabricated materials looks very complicated. Dislocations generated during plastic deformation, deposition, etc., can be of edge, screw, and mixed types. Their Burgers vectors can be determined by deformation mode and texture, and their array can be approximated by a stable or low energy arrangement of edge dislocations after recovery. Figure 1 shows a schematic dislocation array after recovery and principal stress distributions around stable and low energy configurations of edge dislocations, which were calculated using superposition of the stress fields around isolated dislocations, or, more specifically, were obtained by a summation of the components of stress field of the individual dislocations sited in the array. It can be seen that AMSD is along the Burgers vector of dislocations that are responsible for the long-range stress field. The volume of crystal changes little after heavy deformation because contraction in the compressive field and expansion in the tensile fields around dislocations generated during deformation compensate each other. That is, this process takes place in a displacement controlled system. The uniaxial specimen in Figure 2 makes an example of the displacement controlled system. When a stress-free specimen S_0 is elastically elongated by ΔL by force F_A (Figure 2a), the elongated specimen S_F has an elastic strain energy represented by triangle *OAC* (Figure 2b). When V in S_F is replaced by a stress-free volume V, S_R having the stress free V has the strain energy of *OBC* (Figure 2b.) Transformation from the S_F state to the S_R state results in a strain-energy-release represented by *OAB* (Figure 2b). The strain-energy-release can be maximized when the S_F and S_R states have the maximum and minimum strain energies, respectively. In this case, AMSD is the axial direction of S_F, and the S_R state has the minimum energy when MYMD of the stress-free V is along the axial direction that is AMSD. In summary, the strain energy release is maximized when AMSD in the high dislocation density matrix is along MYMD of the stress free crystal, or nucleus. That is, when a volume of *V* in the stress field is replaced by a stress-free single crystal of the volume *V*, the strain energy release of the system occurs. The strain energy release can change depending on the orientation of the stress-free crystal. The strain energy release is maximized when AMSD in the high energy matrix is along MYMD of the stress-free crystal. The stress-free grains formed in the early stage are referred to as nuclei, if they can grow. The orientation of a nucleus is determined such that its strain energy release per unit volume during Rex becomes maximized.

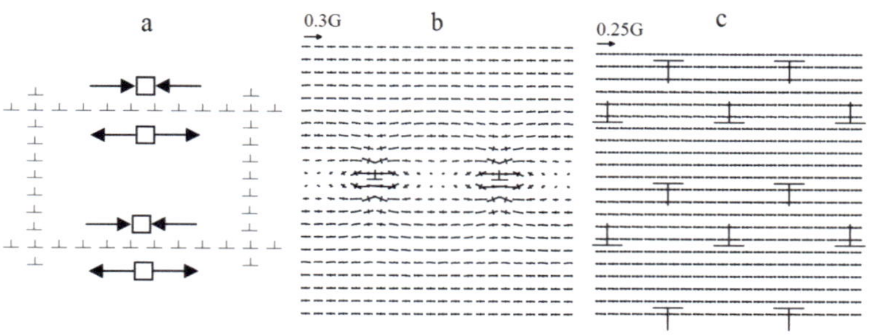

Figure 1. (a) Schematic dislocation array after recovery, where horizontal arrays give rise to long-range stress field, and vertical arrays give rise to short-range stress field [7]. Principal stress distributions around parallel edge dislocations calculated based on (b) 100 linearly arrayed dislocations with dislocation spacing of 10**b**, and (c) low energy array of 100 x 100 dislocations. **b** is Burgers vector and G is shear modulus [8].

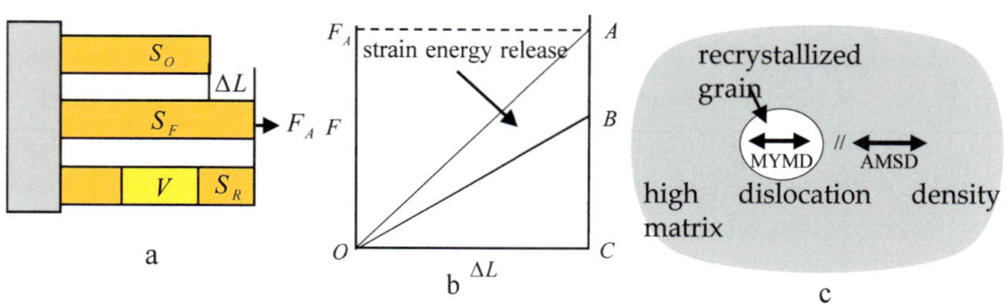

Figure 2. Displacement controlled uniaxial specimen for explaining strain-energy-release being maximized when AMSD in high dislocation density matrix is along MYMD in recrystallized grain.

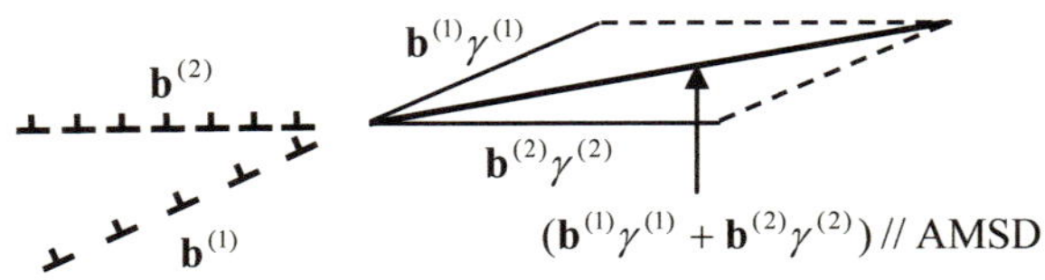

Figure 3. AMSD for active slip systems *i* whose Burgers vectors are $\mathbf{b}^{(i)}$ and activities are $\gamma^{(i)}$.

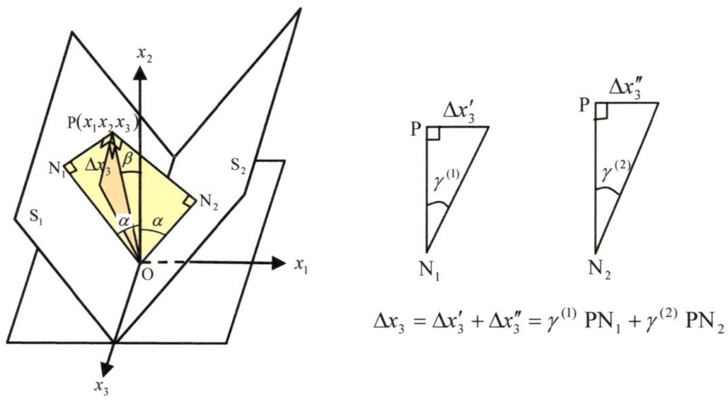

Figure 4. Schematic of two slip planes S_1 and S_2 that share common slip direction along x_3 axis.

We first calculate AMSD in an fcc crystal deformed by a duplex slip of (111)[-101] and (111) [-110] that are equally active. The duplex slip can be taken as a single slip of (111)[-211], which is obtained by the sum of the two slip directions. In this case, the maximum stress direction is [-211]. However, some complication can occur. One slip system has two opposite directions. The maximum stress direction for the (111)[-101] slip system represents the [-101] direction and its opposite direction, [1 0-1]. The maximum stress direction for the (111)[-110] slip system represents the [-110] and [1-1 0] directions. Therefore, there are four possible combinations to calculate the maximum stress direction, [-101] + [-110] = [-211], [-101] + [1-1 0] = [0-1 1], [1 0-1] + [-110] = [0 1-1], and [1 0-1] + [1-1 0] = [2-1-1], among which [-211]//[2-1-1] and [0-1 1]//[0 1-1]. The correct combinations are such that two directions make an acute angle. If the two slip systems are not equally active, the activity of each slip system should be taken into account. If the (111)[-101] slip system is two times more active than the (111)[-110] system, the maximum stress direction becomes 2[-101] + [-110] = [-312]. This can be generalized to multiple slip. For multiple slip, AMSD is calculated by the sum of active slip directions of the same sense and their activities, as shown in Figure 3. It is convenient to choose slip directions so that they can be at acute angles with the highest strain direction of the specimen, e.g., RD in rolled sheets, the axial direction in drawn wires, etc.

When two slip systems share the same slip direction, their contributions to AMSD are reduced by 0.5 for bcc metals and 0.577 for fcc metals as follows. Figure 4 shows two slip planes, S_1 and S_2, intersecting along the common slip direction, the x_3 axis; the x_2 axis bisects the angle between the poles of these planes. The loading direction lies within the quadrant drawn between S_1 and

S_2, and the displacement Δx_3 along the x_3 axis at any point P with coordinates (x_1,x_2,x_3) is considered. If shear strains $\gamma^{(1)}$ and $\gamma^{(2)}$ occur on the slip system 1 (the slip plane S_1 and the slip direction x_3) and the slip system 2 (the slip plane S_2 and the slip direction x_3), respectively, then

$$\Delta x_3 = \gamma^{(1)}\ \mathrm{PN}_1 + \gamma^{(2)}\ \mathrm{PN}_2 \tag{1}$$

where PN_1 and PN_2 are normal to the planes S_1 and S_2, respectively. Therefore,

$$\mathrm{PN}_1 = \mathrm{OP}\ \sin(\alpha - \beta) \text{and } \ \mathrm{PN}_2 = \mathrm{OP}\ \sin(\alpha + \beta) \tag{2}$$

where OP, α, and β are defined in Figure 4. Therefore,

$$\Delta x_3 = \ (\gamma^{(1)} \ + \gamma^{(2)})\ \mathrm{OP}\ \sin\alpha\cos\beta + \ (\gamma^{(2)} \ - \gamma^{(1)})\ \mathrm{OP}\ \cos\alpha\sin\beta \tag{3}$$

Because $\alpha > \beta$ and $(\gamma^{(1)} + \gamma^{(2)}) > (\gamma^{(2)} - \gamma^{(1)})$, the second term of the right hand side is negligible compared with the first term. It follows from OP $\cos\beta = x_2$ that $\Delta x_3 \approx (\gamma^{(1)} + \gamma^{(2)})\, x_2 \sin\alpha$. Therefore, the displacement Δx_3 is linear with the x_2 coordinate, and the deformation is equivalent to single slip in the x_3 direction on the $(\gamma^{(1)}S_1 + \gamma^{(2)}S_2)$ plane. The apparent shear strain γ_a is

$$\gamma_a \ = \Delta x_3 \,/\, x_2 \approx (\gamma^{(1)} \ + \gamma^{(2)})\ \sin\alpha \tag{4}$$

The apparent shear strains $\gamma_a^{(i)}$ on the slip systems i are

$$\gamma_a^{(i)} = \gamma^{(i)} \sin\alpha \tag{5}$$

For bcc metals, $\sin\alpha = 0.5$ (e.g. a duplex slip of (101)[1 1-1] and (011)[1 1-1]) and hence

$$\gamma_a^{(i)}(\mathrm{bcc}) = 0.5\gamma^{(i)} \tag{6}$$

For fcc metals, $\sin\alpha = 0.577$ (e.g. a duplex slip of (-1 1-1)[110] and (1-1-1)[110]) and hence

$$\gamma_a^{(i)}(\mathrm{fcc}) = 0.577\gamma^{(i)} \tag{7}$$

The activity of each slip direction is linearly proportional to the dislocation density ρ on the corresponding slip system, which is roughly proportional to the shear strain on the slip system. Experimental results on the relation between shear strain γ and ρ are available for Cu and Al [9].

If a crystal is plastically deformed by $\delta\varepsilon$ (often about 0.01), then we can calculate active slip systems i and shear strains $\gamma^{(i)}$ on them using a crystal plasticity model, resulting in the shear strain rate with respect to strain of specimen, $d\gamma^{(i)}/d\varepsilon$. During this deformation, the crystal can rotate, and active slip systems and shear strains on them change during $\delta\varepsilon$. When a crystal rotates during deformation, the absolute value of shear strain rates $|d\gamma^{(i)}/d\varepsilon|$ on slip systems i can vary with strain ε of specimen. For a strain up to $\varepsilon = e$, the contribution of each slip system to AMSD is proportional to

$$\gamma^{(i)} = \int_0^e \left| d\gamma^{(i)} / d\varepsilon \right| d\varepsilon \tag{8}$$

The above equation is illustrated in Figure 5. If a deformation texture is stable, the shear strain rates on the slip systems are independent of deformation.

So far methods of obtaining AMSD have been discussed. This is good enough for prediction of fiber textures. However, the stress states around dislocation arrays are not uniaxial but triaxial. Unfortunately we do not know the stress fields of individual dislocations in real crystals, but know Burgers vectors. Therefore, AMSD obtained above applies to real crystals. Any stress state has three principal stresses and hence three principal stress directions which are perpendicular to each other. Once we know the three principal stress directions, the Rex textures are determined such that the three directions in the deformed matrix are parallel to three <100> directions in the Rexed grain, when MYMDs are <100>. In figure 6, let the unit vectors of **A**, **B**, and **C** be $\boldsymbol{a}$ $[a_1\ a_2\ a_3]$, $\boldsymbol{b}$ $[b_1\ b_2\ b_3]$, and $\boldsymbol{c}$ $[c_1\ c_2\ c_3]$, where a_i are direction cosines of the unit vector $\boldsymbol{a}$ referred to the crystal coordinate system. AMSD is one of three principal stress directions. Two other principal stresses are obtained as explained in Figure 6.

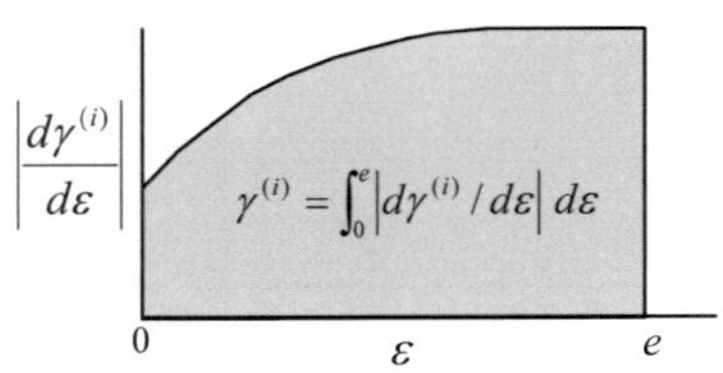

Figure 5. Calculation of $\gamma^{(i)}$ for crystal rotation during deformation up to $\varepsilon = e$.

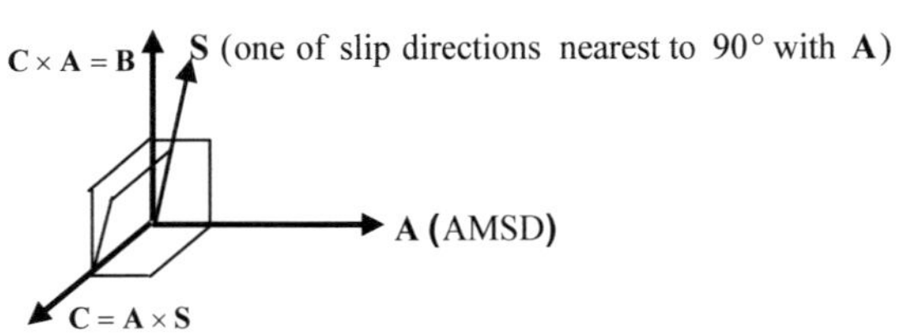

Figure 6. Relationship between three principal stress directions **A**, **B**, and **C**.

If the unit vectors $\boldsymbol{a}$, $\boldsymbol{b}$, and $\boldsymbol{c}$ are set to be along [100], [010], and [001] after Rex, components of the unit vectors are direction cosines relating the deformed and Rexed crystal coordinate systems, when MYMDs are <100>. That is, the $(hkl)[uvw]$ deformation orientation is calculated to transform to the $(h_r\ k_r\ l_r)[u_r\ v_r\ w_r]$ Rex orientation using the following equation.

$$\begin{pmatrix} h_r \\ k_r \\ l_r \end{pmatrix} = \begin{pmatrix} a_1 & a_2 & a_3 \\ b_1 & b_2 & b_3 \\ c_1 & c_2 & c_3 \end{pmatrix} \begin{pmatrix} h \\ k \\ l \end{pmatrix} \quad \begin{pmatrix} u_r \\ v_r \\ w_r \end{pmatrix} = \begin{pmatrix} a_1 & a_2 & a_3 \\ b_1 & b_2 & b_3 \\ c_1 & c_2 & c_3 \end{pmatrix} \begin{pmatrix} u \\ v \\ w \end{pmatrix} \tag{9}$$

It should be mentioned that ***a*** is set to be along [100], but ***b*** is along [010] or [001] depending on physical situations and c is consequently along [001] or [010]. The Rex texture can often be obtained without resorting to the above process because the AMSD//MYMD condition is so dominant that the Rex texture can be obtained by the following priority order.

The 1st priority: When AMSD is cristallographically the same as MYMD, No texture changes after Rex [10].

The 2nd priority: When AMSD crystallographically differs from MYMD, the Rex texture is determined such that AMSD in the matrix is parallel to MYMD in the Rexed grain, with one common axis of rotation between the deformed and Rexed states. The common axis can be ND, TD, or other direction (e.g. <110> for bcc metals). This may be related to minimum atomic movement at the AMSD//MYMD constraints. However, we do not know the exact physical picture of this.

The 3rd priority: When the first two conditions are not met, the method explained to obtain Eq. 9 is used.

3. Electrodeposits and vapor-deposits

When the density of dislocations in electrodeposits and vapor deposits is high, the deposits undergo Rex when annealed. AMSD in the deposits can be determined by their textures. The density of dislocations whose Burgers vectors are directed away from the growth direction (GD)of deposits was supposed to be higher than when the Burgers vector is nearly parallel to GD because dislocations whose Burgers vector is close to GD are easy to glide out from the deposits by the image force during their growth [11]. This was experimentally proved in a Cu electrodeposit with the <111> orientation [12]. Therefore, AMSDs are along the Burgers vectors nearly normal to GD.

3.1. Copper, nickel, and silver electrodeposits

Lee et al. found that the <100>, <111>, and <110> textures (inverse pole figures: IPFs) of Cu electrodeposits which were obtained from Cu sulfate and Cu fluoborate baths [13,14], and a cyanide bath [15] changed to the <100>, <100>, and <√310> textures, respectively, after Rex as shown in Figure 7. The texture fraction (TF) of the (*hkl*) reflection plane is defined as follows:

$$\mathrm{TF}(hkl) = \frac{\mathrm{I}(hkl)/\mathrm{I_o}(hkl)}{\Sigma\left[\mathrm{I}(hkl)/\mathrm{I_o}(hkl)\right]} \tag{10}$$

where I(*hkl*) and I_o(*hkl*) are the integrated intensities of (*hkl*) reflections measured by x-ray diffraction for an experimental specimen and a standard powder sample, respectively, and Σ means the summation. When TF of any (*hkl*) plane is larger than the mean value of TFs, a preferred orientation or a texture exists in which grains are oriented with their (*hkl*) planes parallel to the surface, or with their <*hkl*> directions normal to the surface. When TFs of all reflections are the same, the distribution of crystal orientation is random. TFs of all the reflections sum up to unity. Figure 7 indicates that the deposition texture of <100> remains unchanged after Rex. This is expressed as $<100>_D \rightarrow <100>_R$. All the samples were freestanding and so subjected to no external external stresses during annnealing. The results are explained by SERM in Section 2. We have to know MYMD of Cu and AMSDs of Cu electrodeposits. Young's modulus *E* of cubic crystals can be calculated using Eq. 11 [16].

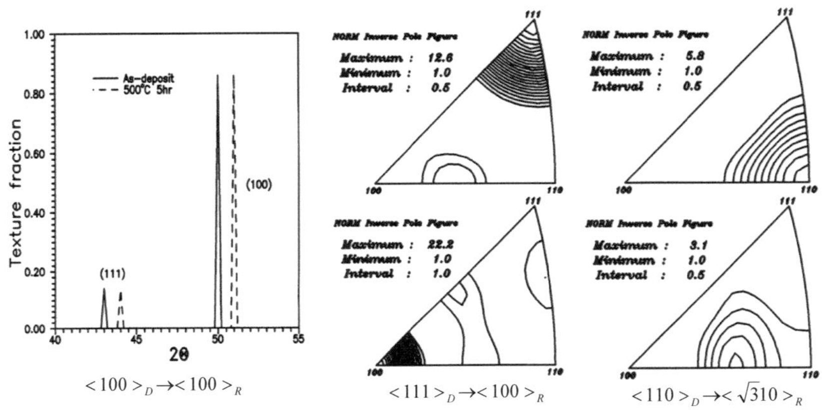

Figure 7. Deposition and Rex textures of Cu electrodeposits. $<hkl>_D \rightarrow <uvw>_R$ means that <*hkl*> deposition texture changes to <*uvw*> Rex texture. For $<100>_D$, Rex peaks are shifted rightward by 1° from their original positions to be distinguished from deposition peaks. TF data [13] and IPFs [14].

$$1/E = S_{11} + [S_{44} - 2(S_{11} - S_{12})](a_{11}^2 a_{12}^2 + a_{12}^2 a_{13}^2 + a_{13}^2 a_{11}^2) \tag{11}$$

where S_{ij} are compliances and a_{1i} are the direction cosines relating the uniaxial stress direction x'_1 to the symmetry axes x_i. When $[S_{44}-2(S_{11}-S_{12})] < 0$, or $A=2(S_{11}-S_{12})/S_{44} > 1$, $(a_{11}^2 a_{12}^2 + a_{12}^2 a_{13}^2 + a_{13}^2 a_{11}^2) = 0$ yields the minimum Young's modulus, which is obtained at $a_{11} = a_{12} = a_{13} = 0$. Therefore, MYMDs are parallel to <100>. When $[S_{44}-2(S_{11}-S_{12})]>0$, or $A< 1$, the maximum value of $(a_{11}^2 a_{12}^2 + a_{12}^2 a_{13}^2 + a_{13}^2 a_{11}^2)$ yields the minimum Young's modulus, which is obtained at $a_{11}^2 = a_{12}^2 = a_{13}^2 = 1/3$. Therefore, MYMDs are parallel to <111>. When $[S_{44}-2(S_{11}-S_{12})] = 0$, or $A = 1$, *E* is independent of direction, in other words, the elastic properties are isotropic. *A* is usually referred to as Zener's anisotropy factor. Summarizing, MYMDs // <100> for $A>1$, MYMDs// <111> for $A<1$, and elastic isotropy for $A=1$.

For fcc Cu, S_{11}=0.018908, S_{44} =0.016051, S_{12} = -0.008119 GPa^{-1} at 800 K [17], which in turn gives rise to $[S_{44}-2(S_{11}-S_{12})] < 0$, and so MYMDs are <100>. MYMDs and the Burgers vectors of Cu are along the <100> directions and the <110> directions, respectively. There are six equivalent directions in the <110> directions, with opposite directions being taken as the same. As already explained, AMSD is along the Burgers vector which is approximately normal to GD.

For the <100> oriented Cu (simply <100> Cu) deposit, two of the six <110> directions are at 90° and the remaining four are at 45° with GD, as shown in Figure 8. The two <110> directions, which is AMSD, change to the <100> directions after Rex, resulting in the <100> Rex texture (Figure 8b) in agreement with the experimental result.

For the <111> Cu deposit, three of the six <110> directions are at right angles with the [111] GD; the remaining three <110> directions are at 35.26° with GD, as shown in Figure 9 a. The former three <110> directions, AMSD, can change to <100> after Rex, but angles between the <110> directions are 60° and the angle between the <100> directions is 90°. Correspondence between the <110> directions in as-deposited grains and the <100> directions in Rexed grains is therefore impossible in a grain. Two of the <110> directions in neighboring grains, which are at right angles with each other, can change to the <100> directions to form the <100> nuclei in grain boundaries, which grow at the expense of high energy region, as shown in Figure 9b. Thus, the <111> deposition texture change to the <100> Rex texture, in agreement with the measured result.

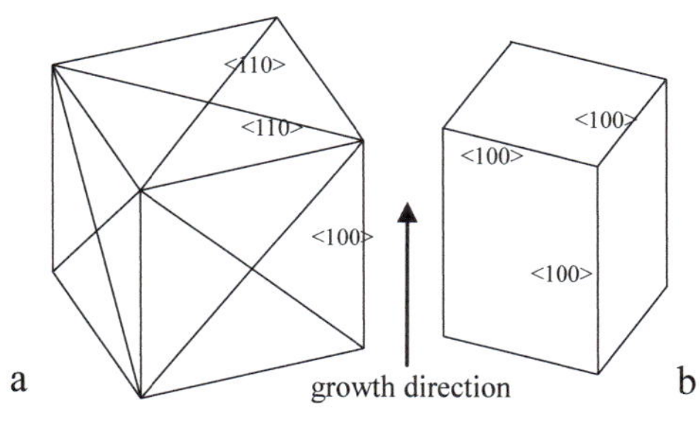

Figure 8. Drawings explaining that <100> deposition texture (a) remains unchanged after Rex (b).

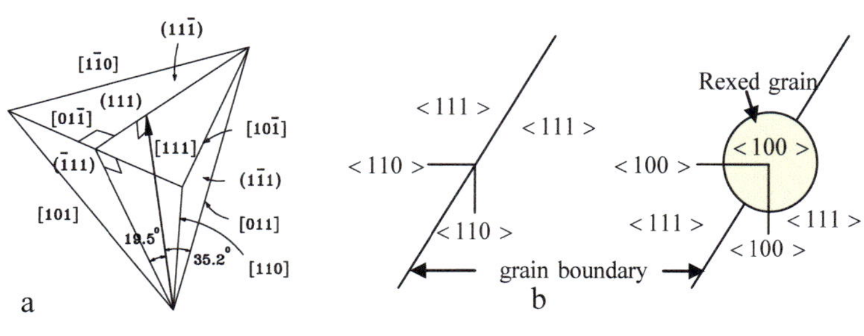

Figure 9. (a) <110> directions in <111> oriented fcc crystal in which arrow indicates [111] growth direction. (b) Drawings for explanation of <111> deposition to <100> Rex texture transformation.

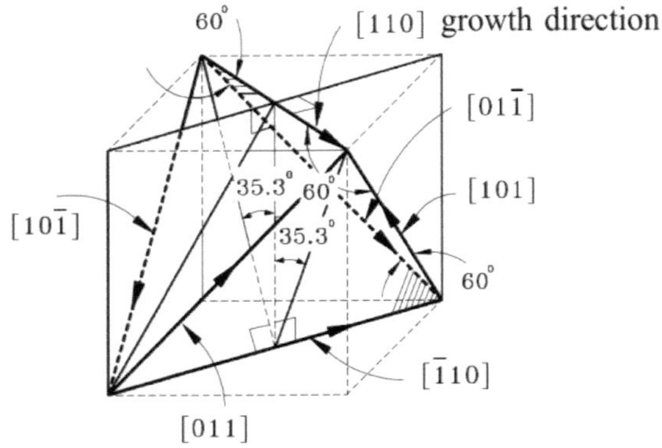

Figure 10. directions in [110] oriented fcc crystal.

For the <110> Cu deposit, one <110> direction is normal to the <110> GD and the remaining four <110> directions are at 60° with the <110> GD, as shown in Figure 10. The first one of the <110> directions and the last four <110> directions are likely to determine the Rex texture because the last four directions are closer to the deposit surface than to GD. Recalling that the <110> directions change to <100> directions after Rex, GD of Rexed grains should be at 60° and 90° with the <100> directions, MYMD, at the same time. GD satisfying the condition is <√310>, in agreement with the experimental results.

So far we have discussed the evolution of the Rex textures from simple deposition textures. A Cu deposit whose texture can be be approximated by a weak duplex texture consisting of the <111> and <110> orientations developed the Rex texture which is approximated by a weak <√310> orientation rather than <100> + <√310> [18]. For the duplex deposition texture, the Rex texture may not consist of the Rex orientation components from the deposition orientation components because differently oriented grains can have different energies. The tensile strengths of copper electrodeposits showed that the tensile strength of the specimens with the <110> texture was higher than those with the <111> texture obtained from the similar electrodeposition condition. This implies that the <110> specimen has the higher defect densities than the <111> specimen [18,19]. Therefore, the <110> grains are likely to have higher driving force for Rex than the <111> grains, resulting in the <√310> texture after Rex, in agreement with experimental result [18].

For Ni, S_{11}= 0.009327, S_{44} = 0.009452, S_{12} = -0.003694 GPa^{-1} at 760 K [20], which in turn gives rise to $[S_{44}-2(S_{11}-S_{12})] < 0$, and so MYMDs are <100>. Therefore, the deposition to Rex texture transformation of Ni electrodeposits is expected to be similar to that of Cu electrodeposits. As expected, freestanding Ni electrodeposits of 30-50 μm in thickness showed that the <100> deposition texture remained unchanged after Rex, and the <110> deposition texture changed to <√310> after Rex [21].

For Ag, S_{11}= 0.03018, S_{44} = 0.02639, S_{12} = -0.0133 GPa^{-1} at 750 K [17], which in turn gives $[S_{44}-2(S_{11}-S_{12})] < 0$, and so MYMDs are <100>. Therefore, the deposition to Rex texture transformation of freestanding Ag electrodeposits is expected to be similar to that of Cu electrodeposits. Figure 11 shows four different deposition and corresponding Rex textures of Ag electrodeposits. Samples a, b, and c shows results similar to Cu electrodeposits, except that minor <221> component, which is the primary twin component of the <100> component in the Rex textures, is stronger than that of Cu deposits. The strong development of twins in Ag is due to its lower stacking fault energy (~22 mJm^{-2}) than that of Cu (~80 mJm^{-2}).

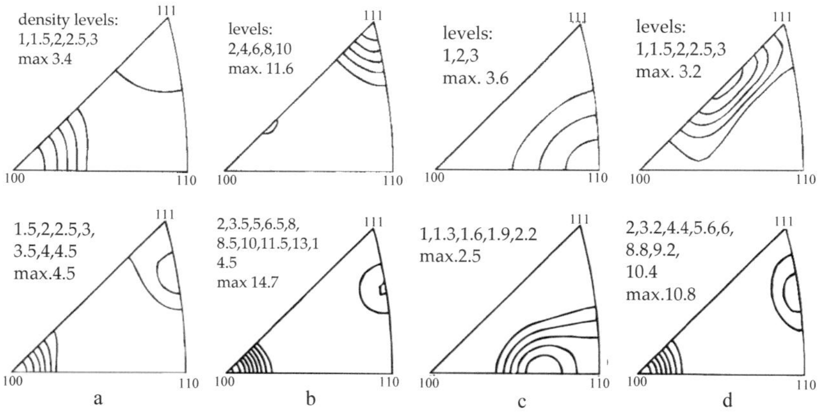

Figure 11. Deposition (top) and Rex (bottom) textures (IPFs) of Ag electrodeposits [22].

The deposition texture of Sample d was well described by 0.32<112> + 0.14$<127>_T$ + 0.25<113> + 0.23$<557>_T$ + 0.06$<19\ 19\ 13>_{TT}$ with each of individual orientations being superimposed with a Gaussian peak of 8°. Here $<127>_T$ indicates the twin orientation of its preceeding <112> orientation, and TT indicates secondary twin. Thus, the main components in deposition texture of Sample d are <112>, <113>, and <557>. The <110> directions that are nearly normal to GD will be AMSD and in turn determine the Rex texture. Table 1 gives angles between <110> and [11w]. Table 1 shows that the probability of <110> directions being normal to GD is the highest. The <110> directions normal to GD will become parallel to the <100> directions (MYMS) after Rex. Therefore, the Rex texture will be the <100> orientation for the same reason as in the <111> orientation of the deposit [22].

3.2. Chromium electrodeposits

Table 2 shows TFs (Eq. 10) of Cr electrodeposits obtained under three electrodeposition conditions. Specimen Cr-A has a strong <111> fiber texture. The texture of Cr-B is characterized by weak <111>, and that of Cr-C is by weak <100>. The optical microstructure and hardness test results and others indicated that all the specimens were fully Rexed at 1173 K. TFs as functions of annealing temperature and time in Figure 12 indicate that the deposition texture of Cr-A little change after Rex. The pole figures in Figures 13 and 14 indicate the deposition textures of Cr-B and Cr-C little change after Rex. In conclusion, the <100> and <111> deposition textures of Cr electrodeposits little change after Rex. These results are compatible with SERM as discussed in what follows. There are four equivalent <111> directions in bcc Cr crystal, with opposite directions being taken as the same. For the <111> Cr deposit, one of four <111>

directions is along GD and the remaining three <111> directions are at an angle of 70.5° with GD (Figure 15). The remaining three <111> directions can be AMSDs. They will become parallel to MYMDs of Rexed grains. The compliances of Cr are S_{11} =0.00314, S_{44} = 0.0101, S_{12} = -0.000567 GP^{-1} at 500 K [23], which lead to $[S_{44}\text{-}2(S_{11}\text{-}S_{12})] > 0$. Therefore, MYMDs of Cr are <111>, which are also AMSDs of the deposit. Therefore, the <111> and <100> textures of Cr deposits do not change after Rex, as can be seen from Figure 15, in agreement with experimental results.

	110	**-110**	**101**	**-101**	**011**	**0-1 1**
557	44.7	90	31.5	81.8	31.5	81.8
112	54.7	90	30	73.2	30	73.2
113	64.8	90	31.5	64.8	31.5	64.8

Table 1. Angles between <110> and [11w] directions (°)

	(110)	**(200)**	**(211)**	**(220)**	**(310)**	**(222)**	**Texture**
Cr-A	0.02	0.05	0	0	0	**0.93**	Strong <111>
Cr-B	0.03	0.15	0.28	0	0.01	**0.53**	<111>
Cr-C	0.19	**0.47**	0.13	0.05	0.13	0.03	<100>

Table 2. Texture fractions (TF) of reflection planes of Cr electrodeposits A, B, and C [14]. Bold-faced numbers indicate highest TFs in corresponding deposits.

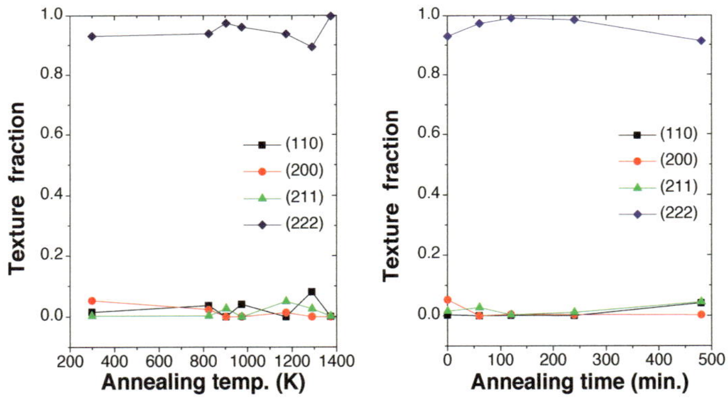

Figure 12. TFs of Cr-A as functions of annealing (a) temperature for 1 h and (b) time at 903 K [14].

3.3. Copper and silver vapor-deposits

Patten et al. [24] formed deposits of Cu up to 1mm in thickness at room temperature in a triode sputtering apparatus using a krypton discharge under various conditions of sputtering rate,

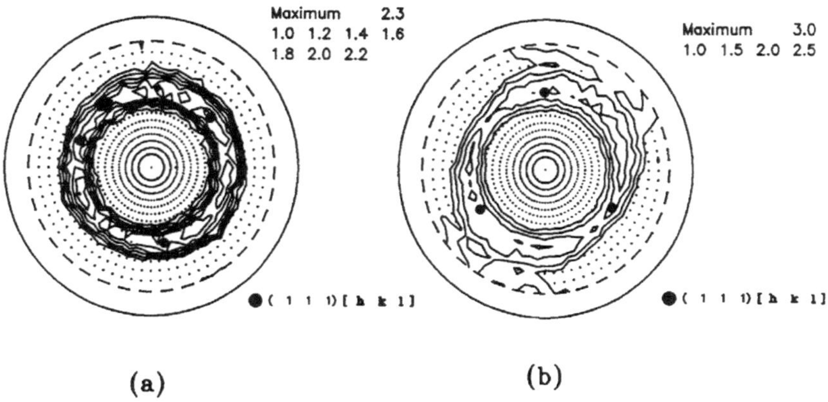

Figure 13. (200) pole figures of Cr-B (left) before and (right) after annealing at 1173 K for 1 h [14].

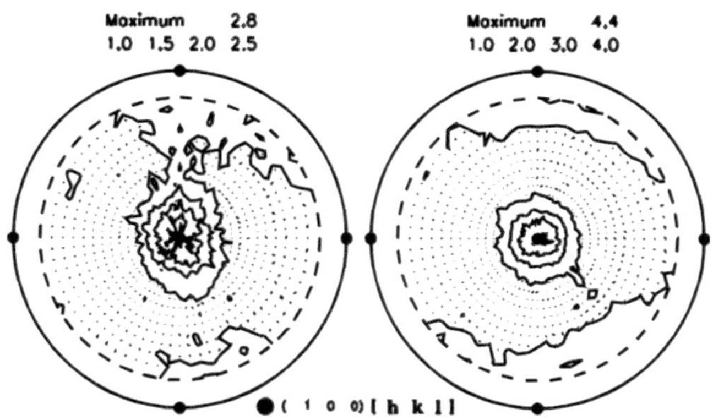

Figure 14. (200) pole figures of Cr-C (left) before and (right) after annealing at 1173 K for 1 h [14].

gas purity, and substrate bias. The 3.81 cm diameter target was made from commercial grade OFHC forged Cu-bar stock containing approximately 100 ppm oxygen by weight with only traces of other elements. The substrates were 2.54 cm diameter by 6.2 mm thick disks made of OFHC Cu. These disks were electron beam welded to a stainless-steel tube to provide direct water-cooling for temperature control during sputtering. As-deposited grains were approximately 100 nm in diameter. Room-temperature Rex and grain growth displaying no twins were observed approximately 9 h after removal from the sputtering apparatus. Nucleation sites were almost randomly distributed. Hardness of the unrecrystallized matrix remained at ~230 DPH from the time it was sputtered until Rex, when it abruptly dropped to approximately 60 DPH in the Rexed grains. Rex resulted in a texture transformation from the <111> deposition texture to the <100> Rex texture. Since the substrate is also Cu, the orientation transition from <111> to <100> cannot be attributed to thermal strains. The driving force for Rex must be the internal stress due to defects such as vacancies and dislocations. Therefore, the texture transition is consistent with the prediction of SERM.

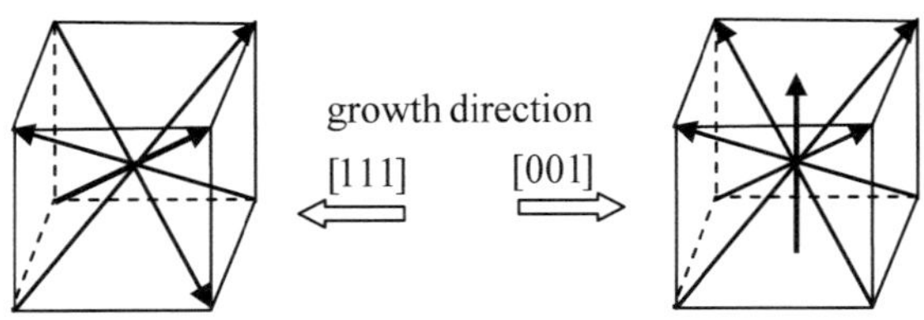

Figure 15. Thin arrows (AMSDs) and thick arrows (GD) in [111] and [001] Cr crystals.

Greiser et al. [25] measured the microstructure and texture of Ag thin films deposited on different substrates using DC magnetron sputtering under high vacuum conditions (base pressure: 10^{-8} mbar, partial Ar pressure during deposition: 10^{-3} mbar). A weak <111> texture in a 0.6 μm thick Ag film deposited on a (001) Si wafer with a 50 nm thermal SiO_2 layer at room temperature becomes stronger with increasing thickness. It is generally accepted that a random polycrystalline structure is obtained up to a critical film thickness unless an epitaxial growth condition is satisfied. Therefore, the <111> texture developed in the 0.6 μm film was weak and became stronger with increasing thickness. This is consistent with the preferred growth model [26]. They also found that the texture of the film deposited at room temperature was "high <111>", whereas the texture of the film deposited at 200 °C was characterized by a low amount of the <111> component and a high amount of the random component. This is also consistent with the preferred growth model.

Post-deposition annealing was carried out in a vacuum furnace at 400 °C with a base pressure of 10^{-6} mbar, a partial H_2 pressure of 10 mbar, and under environmental conditions. The post-deposition grain growth was the same for annealing in high vacuum and in environmental conditions. A dramatic difference in the extent of growth was recognized in the micrographs of the 0.6 and 2.4 μm thick films. The 0.6 μm thick film showed normally grown grains with the <111> orientation; the average grain size was about 1 to 2 μm. This can be understood in light of the surface energy minimization. In contrast, in 2.4 μm thick films, abnormally large grains with the <001> orientation were found. These grains grew into the matrix of <111> grains. The grain boundaries between the abnormally grown grains have a meander-like shape unlike the usual polygonal shape. They could not explain the results by the model of Carel, Thomson, and Frost [27]. According to the model, the strain energy minimization favors the growth of <100> grains. The growth mode should be affected by strain and should not be sensitive to the initial texture. These predictions are at variance with the experimental results in which freestanding, stress-free films also showed abnormal growth of giant grains with <001> texture. The 2.4 μm thick films deposited at 100 °C or below could have dislocations whose density was high enough to cause Rex, which in turn gave rise to the texture change from <111> to <001> regardless of the existence of substrate when annealed, as explained in the previous section. Thus, the <111> to <100> texture change in the 2.4 μm thick films is compatible with SERM [28].

4. Axisymmetrically drawn fcc metals

It is known that the texture of axisymmetrically drawn fcc metals is characterized by major <111> + minor <100> components, and the drawing texture changes to the <100> texture after Rex [29,30]. Figure 16 shows calculated textures in the center region of 90% drawn copper wire taking work hardening per pass into account. The drawing to Rex texture transition was explained by SERM [4]. Since the drawing texture is stable, we consider the [111] and [100] fcc crystals representing the <111> and <100> fiber orientations constituting the texture. Figure 17 shows tetrahedron and octahedron consisting of slip planes (triangles) and slip directions (edges) for the [111] and [100] fcc crystals. The slip planes are not indexed to avoid complication. The slip-plane index can be calculated by the vector product of two of three slip directions (edges) of a triangle constituting the slip-plane triangle. It follows from Figure 17a that three active slip directions that are skew to the [111] axial direction are [101], [110], and [011]. It should be noted that these directions are chosen to be at acute angles with the [111] direction (Section 2). Therefore, AMSD // ([101] + [110] + [011]) = [222] // [111]. That is, AMSD is along the axial direction. According to SERM, AMSD in the deformed matrix is along MYMD in the Rexed grain. MYMDs of most of fcc metals are <100>. Therefore, the <111> drawing texture changes to the <100> Rex texture. Now, the evolution of <100> Rex texture in the <100> deformed matrix is explained. Eight active slip systems in fcc crystal elongated along the [100] direction are calculated to be (111)[1 0-1], (-111)[101], (1-1 1)[110], (1 1-1)[1-1 0], (111)[1-1 0], (-111)[110], (1-1 1)[10-1], and (1 1-1)[101], if the slip systems are {111}<110> [32]. It is noted that the slip directions are chosen to be at acute angles with the [100] axial direction. These slip systems are shown in Figure 17 b. AMSD is obtained, from the vector sum of the active slip directions, to be parallel to [100], which is also MYMD of fcc metals. Therefore, the <100> drawing texture remains unchanged after Rex (1st priority in Section 2), and the <111> + <100> orientation changes to <100> after Rex, regardless of relative intensity of <111> to <100> in the deformation texture. The <100> grains in deformed fcc wires are likely to act as nuclei for Rex. The texture change during annealing might take place by the following process. The <100> grains retain their deformation texture during annealing by continuous Rex, or by recovery-controlled processes, without long-range high-angle boundary migration. The <100> grains grow at the expense of their neighboring <111> grains that are destined to assume the <100> orientation during annealing.

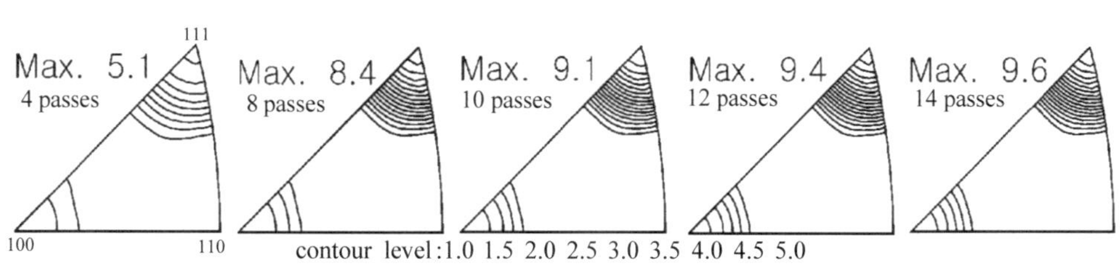

Figure 16. Calculated IPFs in centeral axis zone of Cu wire drawn by 90% in 14 passes (~15% per pass) through conical-dies of 9° in half-die angle, taking strain-hardening per pass into count [31].

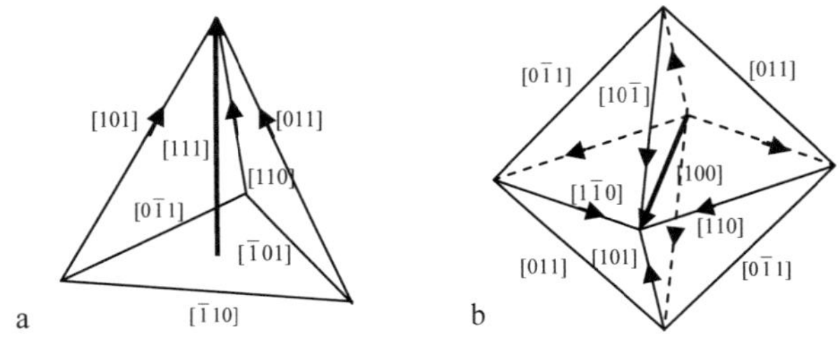

Figure 17. Tetrahedron and octahedron representing slip planes (triangles) and directions (edges) in [111] and [100] fiber oriented fcc crystals. Thick arrows show (a) [111] and (b) [100] axial directions.

4.1. Silver

Cold drawn Ag wires develop major <111> + minor <100> at low reductions (less than about 90%) as do other fcc metals, whereas they exhibit major <100> + minor <111> at high reductions (99%) as shown in Figure 18 [32]. This result is in qualitative agreement with that of Ahlborn and Wassermann [33], which shows that the ratio of <100> to <111> of Ag wires was higher at 100 and -196°C than at room temperature. They attributed the higher <100> orientation to Rex and mechanical twinning, because Ag has low stacking fault energy. They suggested that the <111> orientation transformed to the <115> orientation by twinning, which rotated to the <100> orientation by further deformation.

The hardness of deformed Ag wires as a function of annealing time at 250 and 300 °C indicated that Rex was completed after a few min. This was also confirmed by microstructure studies [32]. Figure 18 shows the annealing textures of drawn Ag wires of 99.95% in purity, which shows that drawing by 61 and 84% and subsequent annealing at 250 °C for 1 h gives rise to nearly random orientation. Ag wires with the <111> + <100> deformation texture develop Rex textures of major <100> and minor <111>, or major <100> + its twin component <122> and minor <111>. The almost random orientation can be seen in Figures 19 d. Figure 20 shows the IPFs of 99% drawn 99.99% Ag wire annealed at 600 ℃ for 1 min to 200 h. Their microstructures showed that the specimen annealed at 600 °C for 1min is almost completely Rexed. The specimen has major <100> + minor <111> as the specimens annealed at 300 °C. After annealing at 600 °C for 3min, some grains showed abnormal grain growth (AGG), indicating complete Rex, and the intensity of <100> component increased. However, as the annealing time incresed, the orientation density ratio (ODR) of <111> to <100> increased, accompanied by grain growth. It is noted that the annealing texture is diffuse at the transient stage from <100> to <111> (5 min in Figure 20 and Figure 19d). The <100> to <111> transition is associated with AGG in low dislocation-density fcc metals, which has been discussed in [31,32]. The Rex results before AGG lead to the conclusion that the Rex texture of the heavily drawn Ag wires is <100> regardless of relative intensity of <111> and <100>, as expected from SERM.

4.2. Aluminum, copper, and gold

Axisymmetrically extruded Al alloy rod [34], drawn Al wire [30] and Cu and some Cu alloy wires [29] generally have major <111> + minor <001> double fiber textures in the deformed

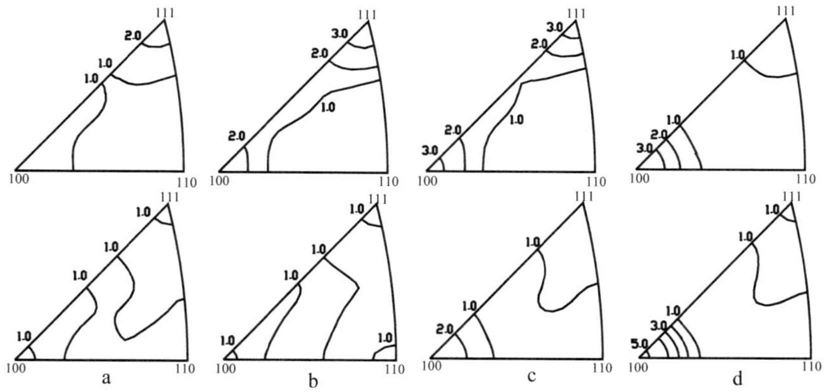

Figure 18. IPFs of (a) 61, (b) 84, (c) 91, and (d) 99% drawn Ag wires (initial texture: random) of 99.95% in purity (top) before and (bottom) after annealing at 250 °C for 1 h [32].

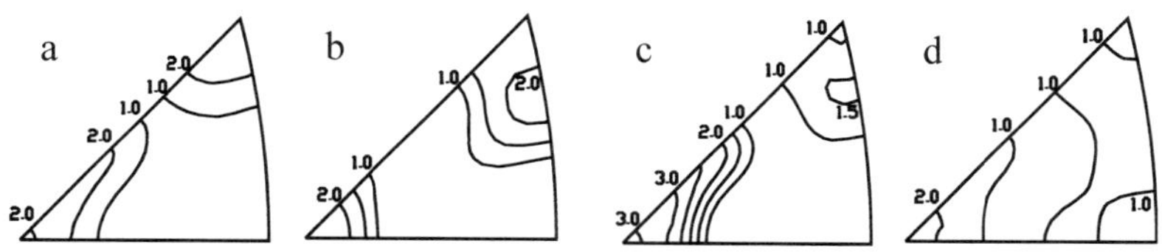

Figure 19. IPFs of 99.99% pure Ag wires (a) drawn by 90% and (b) annealed at 300 °C for 1 h; (c) drawn by 99% and (d) annealed at 300 °C for 1 h [32].

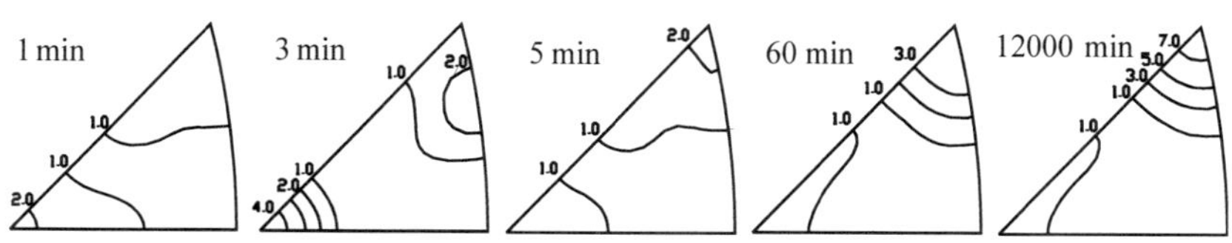

Figure 20. IPFs of 99.99% Ag wire drawn by 99% and annealed at 600 °C for 1-12000 min [32].

state. Park and Lee [35] studied drawing and annealing textures of a commercial electrolytic tough-pitch Cu of 99.97% in purity. A rod of 8mm in diameter, whose microstructure was characterized by equiaxed grains having a homogeneous size distribution, was cold drawn by 90% reduction in area in 14 passes through conical dies of 9° in half-die-angle with about 15% reduction per pass. The drawing speed was 10 m/min. The drawn wire was annealed in a salt bath at 300 or 600 °C and in air, argon, hydrogen or vacuum (< 1×10^{-4} torr) at 700 °C for various periods of time. Figure 21 shows orientation distribution functions (ODFs) for the 90% drawn Cu wire. The drawing texture can be approximated by a major <111> + minor <100> duplex fiber texture. The orientation density ratio of the <111> to <100> components is about 2.6. The orientation densities were obtained by averaging the f(g) values on the [φ_1=0-90°, Φ=0°,

φ_2=45°] line representing the <100> fiber texture and the [0-90°,55°,45°] line representing <111> in the φ_2=45° section of ODF. When annealed at 300 and 600 °C, the specimen developed textures of major <100> + minor <111> as expected from SERM. However, after annealing at 700 °C for 3 h, the grain size is so large that the ODF data consist of discrete orientations and the density of the <100> orientation is reduced while the density around the <1 1 1.7> orientation increases drastically. This is due to AGG and not discussed here. Wire drawing undergoes homogeneous deformation only in the axial center region, textures of the center regions were measured using electron backscatter diffraction (EBSD). The EBSD results are shown in Figure 22. The center region of the as-drawn specimen develops the major <111> + minor <100> fiber duplex texture as expected for axisymmetric deformation. The texture of the center region is similar to the gloval texure in Figure 21 because the deformation in wire drawing is relatively homogeneous. The annealing textures obtained at 700 °C is not the primary Rex texture.

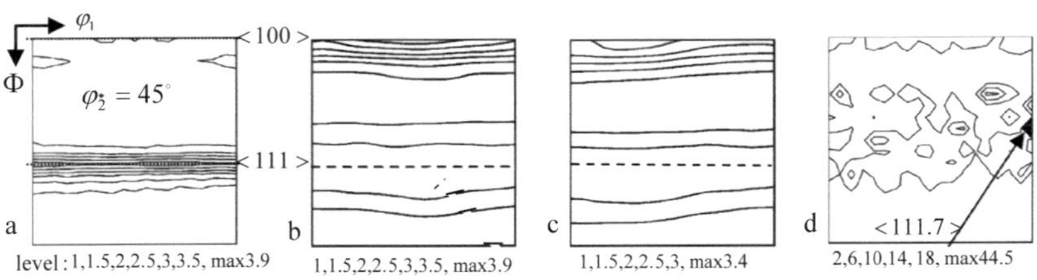

Figure 21. ODFs of 90% drawn Cu wire (a) before and after annealing at (b) 300, (c) 600, (d) 700 °C for 3 h, measured by X-ray [35].

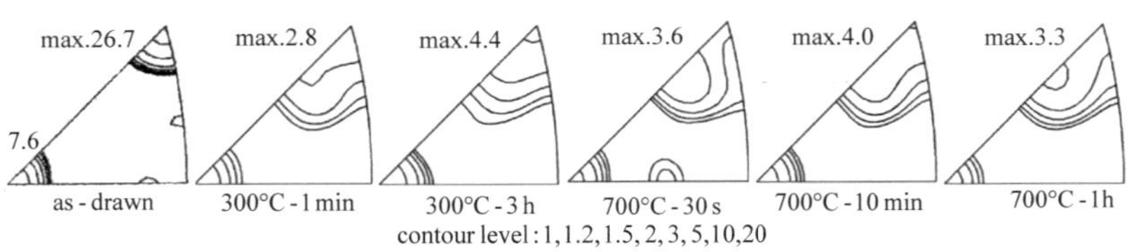

Figure 22. IPFs for center regions of 90% drawn Cu wires after annealing at 300 and 700°C [35].

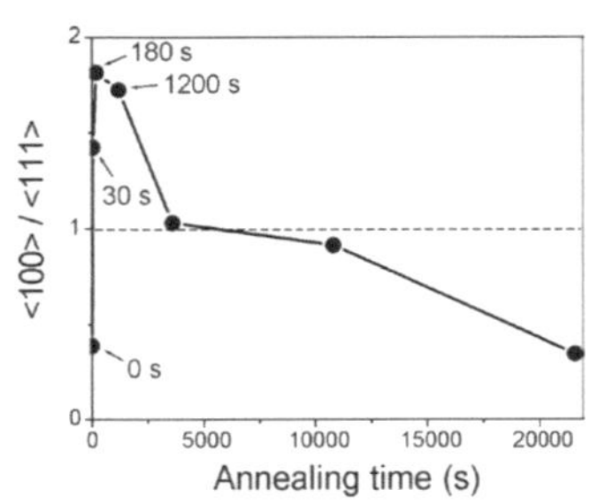

Figure 23. ODR of <100> to <111> of 90% drawn Cu wire vs. annealing time at 700°C [35].

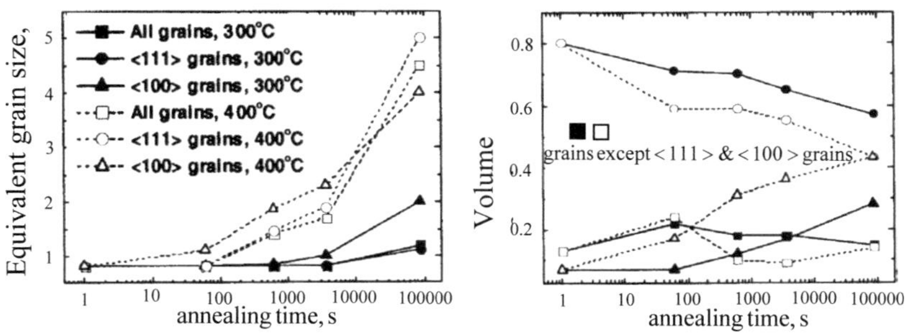

Figure 24. Grain size and volume fraction of ● ○ <111> and ▲△ <100> grains in Au wire vs. annealing time at 300 °C (solid symbols) and 400 °C (open symbols) [36].

Figure 23 shows ODR of <100> to <111> of the 90% drawn Cu wire as a function of annealing time at 700 °C. The ratio increases very rapidly up to about 1.8 after annealing for 180 s, wherefrom it decreases and reaches to about 0.3 after 6 h. The increase in the ratio indicates the occurrence of Rex and the decrease indicates the texture change during subsequent grain growth, that is, AGG. A similar phenomenon is observed in drawn Ag wire during annealing (Figure 20).

Cho et al. [36] measured the drawing and Rex textures of 25 and 30 μm diameter Au wires of over 99.99% in purity, which had dopants such as Ca and Be that total less than 50 ppm by weight. The Au wires were made by drawing through a series of diamond dies to an effective strain of 11.4.

Figure 24 shows the grain size and the volume fraction of the <111> and <100> grains as a function of annealing time at 300 and 400 °C. These values are based on EBSD measurements. The aspect ratio of grain shape was in the range of 1.5 - 2, which is little influenced by annealing time and temperature [36]. The grain growth occurs in whole area of the wire and is more rapid at 400 °C than at 300 °C as expected for thermally activated motion of grain boundaries. The volume fraction of the <111> grains decreases and that of the <100> grains increases with annealing time when Rex takes place, as expected from SERM.

5. Plane-strain compressed fcc metallic single crystals

5.1. Channel-die compressed {110}<001> aluminum single crystal

The annealing texture of single-phase crystals of Al-0.05% Si of the Goss orientation {110}<001> deformed in channel-die compression was studied by Ferry et al. [37]. In the channel-die compression, the compression and extension directions were <110> and <001> directions, respectively. Their experimental results showed that, even after deformation to a true strain of 3.0 which is equivalent to a compressive reduction of 95%, the original orientation was maintained as shown in Figure 25a. Figure 25b shows one (110) pole figure typical of a deformed crystal after annealing at 300 °C for 4 h. The comparison of Figures 25a and 25b suggests that the annealing texture is essentially the same as the deformation texture.

They also reported that even after 90% reduction and annealing for up to 235 h, the orientation was the same as that of the as-deformed crystal. For deformed specimens electropolished and annealed for various temperatures between 250 and 350 °C, no texture change took place before and after annealing, although grains which had different orientations were sometimes found to grow from the crystal surface after very long annealing treatments. For samples deformed over the true strain range of 0.5 to 3.0 in their work, annealing at a given temperature resulted in similar microstructural evolution. They called the phenomenon discontinuous subgrain growth during recovery. They stated that crystals of an orientation which was stable during deformation were generally resistant to Rex. This statement cannot be justified in light of single crystal examples in Sections 5.2 to 5.4.

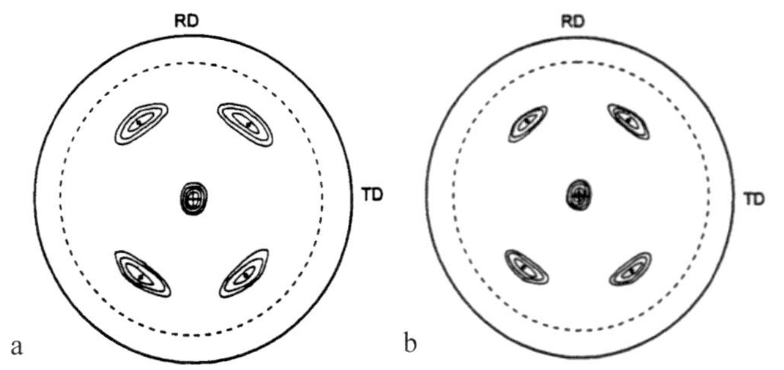

Figure 25. pole figures for 95% channel-die compressed Al single crystal (a) before and (b) after annealing at 300 ˚C for 4 h. (Contour levels: 2, 5, 11, 20, 35, 70 x random) [37].

The result was discussed based on SERM [38]. The (110)[001] orientation is calculated by the full constraints Taylor-Bishop-Hill model to be stable when subjected to plane strain compression. The active slip systems for the (110)[001] crystal are calculated to be (111)[0-1 1], (111)[-101], (-1-1 1)[011], and (-1-1 1)[101], whose activities are the same. It is noted that all the slip directions are chosen so that they can be at acute angle with the maximum strain direction [001]. AMSD is [0-1 1] + [-101] + [011] + [101] = [004]//[001], which is MYMD because $[S_{44}\text{-}2(S_{11}\text{-}S_{12})] < 0$ from compliances of Al [39]. When AMSD in the deformed state is parallel to MYMD in Rexed grains, the deformation texture remains unchanged after Rex (1st priority in Section 2).

5.2. Aluminum crystals of {123}<412> orientations

Blicharski et al. [40] studied the microstructural and texture changes during recovery and Rex in high purity Al bicrystals with S orientations, e.g. (123)[4 1-2]/(123)[-4-1 2] and (123)[4 1-2]/(-1-2-3)[4 1-2], which had been channel-die compressed by 90 to 97.5% reduction in thickness. The geometry of deformation for these bicrystals was such that the bicrystal boundary, which separates the top and bottom crystals at the midthickness of the specimen, lies parallel to the plane of compression, i.e. {123} and the <412> directions are aligned with the channel, and the die constrains deformation in the <121> directions. The annealing of the deformed bicrystals was conducted for 5 min in a fused quartz tube furnace with He + 5%H_2 atmosphere. The textures of the fully Rexed specimens were examined by determining the {111} and {200} pole figures from sectioned planes at 1/4, 1/2 and 3/4 specimen thickness. This roughly corresponds to the positions at the midthickness of the top crystal, the bicrystal boundary, and the midthickness of the bottom crystal, respectively. The deformation textures of the two bicrystals, (123)[4 1-2]/(123)[-4-1 2] and (123)[4 1-2]/(-1-2-3)[4 1-2], channel-die compressed by 90%, are reproduced in Figure 26. The initial orientation of the component crystals is also indicated in these pole figures. The annealing textures are shown in Figure 27. As Bricharski *et al.* pointed out, the Rex textures of the fully annealed bicrystal specimens do not have 40° <111> rotational orientation relationship with the deformation textures (compare Figures 26 and 27). Lee and Jeong [41] dicussed the Rex textures based on SERM. The slip systems activated during deformation and their activities (shear strains on the slip systems) must be known. Figure 28 shows the orientation change of crystal {123}<412> during the plane strain compression. Comparing the calculated results with the measured values in Figure 28, the measured orientation change during deformation seems to be best simulated by the full constraints strain rate sensitivity model. Figure 29 shows the calculated shear strain increments on active slip systems of the (123)[4 1-2] crystal as a function of true thickness strain, when subjected to the plane strain compression. The experimental deformation texture is well described by (0.1534 0.5101 0.8463)[0.8111 0.4242 -0.4027], or (135)[2 1-1], which is calculated based on the full constraints strain rate sensitivity model with $m = 0.01$. The reason why the measured deformation texture is simulated at the reduction slightly lower than experimental reduction may be localized deformation like shear band formation occurring in real deformation. The localized deformation might not be reflected in X-ray measurements. The scattered experimental Rex textures may be related to the nonuniform deformation. Now that the shear strains on active slip systems are known, we are in position to calculate AMSD. For a true thickness strain of 2.3, or 90% reduction, the γ values (Eq. 8) of the (111)[1 0-1], (111)[0 1-1], (1-1 1)[110], (1-1-1)[110], (1-1 1)[011], and (1-1-1)[101] slip systems calculated using the data in Figure 29 are proportional to 2091, 776, 1424, 2938, 76, and 139, respectively. The contributions of the (1-1 1)[011] and (1-1-1)[101] slip systems are negligible compared with others. Therefore, the (111)[1 0-1], (111)[0 1-1], (1-1 1) [110], and (1-1-1)[110] systems are considered in calculating AMSD. It is noted that all the slip directions are chosen so that they can be at acute angle with RD, [0.8111 0.4242 -0.4027]. AMSD is calculated as follows:

$$209[10-1]+776[0\,1-1]+1424\times 0.577[110]+2938\times 0.577[110]=[4608\ 3293-2867]//[0.7259\ 0.5187-0.4516]\text{unit vector} \quad (12)$$

where the factor 0.577 originates from the fact that the slip systems of (1-1 1)[110] and (-111) [110] share the same slip direction [110] (Eq. 7). Two other principal stress directions are obtained as explained in Figure 6. Possible candidates for the direction equivalent to **S** in Figure 6 are the [011], [101], and [1-1 0] directions, which are not used in calculation of AMSD among six possible Burgers vector directions. The [011], [101], and [1-1 0] directions are at 87.3, 78.8 and 81.6°, respectively, with AMSD. The [011] direction is closest to 90° (Figure 30). The directions equivalent to **B** and **C** in Figure 6 are calculated to be [-0.0345 0.6833 0.7294] and [0.6869 -0.5139 0.5139] unit vectors, respectively. In summary, OA, OB, and OC in Figure 30, which are equivalent to **A**, **B**, and **C**, are to be parallel to the <100> directions in the Rexed grain. If the [0.7259 0.5187 -0.4516], [0.6869 -0.5139 0.5139] and [-0.0345 0.6833 0.7294] unit vectors are set to be parallel to [100], [010] and [001] directions after Rex (Figure 30), components of the unit vectors are direction cosines relating the deformed and Rexed crystal coordinate axes. Therefore, ND, [0.1534 0.5101 0.8463], and RD, [0.8111 0.4242 -0.4027], in the deformed crystal coordinate system can be transformed to the expressions in the Rexed crystal coordinate system using the following calculations (refer to Eq. 9):

$$\begin{bmatrix} 0.7259 & 0.5187 & -0.4516 \\ 0.6869 & -0.5139 & 0.5139 \\ -0.0345 & 0.6833 & 0.7294 \end{bmatrix} \begin{bmatrix} 0.1534 \\ 0.5101 \\ 0.8463 \end{bmatrix} = \begin{bmatrix} -0.0062 \\ 0.2781 \\ 0.9606 \end{bmatrix} \tag{13}$$

$$\begin{bmatrix} 0.7259 & 0.5187 & -0.4516 \\ 0.6869 & -0.5139 & 0.5139 \\ -0.0345 & 0.6833 & 0.7294 \end{bmatrix} \begin{bmatrix} 0.8111 \\ 0.4242 \\ -0.4027 \end{bmatrix} = \begin{bmatrix} 0.9907 \\ 0.1322 \\ -0.0319 \end{bmatrix} \tag{14}$$

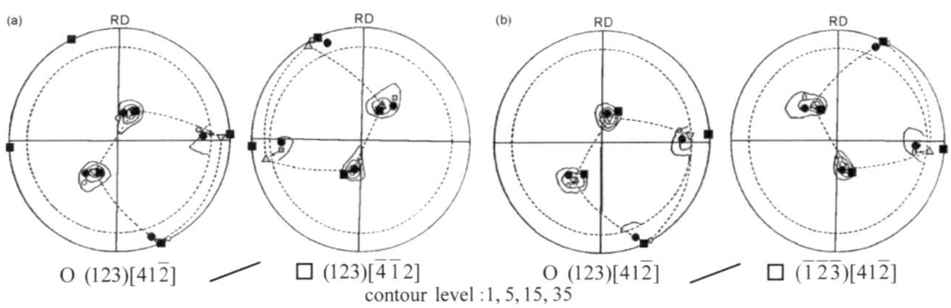

Figure 26. pole figures for 90% channel-die compressed Al crystals of {123}<412> orientations [40]. A/B indicates bi-crystal composed of A and B crystals. ● ~{135}<211>; ■ ~{011}<522>.

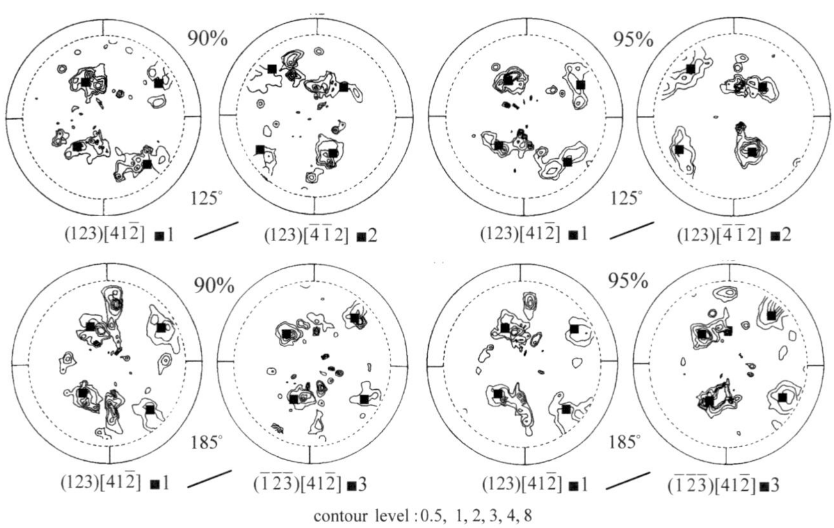

Figure 27. pole figures of 90 and 95% channel-die compressed Al bicrystals after annealing at 125 and 185 °C for 5 min [40]. ■ SERM-calculated Rex orientations: ■1 (-0.0062 0.2781 0.9606)[0.9907 0.1322 -0.0319]; ■2 (-0.0062 0.2781 0.9606)[-0.9907 -0.1322 0.0319]; ■3 (0.0062 -0.2781 -0.9606)[0.9907 0.1322 -0.0319] [41].

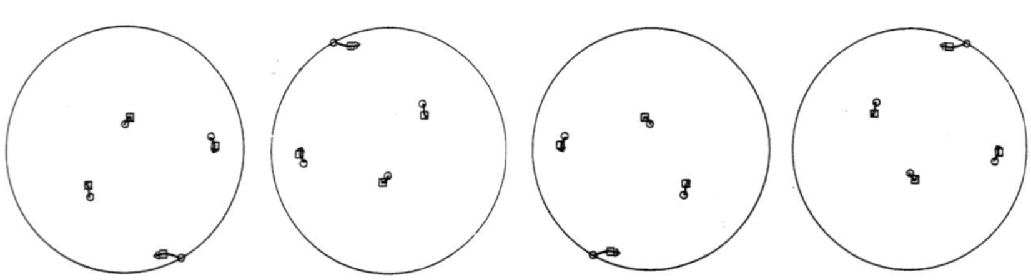

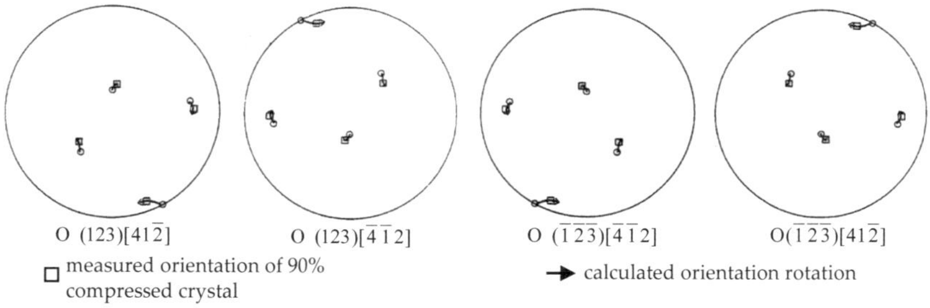

Figure 28. Orientation rotations of {123}<412> crystals during plain strain compression by 90% [41].

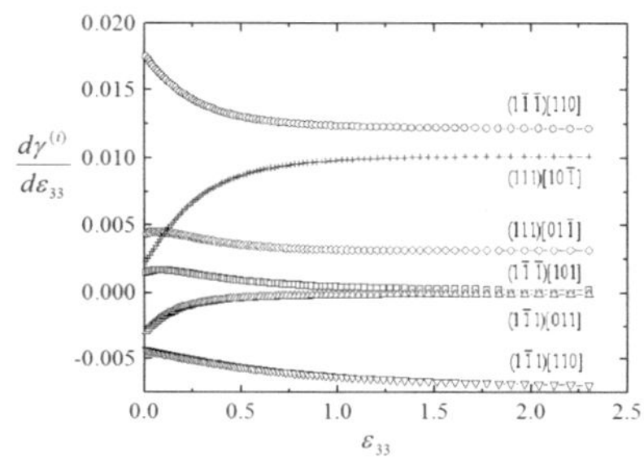

Figure 29. Calculated shear strain rate with respect to thickness reduction of 0.01, $d\gamma^{(i)}/d\varepsilon_{33}$, on active slip systems of (123)[4 1-2] crystal as a function of true thickness strain, ε_{33}, [41].

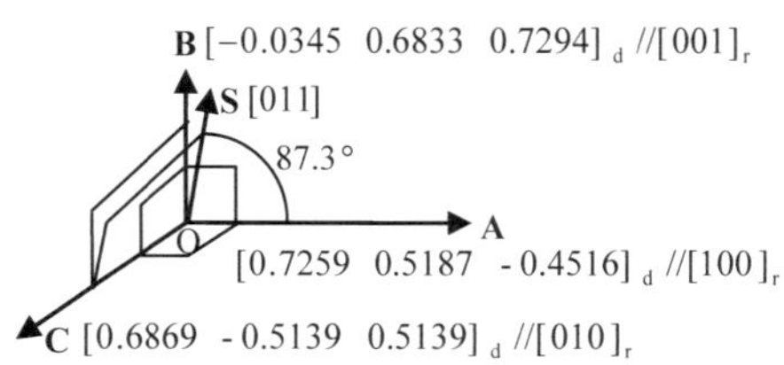

Figure 30. Orientation relations in deformed and Rexed states. Subscripts d and r indicate deformed state and Rexed state, respectively.

The calculated result means that the (0.1534 0.5101 0.8463)[0.8111 0.4242 -0.4027] crystal, which is obtained by the channel die compression by 90% reduction, transforms to the Rex texture (-0.0062 0.2781 0.9606)[0.9907 0.1322 -0.0319]. Similarly, crystals deformed by channel die compression from (123)[-4-1 2] and (-1-2-3)[4 1-2] orientations transform to (-0.0062 0.2781 0.9606)[-0.9907 -0.1322 0.0319] and (0.0062 -0.2781 -0.9606)[0.9907 0.1322 -0.0319], respectively, after Rex. The results are plotted in Figure 27 superimposed on the experimental data. It can be seen that the calculated Rex textures are in good agreement with the measured data.

5.3. Aluminum crystal of {112}<111> obtained by channel-die compression of (001)[110] crystal

Butler et al. [42] obtained a {112}<111> Al crystal by channel-die compression of the (001)[110] single crystal. The (001)[110]orientation is unstable with respect to plane strain compression, to form the (112)[1 1-1] and (112)[-1-1 1] orientations as shown in Figure 31a. The Rex texture produced after annealing at 200 °C was a rotated cube texture (Figure 31b). Lee [43] analyzed the result based on SERM. Figure 32 shows shear strains/extension strain on slip systems of 1 to 6 as a function of rotation angle about TD [-110] of the (001)[110] fcc crystal obtained from the Taylor-Bishop-Hill theory. The contribution of the slip systems to the deformation is approximated to be proportional to the area under the shear strains $\gamma^{(i)}$ on slip systems i / extension strain - rotation angle θ curve in Figure 32. The area ratio becomes

$$\int_0^{35} \gamma^{(1)} d\theta : \int_0^{10} \gamma^{(3)} d\theta : \int_{10}^{35} \gamma^{(5)} d\theta = 30 : 3 : 20.6 \qquad (15)$$

All the slips may not occur on the related slip systems uniformly in a large single crystal. Some regions of the crystal may be deformed by 1, 3, and 5 slip systems, while some other regions by 2, 4, and 6 slip systems.

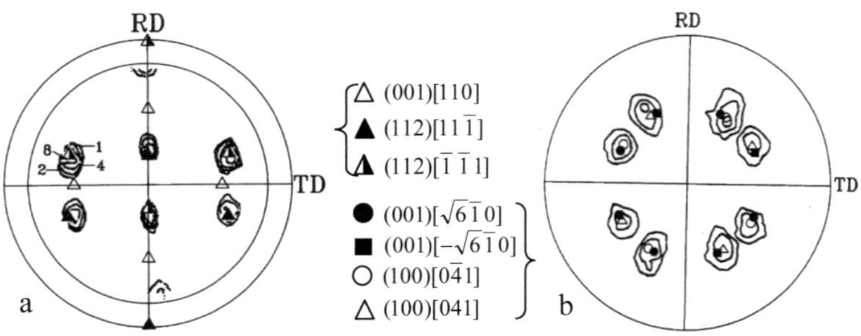

Figure 31. (a) (111) pole figure of Al single crystal with initial orientation (001)[110] after 70% reduction by channel-die compression; (b) (111) pole figure of measured Rex texture (contours), (100)[0-4 1], and (100)[041] [42]. (001)[√6 -1 0] and (001)[-√6 -1 0] are calculated by SERM [43].

For the contribution of the former three slip systems to the crystal deformation, AMSD is obtained by the vector sum of the [1 0-1], [0-1-1], and [110] directions whose contributions are assumed to be proportional to the area ratio obtained earlier (30 : 3 : 20.6). The vector sum is shown in Figure 33. The resultant direction passes through point E, which divides line BC by a ratio of 1 to 2. Thus, AMSD // AE // [3 1-2]. Another high stress direction equivalent to **S** in Figure 6 is BD, or [-110] which is not used in calculation of AMSD among possible Burgers vectors. The [-110] direction is not normal to AMSD. The direction that is at the smallest possible angle with the [-110] direction and normal to AMSD must be on a plane made of AMSD and the [-110] direction. The plane normal is obtained to be the [112] direction by the vector product of AMSD and the [-110] direction, which is equivalent to **C** in Figure 6. The

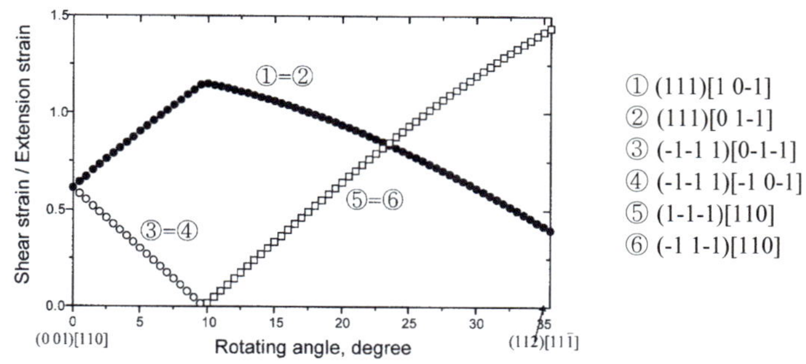

Figure 32. Shear strains on slip systems of 1 to 6 as a function of rotation angle about TD [-110] of (001)[110] crystal [43].

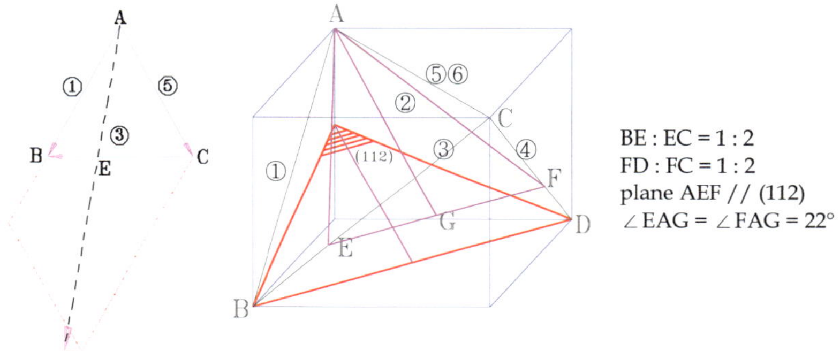

Figure 33. Vector sum of slip directions ① [1 0-1], ③ [0-1-1], and ⑤ [110] assuming that their activities are proportional to 30:3:20.6 (Eq. 15).

direction that is equivalent to **B** in Figure 6 is calculated to be [2-4 1] by the vector product of AMSD [3 1-2] and [112]. Thus, the [3 1-2], [2-4 1], and [112] directions become parallel to <100> in the Rexed grains.

If the directions [3 1-2], [2-4 1], and [112], whose unit vectors are [3/√14 1/√14 -2/√14], [2/√21 -4/√21 1/√21], and [1/√6 1/√6 2/√6], respectively, are set to be parallel to [100], [010] and [001] directions in the Rexed crystal, components of the unit vectors are direction cosines relating the deformed and Rexed crystal coordinate axes (Eq. 9). Therefore, ND, [112], and RD, [1 1-1], in the deformed crystal coordinate system can be transformed to the expressions in the Rexed crystal coordinate system using the following calculation:

$$\begin{bmatrix} 3/\sqrt{14} & 1/\sqrt{14} & -2/\sqrt{14} \\ 2/\sqrt{21} & -4/\sqrt{21} & 1/\sqrt{21} \\ 1/\sqrt{6} & 1/\sqrt{6} & 2/\sqrt{6} \end{bmatrix} \begin{bmatrix} 1 \\ 1 \\ 2 \end{bmatrix} // \begin{bmatrix} 0 \\ 0 \\ 1 \end{bmatrix} \text{ and } \begin{bmatrix} 3/\sqrt{14} & 1/\sqrt{14} & -2/\sqrt{14} \\ 2/\sqrt{21} & -4/\sqrt{21} & 1/\sqrt{21} \\ 1/\sqrt{6} & 1/\sqrt{6} & 2/\sqrt{6} \end{bmatrix} \begin{bmatrix} 1 \\ 1 \\ -1 \end{bmatrix} // \begin{bmatrix} \sqrt{6} \\ -1 \\ 0 \end{bmatrix}$$

Therefore, the (112)[11-1] deformation texture transforms to the (001)[√6-1 0] Rex texture. Similarly, from the (111)[0 1-1], (-1-1 1)[-1 0-1], and (-1 1-1) [110] slip systems, another AMSD AF, or the [1 3-2] direction, can be obtained. In this case, the (112)[1 1-1] deformation texture transforms into the (001)[-√6 -1 0] Rex texture. The {001}<√6 1 0> orientation has a rotational relation with the {001}<100> orientation through 22° about the plane normal. The calculated Rex texture is superimposed on the measured data in Figure 31b. The calculated results are in relatively good agreement with the measured data. It is noted that Figure 32 does not represent the correct strain path during deformation. Therefore, there is a room to improve the calculated Rex texture. The Rex texture is at variance with the {001}<100> Rex texture in polycrystalline Al and Cu.

5.4. Copper crystal of (123)[-6-3 4] in orientation rolled by 99.5% reduction in thickness

Kamijo et al. [44] rolled a (123)[-6-3 4] Cu single crystal reversibly by 99.5% under oil lubrication. The (123)[-6-3 4] orientation was relatively well preserved up to 95%, even though the orientation spread occurred as shown in Figure 34a. However, the crystal rotation proceeded with increasing reduction. A new (321)[-4 3 6] component, which is symmetrically oriented to the initial (123)[-6-3 4] with respect to TD, developed after 99.5% rolling as shown in Figure 34b. It is noted that other two equivalent components are not observed. The rolled specimens were annealed at 538 K for 100 s to obtain Rex textures. In the Rex textures of the crystals rolled less than 90%, any fairly developed texture could not be observed, except for the retained rolling texture component. They could observe a cube texture with large scatter in the 95% rolled crystal and the fairly well developed cube orientation in the 99.5% rolled crystal after Rex as shown in Figure 34c. They concluded that the development of cube texture in the single crystal of the (123)[-6-3 4] orientation was mainly attributed to the preferential nucleation from the (001)[100] deformation structure. The cube deformation structure was proposed to form due to the inhomogeneity of deformation. Lee and Shin [45] explained the textures in Figure 34 based on SERM. Figure 35 shows $d\gamma^{(i)}/d\varepsilon_{11,}$ with $d\varepsilon_{11} = 0.01$, on active slip systems i as a function of ε_{11} for the (123)[-6-3 4] crystal, which was calculated by the $d\varepsilon_{13}$ and $d\varepsilon_{23}$ relaxed strain rate sensitive model (m = 0.02) with the subscripts 1, 2, and 3 indicating RD, TD, and ND. The changes in $d\gamma^{(i)}/d\varepsilon_{11}$ depending on ε_{11} indicate that the (123)[-6-3 4] orientation is unstable with respect to the plane strain compression. A part of the (123)[-6-3 4] crystal, particularly the surface layers where $d\varepsilon_{23}$ is negligible due to friction between rolls and sheet, seems to rotate to the {112}<111> orientation. A part of the {112}<111> crystal further rotates to (321)[-436] with increasing reduction. This is why (321)[-436] has lower density than (123)[-6-3 4] along with weak {112}<111>. The orientation rotation is shown in Figure 35b. Since important components in the deformation texture are (123)[-6-3 4], (321)[-436], and {112}<111>, their Rex textures are calculated using SERM. If the (123)[-6-3 4] orientation is stable, $d\gamma^{(i)}/d\varepsilon_{11}$ on the active slip systems do not vary with strain. It follows from Fig. 35a that $d\gamma^{(i)}/d\varepsilon_{11}$ on C, J, M, and B are 0.014, 0.01, 0.007, and 0.003, respectively, at zero strain. Therefore, AMSD is 0.014 [-101] + 0.01×0.577 [-1-1 0] + 0.007 ×0.577 [-1-1 0] + 0.003 [0-1 1] = [-0.02381 -0.01281 0.017], where the factor 0.577 originates from the fact that the duplex slip systems of (1-1 1)[-1-10] and (-111) [-1-10] share the same slip direction (Figure 4). The [-0.02381 -0.01281 0.017] direction, or the [-0.745 -0.401 0.532] unit vector, will be parallel to one of the <100> directions, MYMDs of Cu, after Rex. Orientation relationship between the matrix and Rexed state is shown in Figure 36a, which is obtained as explained in Figure 6. The Rex orientation of the (123)[-6-3 4] matrix is calculated as follows:

$$\begin{pmatrix} -0.745 & -0.401 & 0.532 \\ 0.069 & 0.747 & 0.660 \\ -0.663 & 0.529 & -0.529 \end{pmatrix}\begin{pmatrix} 1 \\ 2 \\ 3 \end{pmatrix} = \begin{pmatrix} 0.049 \\ 3.543 \\ -1.192 \end{pmatrix} \text{ and } \begin{pmatrix} -0.745 & -0.401 & 0.532 \\ 0.069 & 0.747 & 0.660 \\ -0.663 & 0.529 & -0.529 \end{pmatrix}\begin{pmatrix} -6 \\ -3 \\ 4 \end{pmatrix} = \begin{pmatrix} 7.801 \\ -0.017 \\ -0.275 \end{pmatrix}$$

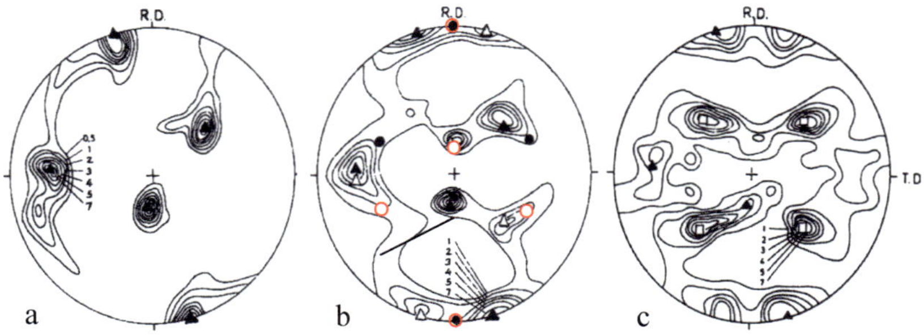

Figure 34. pole figures for (123)[-6-3 4] Cu single crystal after rolling by (a) 95%, (b) 99.5%, and (c) 99.5% and subsequent annealing at 538 K for 100 s [44]. ▲(123)[-6-3 4]; Δ(321)[-436]; ●(112)[-1-1 1]; ○ (112)[11-1]; □(001)[100].

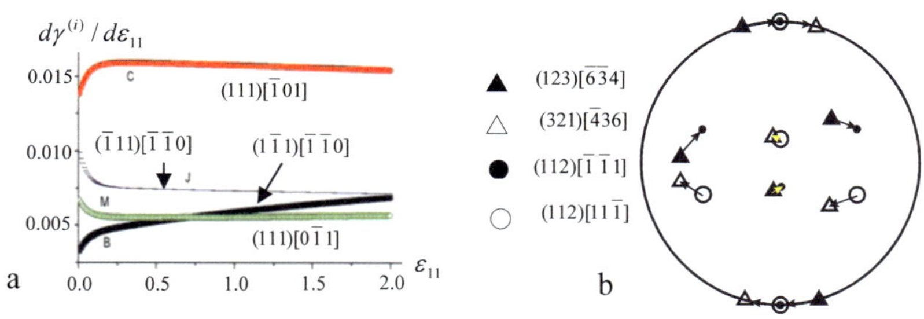

Figure 35. (a) Shear strain rates $d\gamma/d\varepsilon_{11}$ with $d\varepsilon_{11}$=0.01 vs. ε_{11} on active slip systems (C, J, M, B) and (b) (111) pole figure showing orientation rotation from {123}<634> to {112}<111> [45].

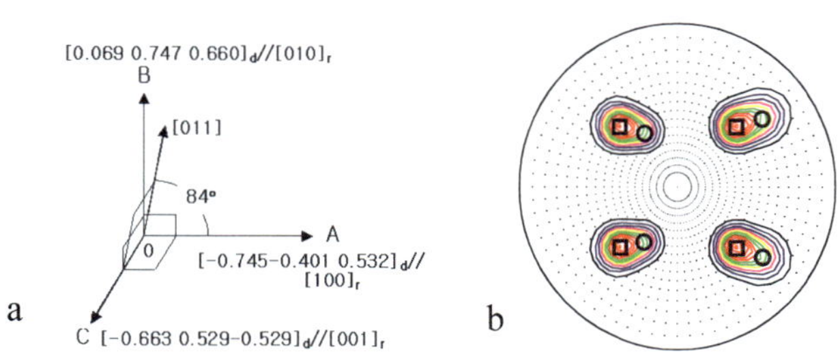

Figure 36. a) Orientation relationship between deformed ($_d$) and Rexed ($_r$) states and (b) (111) pole figures of ○ (0 3-1)[100] and □ (001)[100] orientations. Contours were calculated assuming Gaussian scattering (10°) of (0 3-1)[100] and (001)[100] components with their density ratio being 2:1 [45].

The calculated Rex orientation is (0.049 3.543–1.192)[7.801-0.017-0.275] ≈ (0 3-1)[100]. Similarly the (321)[-436] crystal is calculated to have slip systems of (111)[-101], (111)[-110], (-1-1 1)[011], and (-1 1-1)[011], on which the shear strain rates at $d\varepsilon_{11}=0$ are 0.014, 0.003, 0.01, and 0.007, respectively. The (321)[-436] is calculated to transform to the (-0.049 3.543-1.192)[7.801 0.017 -0.275] ≈ the (0 3-1)[100] Rex texture. This result is understandable from the fact that the (0 3-1)[100] orientation is symmetrical with respect to TD as shown in Figure 36b and the deformation orientations, (123)[-6-3 4] and (321)[-436], are also symmetrical with respect to TD as shown in Figure 35b. The {112}<111> rolling orientation to the {001}<100> Rex orientation transformation is discussed based on SERM in Section 6.1.

According to the discussion in Section 6.1, if the cube oriented regions are generated during rolling, they are likely to survive and act as nuclei and grow at the expense of neighboring {112}<111> region during annealing because the region tend to transform to the {001}<100> orientation to reduce energy. The grown-up cube grains will grow at the expense of grains having other orientations such as the {123}<634> orientation, resulting in the {001}<100> texture after Rex, even though the Cu orientation is a minor component in the deformation texture. Meanwhile, the main S component in the deformation texture can form its own Rex texture, the near (0 3-1)[100] orientation. In this case, the Rex texture may be approximated by main (001)[100] and minor (0 3-1)[001] components. Figure 36b shows the texture calculated assuming Gaussian scattering (half angle=10°) of these components with the intensity ratio of (001)[100]: (0 3-1)[001] = 2 : 1. It is interesting to note that the cube peaks diffuse rightward under the influence of the minor (0 3-1)[100] component in agreement with experimental result in Figure 34c.

6. Cold-rolled polycrystalline fcc metals and alloys

6.1. Cube recrystallization texture

The rolling texture of fcc sheet metals with medium to high stacking fault energies is known to consist of the brass orientation {011}<211>, the Cu orientation {112}<111>, the Goss orientation {011}<100>, the S orientation {123}<634>, and the cube orientation {100}<001>. The fiber connecting the brass, Cu, and S orientations in the Euler space is called the β fiber. Major components of the plane-strain rolling texture of polycrystalline Al and Cu are known to be the Cu and S orientations. The Rex texture of rolled Al and Cu sheets is well known to be the cube texture. The 40°<111> orientation relationship between the S texture and the cube texture has been taken as a proof of OG, and has made one believe that the S orientation is more responsible for the cube Rex texture. OG is claimed to be associated with grain boundary mobility anisotropy. However, experimental data indicate that the Cu texture is responsible for the cube texture. For an experimental result of Table 3, the deformation texture is not strongly developed below a reduction of 73% and its Rex texture is approximately random. At a reduction of 90%, a strong Cu texture is obtained and its Rex texture is a strong cube texture. For 95% cold rolled Al-0 to 9%Mg alloy after annealing at 598K for 0.5 to 96 h, the highest density in the Cu component in the deformation texture and the highest density in the cube component in Rex textures were observed at about 3% Mg (Figure 37). This implies that the Cu component is responsible for the cube component. However, these cannot prove that the Cu texture is responsible for the cube texture because deformation components with the highest density are not always linked with highest Rex components [47].

Changes in orientation densities of 95% rolled Cu during annealing at 400 to 500 °C (Figure 38), 95% rolled AA8011 Al alloy during annealing at 350 °C (Figure 39a), and 95% rolled Fe-50%Ni alloy during annealing at 600°C (Figure 39b), and 95% rolled Cu after heating to 150 to 300°C at a rate of 2.5 K/s followed by quenching showing that the Cu component disappears most rapidly when the cube orientation started to increase [52]. These results imply that the Cu component is responsible for the cube Rex texture. Rex is likely to occur first in high strain energy regions. It is known that the energy stored in highly deformed crystals is proportional to the Taylor factor ($\Sigma d\gamma^{(k)}/d\varepsilon_{ij}$ with γ and ε_{ij} being shear strains on slip systems k and strains of specimen, respectively). The Taylor factor is calculated to be 2.45 for the cube oriented fcc crystal using the full constraints model, 3.64 for the Cu oriented fcc crystal using the ε_{13} relaxed constraints rate sensitive model, 3.24 for the S oriented fcc crystal using the ε_{13} and ε_{23} relaxed constraints rate sensitive model, and 2.45 for the brass orientation using the ε_{12} and ε_{23} relaxed constraints rate sensitive model. In the rate sensitive model calculation, the rate sensitivity index was 0.01 and each strain step in rolling was 0.025. The measured stored energies for 99.99% Al crystals channel-die compressed by a strain of 1.5 showed that the Cu oriented region had higher energies than the S oriented region [53]. The Taylor factors and the measured stored energies indicate that the driving force for Rex is higher in the Cu oriented grains than in the S oriented grains. Therefore, the Cu component in the deformation texture is more responsible for the cube Rex texture than the S component.

	Rolling reduction	Brass	Copper	Goss	S	Cube
Rolling texture	58%	3.6	2.6	1.1	1.4	0.6
	73%	2.8	3.0	0.9	1.1	1.1
	90%	0.7	5.7	0.1	0.7	1.3
Rex texture	58%	2.1	1.4	1.0	1.3	1.2
	73%	1.8	1.5	1.3	1.4	2.1
	90%	0.2	0.8	0.2	0.4	20.0

Table 3. Texture component strength of high purity OFE copper [46]

The copper to cube texture transition was first explained by SERM [4], and elaborated later [54]. The orientations of the (112)[1 1-1] and (123)[6 3-4] Cu single crystals remain stable in the center layer for all degree of rolling [55]. The Cu orientation (112)[1 1-1] is calculated to be stable by the ε_{13} relaxed constraint model [56,57]. For the (112)[1 1-1] crystal, the active slip systems are calculated by the RC model to be (-1 1-1)[110], (1-1-1)[110], (111)[1 0-1], and (111)[0 1-1], on which shear strain rates are the same regardless of reduction ratio. Almost the same

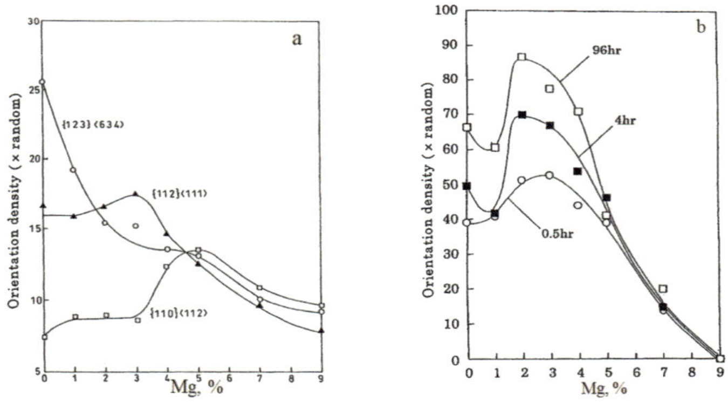

Figure 37. Effect of Mg content on (a) densities of {112}<111>, {123}<634>, and {110}<112> orientations in Al-Mg alloys cold rolled by 95% and on (b) density of {001}<100> orientation in specimens annealed at 598 K for 0.5, 4, and 96 h [48].

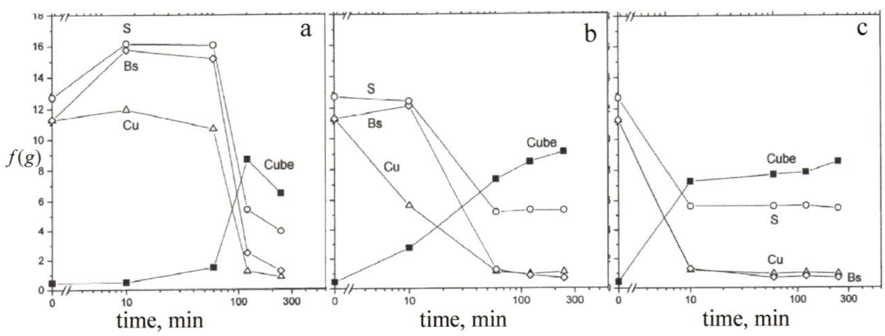

Figure 38. Changes in densities of copper Cu, S, brass Bs, and cube orientations in 95% cold rolled copper during annealing at (a) 400, (b) 450, and (c) 500 °C [49].

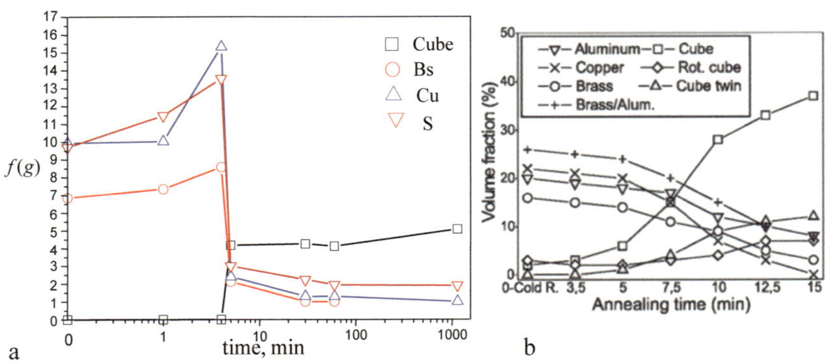

Figure 39. a) Changes in densities of cube, brass, copper and S orientations in 95% cold-rolled AA8011 Al alloy during annealing at 350 °C [50]. (b) Evolution of bulk textures in 90% cold-rolled Fe-50%Ni alloy during annealing at 600 °C [51].

result is obtained by ε_{13}, ε_{23} relaxed constraints rate sensitive model [54]. The slip directions are chosen to be at acute angles with RD (Section 2). To calculate AMSD, the active slip systems are weighted by the shear strain rates on them. AMSD is calculated to be [0 1-1] + [1 0-1] + 0.577 × 2[110] ≈ 2 [1 1-1]. Here the factor 0.577 is related to the fact that the slip direction [110] is shared by the (-1 1-1) and (1-1-1) slip planes (Eq. 7). The [1 1-1] direction is equivalent to **A** in Figure 6. **S** becomes [1-1 0] because it is one of slip directions nearest to 90° with **A**. In fact. **S** is normal to **A**, so **S** becomes **B** and **C** = **A**×**B** = [-1-1-2]. Since MYMDs are <100>, the Rex orientation is calculated using Eq. 9 to be (0 0-1)[100], which is equivalent to (001)[100]. In conclusion, the (112)[1 1-1] deformation texture is calculated to change to the (001)[100] Rex texture.

The above calculation indicates that the Cu orientation tends to turn into the cube orientation during annealing. In order for the transformation to occur, the cube oriented nuclei are needed, whether they may be generated from the deformed matrix or already existing cube bands. In order for the cube bands to be nuclei, they must be stable during annealing. The cube orientation (001)[100] is calculated by the full constrains method to be metastable with respect to plane strain compression, with active slip systems being (111)[1 0-1], (1 1-1)[101], (1-1-1)[101], and (1-1 1)[1 0-1] on which the shear strain rates are the same. If cube oriented grains survive after rolling, they must have undergone the plane strain compression with the slip systems. Therefore, AMSD is [1 0-1] + [101] + [101] + [1 0-1] = [400] // [100]. This is MYMD of Cu. Since AMSD is the same as the MYMD, the cube texture is expected to remain unchanged whether Rex or recovery (1st priority in Section 2).

SERM does not tell us how the cube oriented nuclei form. If the cube oriented grains survived during rolling, they are likely to survive and act as nuclei and grow at the expense of neighboring Cu oriented grains during annealing, because the Cu oriented grains tend to transform to the cube orientation. The grown up cube grains will grow at the expense of grains having other orientations such as the S and brass orientations, resulting in the cube texture after Rex. This discussion applies to other fcc metals with high stacking fault energy (SFE).

6.2. Goss recrystallization texture

The evolution of rolling textures in copper alloys depends strongly on their SFEs. A continuous transition from the copper orientation to the brass orientation tends to occur with increasing content of alloying elements or decreasing SFE. However, Mn can be dissolved in copper up to 12 at.% without significantly changing SFE unlike various Cu alloys [58]. Engler [59,60] studied the influence of Mn on the deformation and Rex behavior of Cu-4 to 16%Mn alloys, as this should yield a clear separation of the effects caused by the changes in SFE from those due to other factors. It is particularly interesting that the alloys develop a deformation texture in which the density of the brass orientation can be higher than the densities of the copper orientation and the S orientation despite the fact that SFEs of the alloys are almost the same as that of pure Cu. The brass orientation is obtained in many Cu alloys with low SFEs, which is well known to transforms to the {236}<385> orientation. However, the Cu-Mn alloys do not develop the {236}<385> orientation after Rex. The texture transformation cannot be well explained by 40° <111> relation between the deformation and Rex textures.

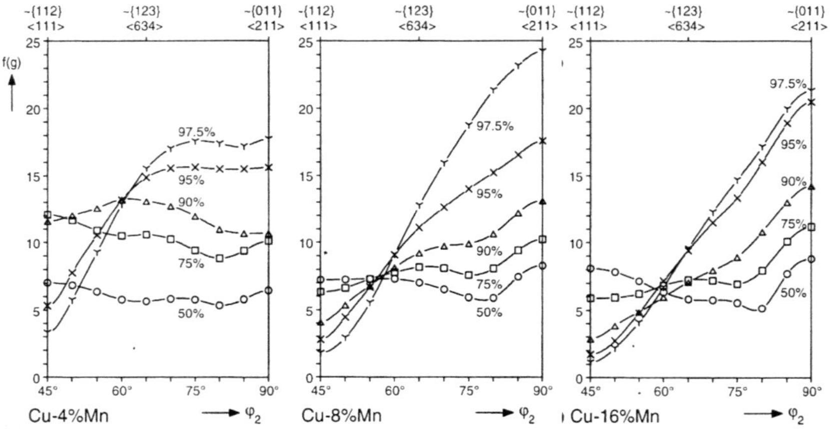

Figure 40. β-fiber intensity lines of Cu-4%Mn, Cu-8%Mn, and Cu-16%Mn alloys after rolling reductions from 50 to 97.5% [59].

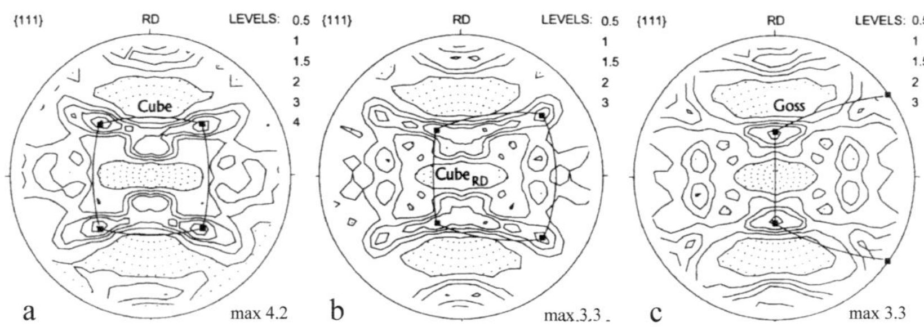

Figure 41. pole figures of (a) Cu-4%Mn, (b) Cu-8%Mn, and (c) Cu-16%Mn alloys after complete Rex (97.5% rolling, annealing for 1000 s at 450 °C) [60].

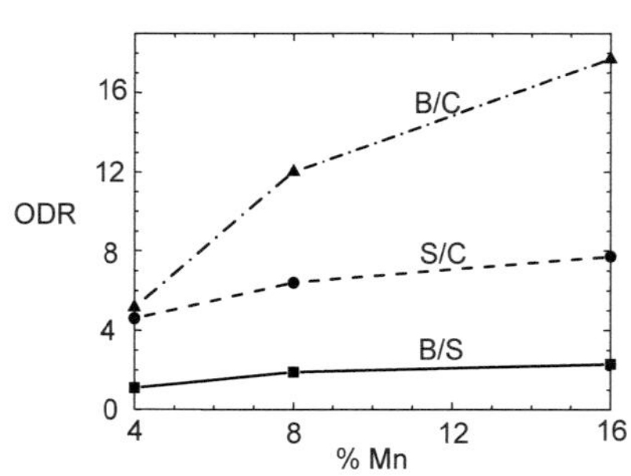

Figure 42. Orientation density ratios (ODR) among brass B, S, and copper C components in rolling texture (Figure 40) as a function of Mn concentration in Cu-Mn alloy [61].

Figure 40 shows the orientation densities f(g) along the β-fiber of Cu-4%Mn, Cu-8%Mn, and Cu-16%Mn alloys after rolling reductions of 50 to 97.5%. The figure indicates that with increasing Mn content and rolling reduction the brass orientation tends to dominate the rolling texture. The brass orientation in the Cu-Mn alloys is particularly interesting because the transformation of the orientation to the Rex texture will not be complicated by twinning as in low SFE alloys. Figure 41 shows {111} pole figures of the three Cu-Mn alloys rolled by 97.5% after complete Rex by annealing for 1000 s at 450 °C. In Cu-4%Mn the texture maximum lies in the cube-orientation. In Cu-8%Mn the texture maximum has shifted from the cube orientation to an orientation which can be approximated by the {013}<100> orientation. In Cu-16%Mn the texture maximum is in the Goss orientation. The orientation density ratios among the copper, S, and brass components in the rolling texture are shown in Figure 42. The density ratio of the brass to S component increases from about 1 to 2, the density ratio of the S to copper component increases from about 5 to 8, and the density ratio of the brass to copper component increases from about 5 to 18 with increasing Mn content from 4 to 16% in the Cu- Mn alloy. The density ratio of the S to copper and that of the brass to copper component are lowest in 4%Mn and highest in 16%Mn.

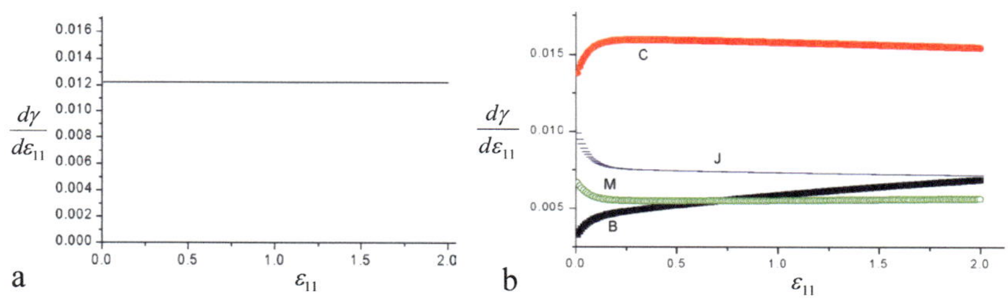

Figure 43. $d\gamma/d\varepsilon_{11}$ ($d\varepsilon_{11}$=0.01) vs. ε_{11} curves on active slip systems of (a) (111)[0-1 1] and (-1-1 1)[101] of (110)[1-1 2] crystal and of (b) J (1-1-1)[110], B (111)[0 1-1], M (-1 1-1)[110], and C (111)[1 0-1] of (123)[6 3-4]. Values of $d\gamma/d\varepsilon_{11}$ on B, M, J, and C are 0.003, 0.006, 0.01, and 0.014, respectively, at ε_{11}=0 [61].

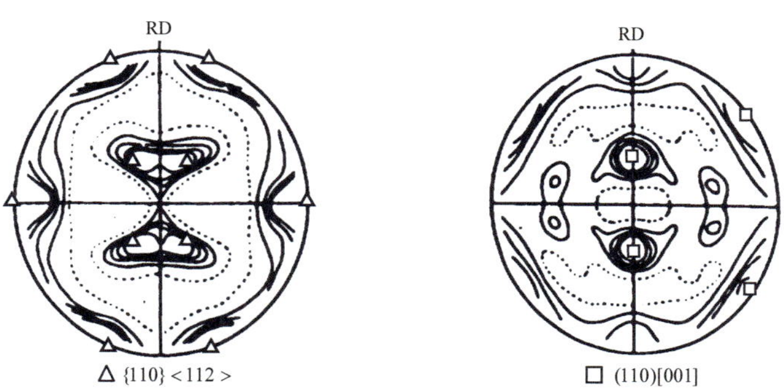

Figure 44. {110}<112> rolling and {110}<001> Rex textures of Cu-1% P alloy [62].

Comparison of the Rex textures with the corresponding deformation textures indicates that the brass component in the deformation texture seems to be responsible for the Goss components in the Rex texture. In what follows, the Rex textures are discussed based on SERM [61]. In order to find which component in the rolling texture is responsible for the Goss Rex texture, the brass rolling texture is first examined because it is the highest component in the deformation texture of Cu-16% Mn alloy, which changed to the Goss texture when annealed. When fcc crystals with the (110)[1-1 2] orientation are plane strain compressed along the [110] direction and elongated along the [1-1 2] direction, the relation between the strain ε_{11} of specimen and shear strain rates $d\gamma/d\varepsilon_{11}$ on active slip systems was calculated by the ε_{13} and ε_{23} relaxed strain rate sensitive model. Figure 43a shows the calculated results, which indicate that active slip systems are (111)[0-1 1] and (-1-1 1)[101] and their shear strain rates do not vary with strain of specimen indicating that the brass orientation is stable with respect to the strain. It is noted that the active slip directions were chosen to be at acute with the [1-1 2] RD. Thus, AMSD = [0-1 1] + [101] = [1-1 2] is the same as RD.

According to SERM, AMSD is parallel to MYMD of Rexed grain, the <100> directions in fcc metals. Therefore, the Rexed grains will have the (hk0)[001] orientation. The 2nd priority in Section 2 gives rise to the (110)[001] orientation because the (110) plane is shared by the deformed and Rexed grains. That is, the (110)[1-1 2] rolling texture transforms to the (110)[001] Rex texture. Similarly, for the (011)[2-1 1] crystal, equally active slip systems of (111)[1-1 0] and (1-1-1)[101] are obtained. Therefore, the (011)[2-1 1] rolling texture is calculated to transform to the (011)[100] Rex texture. It is concluded that the Goss Rex texture is linked with the brass rolling texture. The Goss orientation is stable with respect to plane strain compression and thermally stable (Section 5.1). Therefore, the Goss grains that survived during rolling are likely to act as nuclei during subsequent Rex and will grow at the expense of surrounding brass grains which are destined to change to assume the Goss orientation.

Figure 44 shows the rolling and Rex textures of Cu-1% P alloy sheet. The {110}<112> rolling texture changes to the (110)[001] texture after Rex. This is another example of the transition from the {110}<112> rolling texture to the {110}<001> Rex texture as explained in the Cu-16% Mn alloy.

6.3. {031}<100> recrystallization texture

In Section 5.4 the (123)[-6-3 4] rolling to (031)[100] Rex orientation transformation was discussed. Here we discuss the {123}<634> rolling to {031}<100> Rex orientation transformation. Figure 43b shows the shear strain rates as a function of strain for the (123)[6 3-4] crystal, which was calculated by the ε_{13} and ε_{23} relaxed strain rate sensitive model. The figure indicates that the S orientation is not stable with respect to the strain. Therefore, we calculate AMSD using the shear strain rates of 0.014, 0.01, 0.006 (0.007 was used in Section 5.4), and 0.003 at zero strain on the C, J, M, and B active slip systems (Figure 43b). AMSD is 0.014 [10-1] + 0.01 ×0.577 [110] + 0.006 × 0.577 [110] + 0.003 [0 1-1] = [0.023 0.0122 –0.017], where the factor 0.577 originates from the fact that the (1-1 1) and (1-1-1) slip planes share the [110] slip direction (Eq. 7). The [0.023 0.0122 –0.017] AMSD is parallel to the [0.7397 0.3924 –0.5467] unit vector. Following the method explained in Figure 6, we obtain Figure 45. Therefore, the (123)[6 3-4] rolling orientation is calculated to transform to the Rex texture as explained in Eq. 9. The calculated results are as follows:

$$\begin{pmatrix} 0.7397 & 0.3924 & -0.5467 \\ 0.0812 & 0.7545 & 0.6513 \\ 0.6684 & -0.5263 & 0.5263 \end{pmatrix}\begin{pmatrix} 1 \\ 2 \\ 3 \end{pmatrix}=\begin{pmatrix} -0.1156 \\ 3.5441 \\ 1.1947 \end{pmatrix}\quad \begin{pmatrix} 0.7397 & 0.3924 & -0.5467 \\ 0.0812 & 0.7545 & 0.6513 \\ 0.6684 & -0.5263 & 0.5263 \end{pmatrix}\begin{pmatrix} 6 \\ 3 \\ -4 \end{pmatrix}=\begin{pmatrix} 7.8 \\ 0.1455 \\ 0.3263 \end{pmatrix}$$

The calculated result means that rolled fcc metal with the (123)[6 3-4] orientation transforms to (-0.1156 3.5441 1.1947)[7.8 0.1455 0.3263] ≈ (-1 31 10)[54 1 2] after Rex. For polycrystalline metals, the {123}<634> deformation texture transforms to the {-0.1156 3.5441 1.1947}<7.8 0.1455 0.3263> ≈ {1 31 10}<54 1 2> Rex texture. The Rex texture is shown in Figure 46a. If the {-0.1156 3.5441 1.1947}<7.8 0.1455 0.3263> orientations are expressed as Gaussian peaks with scattering angle of 10°, the Rex texture is very well approximated by the {310}<001> texture as shown in Figure 46b. This texture is similar to Figure 27 which shows the Rex texture of the plane strain compressed {123}<412> crystal.

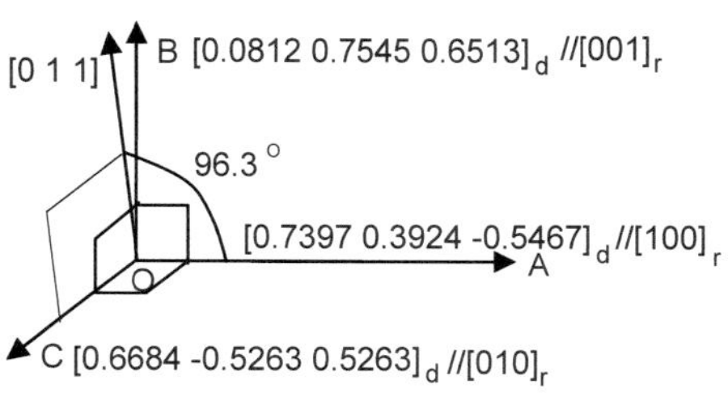

Figure 45. Orientation relationship between deformed and Rexed states [61].

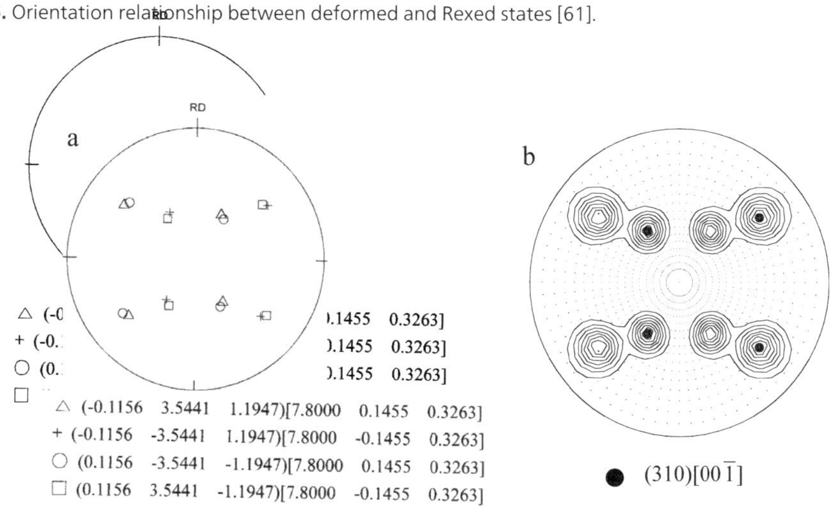

Figure 46. (a) (111) pole figure of {0.1156 3.5441 1.1947}<7.8 0.1455 0.3263> ≈ {1 31 10}<54 1 2>. (b) Sum of {0.1156 3.5441 1.1947}<7.8 0.1455 0.3263> expressed as Gaussian peaks with scattering angle of 10°. Calculated orientation can be approximated by {310}<001> [61].

It is noted that the highest density component in the deformation texture does not always dominate the Rex texture. All the components in the deformation texture are not in equal position to nucleate and grow the corresponding components in the Rex texture. The brass component has the highest density, but has lowest stored energy or the Taylor factor, while the copper component has the lowest density, but has the highest stored energy or the Taylor factor. If grains with the Goss or cube orientation survived during rolling, they must have undergone plane strain compression. They could undergo recovery and act as nuclei for Rex during annealing. This is the reason why the cube Rex texture could be obtained even though the copper component is the least in the deformation texture. When other conditions are the same, the higher relative density component in the deformation texture will give rise to the higher density in the corresponding component in the Rex texture, as shown in the highest relative copper component in the deformation texture yielding the highest cube component in the Rex texture in the Cu-4%Mn alloy among the three Cu-Mn alloys.

7. Plane-strain compressed {110}<001> bcc steel crystal

The Goss orientation {110}<001> in about 3% Si steel has been the subject of speculation due to its scientific and technological points of view. The grain oriented Si steel is made by hot rolling, cold rolling, followed by annealing. The Goss texture is formed near the sheet surface layer rolled in the α phase region at elevated temperatures. The friction between the sheet and rolls tends to increase with increasing temperature, and in turn increases the shear deformation and the Goss texture (Figure 47).

7.1. Thermal stability of Goss orientation formed by shear deformation

During hot rolling, Rex can take place, thereby the Goss orientation may change to a different orientation. Lee and Lee [64] obtained an IF steel specimen with only the shear texture by a multi- layer warm rolling and discussed the evolution of its Rex texture. The material used was a hot rolled 3.2 mm thick IF steel sheet. The hot-rolled sheet was cold-rolled to 1.1 mm in thickness in several passes. Four of the 1.1mm thick sheet were stacked, heated at 700 °C for 30 min and rolled by 70% in the ferrite region without lubrication. The rolled specimen was quenched into 25 °C water. Each layer was separated from the warm rolled sheet. In order to obtain a uniform shear texture, the surface layer was thinned from the inner surface to a half thickness by chemical polishing. The thinned surface and center layers were annealed at 750 °C for 1 h in Ar atmosphere.

The measured (110) pole figures and ODFs of the outer and inner surfaces of the 75% warm-rolled surface layer were similar. The similarity indicates that the texture of the layer is uniform. The texture was approximated by the Goss orientation plus minor {112}<111>. The center layer was similar to the typical texture of cold rolled steel sheet, RD//<110> and ND//{111}(Section 8). The surface texture could also be described as that which is obtained when the center layer texture is rotated through 35° about TD. The measured textures were similar to the calculated textures in Figure 47. The textures of the chemically thinned rolled surface layer and the center layer after annealing at 750 °C for 1 h showed that the texture of the surface layer was almost the same before and after annealing while the center layer underwent a texture change after annealing. Microstructures and hardness tests of the surface layer before and after annealing indicated Rex occurring after annealing [64].

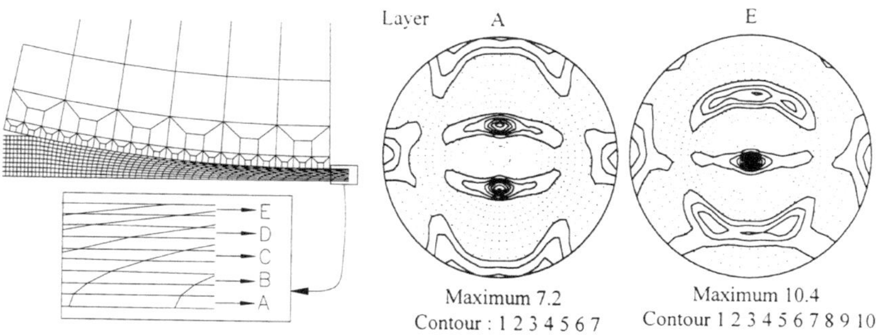

Figure 47. Deformed FEM meshes in rolling and calculated (110) pole figures of layers A and E. In FEM calculation, flow characteristics of IF steel $\sigma = 500\varepsilon^{0.256}$ MPa, roll diameter of 310 mm, initial sheet thickness of 3.4 mm, and reduction of 70% were used [63].

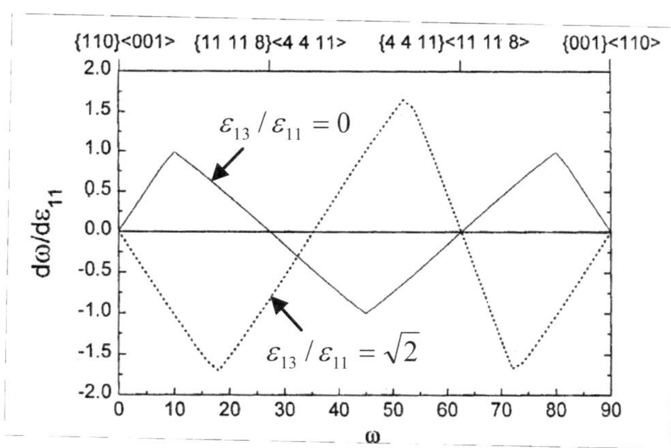

Figure 48. Rotation rate $d\omega/d\varepsilon_{11}$about TD//<110> with respect to ε_{11} for bcc crystal [64].

$d\varepsilon_{13}/d\varepsilon_{11}$	$\gamma^{(1)}$	$\gamma^{(2)}$	$\gamma^{(3)}$	$\gamma^{(4)}$
0.5	1.225	1.225	0.245	0.245
1.0	1.225	1.225	1.120	1.120
1.2	1.225	1.225	1.466	1.466
√2	1.225	1.225	1.837	1.837
1.5	1.225	1.225	1.986	1.986

Table 4. Shear strain on each slip system as a function of $d\varepsilon_{13}/d\varepsilon_{11}$ [64].

The unchanged texture in the surface layer after annealing can be explained based on SERM. AMSD is obtained from the slip systems activated during deformation. On the basis of the Taylor-Bishop-Hill theory, the (110)[001] orientation is calculated to be stable at $\varepsilon_{13}/\varepsilon_{11}=\sqrt{2}$ (Figure 48), and active slip systems for the (110)[001] crystal is calculated to be 1 (0-1 1)[111], 2 (-101)[111], 3 (110)[-111], and 4 (110)[1-1 1]. The shear strain on each slip system calculated as a function of $\varepsilon_{13}/\varepsilon_{11}$ is given in Table 4. The shear strains on slip systems 1 and 2 do not change, but those on slip systems 3 and 4 increase with increasing $\varepsilon_{13}/\varepsilon_{11}$. The slip systems 1 and 2 are effectively equivalent to the (-1-1 2)[111] system, and the slip systems 3 and 4 are equivalent to the (110)[002] system. AMSD may be parallel to $\gamma^{(1,2)}$[111]+$\gamma^{(3,4)}$[002] with $\gamma^{(1,2)}$ and $\gamma^{(3,4)}$ being shear strains on slip systems 1 and 2 and slip systems 3 and 4, respectively. Even though the (110)[001] orientation is stable at $\varepsilon_{13}/\varepsilon_{11}=\sqrt{2}$, in real rolling the $\varepsilon_{13}/\varepsilon_{11}$ value is small in the entrance region, increases very rapidly up to a maximum value just ahead of the neutral point, and then decreases in the exit region. The slip systems having the higher shear strain give dominant contribution to AMSD. As shown in Table 4, slip systems 3 and 4 are likely to give dominant contribution to AMSD. Therefore, AMSD is likely to be parallel to the [001] direction, which is also MYMD. Therefore, the Goss orientation is likely to remain unchanged after annealing in agreement with the experimental result (1st priority in Section 2). Another reason for the thermal stability of shear deformation textures is described in [65].

7.2. Plane-strain compressed {110}<001> bcc metals

The (110)[001] orientation of bcc metals is calculated to be metastable with respect to plane strain compression (Figure 48), with active slip systems being (-1 0-1)[-1-1 1], (1 0-1)[111], (0-1-1) [-1-1 1], and (0 1-1)[111], on which the shear strain rates are the same. It is noted that the slip directions are chosen to be at acute angles with the [001] direction (Section 2). The two slip directions, [-1-1 1] and [111], are on the (-110) plane, which can be a slip plane in bcc crystals. Therefore, AMSD is [-1-1 1] + [111] = [002] // [001]. This is also MYMD of iron. Since the AMSD is the same as MYMD, if the Goss oriented crystal survives the plane-strain compression, the Goss texture is likely to remain unchanged during annealing according to SERM (1st priority in Section 2).

7.3. Evolution of Goss recrystallization texture from {111}<112> rolling texture

The Goss orientation, which is not stable with respect to plane strain deformation, rotates toward the {111}<112> orientation forming a strong maximum [66]. The relaxed constraints Tayor model, in which shear strains parallel to RD may occur, causes the formation of the {111}<112> orientation [67]. The {111}<112> rolling component is known to lead to the Goss orientation after Rex [66, 68]. Dorner et al. [68] attributed the transition from the {111}<112> deformation texture to the Goss Rex texture to the fact that the Taylor factor (2.4) of the Goss grains is lower than that (3.7) of the {111}<112> matrix. Dorner et al. [69], in their study with 3.2% Si-steel single crystals, also found two types of Goss crystal volumes in 89 % cold-rolled specimen. Most of the Goss crystal regions are situated inside of shear bands. The Goss crystal volumes are also observed inside of microbands. These Goss crystals may act as nuclei because they are thermally stable (Section 7.2).

http://dx.doi.org/10.5772/54123

The evolution of the Goss orientation in the (111)[1 1-2] component, a {111}<112>, has been explained by SERM [70]. Slip systems of (-1-1 0)[-1 1-1], (-1-1 0)[1-1-1], (101)[1 1-1], and (011) [1 1-1] are calculated, by the relaxed constraints Taylor model, to be equally active in the (111) [1 1-2] crystal undergoing the plane strain compression. It is noted that the three slip directions are chosen to be at acute angles with RD [1 1-2] of the crystal. Taking the (101)[1 1-1] and (011) [1 1-1] slip systems sharing the same slip direction [1 1-1] into account, AMSD is [-1 1-1] + [1-1-1] + [1 1-1] = [1 1-3]. According to SERM, this AMSD [1 1-3] becomes parallel to MYMD, the <100> directions in bcc iron, in Rexed crystals. Other directional relationships between the matrix and Rexed crystal can be obtained from the 2nd priority in Section 2. Let one of the <100> directions be the [001] direction, then it must be on the (100), (010) or (110) plane, taking the symmetry condition into account. TD of the (111)[1 1-2] crystal is the [1-1 0] direction. These facts give rise to orientation relationship between the deformed and Rexed states (Figure 49). It is noted that the [1-1 0] direction is TD of both the deformed and Rexed states. It follows that the (111)[1 1-2] orientation becomes the (441)[1 1-8] orientation after Rex. The symmetry yields another equivalent orientation, (441)[-1-1 8]. The (110) pole figure of the {441}<118> orientation is shown in Figure 50a along with the Goss orientation {110}<001>. The {441}<118> orientation is deviated from the Goss orientation by 10°. If each {441}<118> orientation is represented by the Gauss type scattering with a half width angle of 12°, the calculated result is as shown in Figure 50b, which is in very good agreement with the measured data in Figure 50c, where the highest intensity poles are the same as those of the Goss orientation, even though it is not real Goss orientation. It is also interesting to note that the rotation angle between (111)[1 1-2] and (441)[1 1-8] about a common pole of [110] is calculated to be 25°and the rotation angle between (111)[1 1-2] and (110)[001] about a common pole of [110] is 35°. Thus the {111}<112> matrix can favor the growth of Goss-oriented crystals or nuclei, which are stable during annealing, if any, or may generate Goss-oriented nuclei, especially in polycrystalline materials.

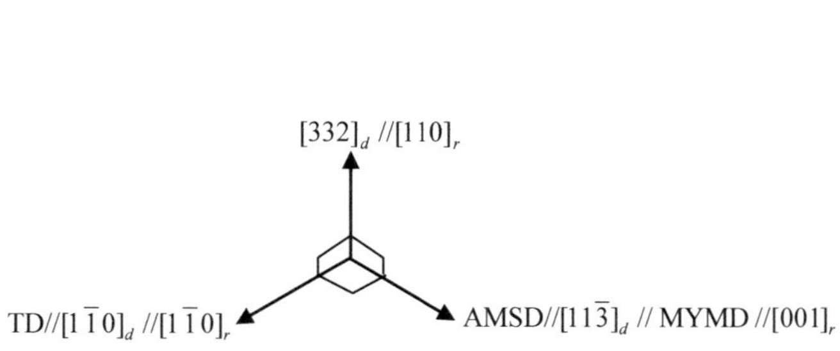

Figure 49. Orientation relationship between deformed ($_d$) and recrystallized ($_r$) states.

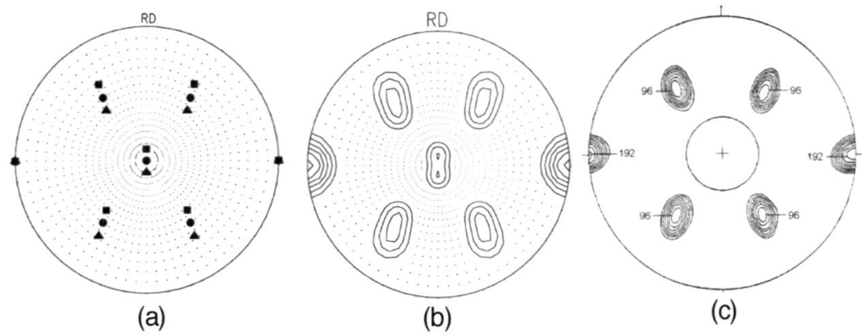

Figure 50. pole figures for (a) {110}<001> and {441}<118> orientations (●(110)[0 0-1], ■(441)[1 1-8], ▲(4 4-1)[-1-1-8]) [70], (b) Gaussian function of {441}<118> with half width of 12° and I_{max}=11 [70], and (c) Si steel specimen Rexed for 1 min at 980 °C [66].

8. Cold-rolled polycrystalline bcc metals

It is well known that the rolling texture of bcc Fe is characterized by the α fiber (<110>//RD) plus the γ-fiber (<111>//ND) and the rolling texture is replaced by the γ-fiber after Rex (Figure 51). This texture transformation will be discussed based on SERM. Figure 52 shows ODFs of 50, 80, and 95% cold-rolled IF steel sheets and their Rex textures, which indicate that the deformation textures are approximated by the α and γ fibers and the Rex texture by the γ fiber, as well known. As the deformation increases, peak type orientations tend to form. For the 80 and 95% cold rolled specimens, the {665}<110>, {558}<110>, and {001}<110> orientations develop as the main components. The {665}<110> and {558}<110> orientations may be approximated by the {111}<110> and {112}<110> orientations, respectively. The {001}<110> component is the principal component inherited from the hot band. It is stable and its intensity increases with deformation [72,73]. The Rex texture is approximated by the γ fiber whose main component is approximated by {111}<112>. The density of this orientation increased with increasing cold rolling reduction.

Figure 53 shows the orientation densities along the α and γ fibers for IF steel rolled by 80% and annealed at 695 °C. Up to 100 s, little change in the orientation density occurs, although appearance of the {111}<112> component in the γ fiber is apparent. For the specimen annealed for 200 s, the orientation density along the γ fiber is almost as high as that of the fully annealed one, while the density along α fiber decreases with increasing annealing time.

8.1. Recrystallization in γ fiber

We want to know if the {111}<112> Rex texture results from the {111}<110> deformation texture. The (111)[1-1 0] orientation is taken as an orientation representing the {111}<110> deformation texture. The (111)[1-1 0] orientation is calculated to be stable using the rate sensitive model with pancake relaxations (ε_{13} and ε_{23} are relaxed.). The active slip systems of the (111)[1-1 0] crystal are calculated to be (101)[1-1-1], (0-1-1)[1-1 1], (211)[1-1-1], and (-1-2-1)[1-1 1], on which the shear strain rates $d\gamma^{(k)} / d\varepsilon_{11}$ with respect to ε_{11} are 0.0612, 0.0612, 0.107 and 0.107, respectively. It is noted that the slip directions are chosen to be at acute angles with RD. The four slip systems can be effectively divided into the following two slip systems. 0.0612(101)[1-1-1] + 0.107(211)[1-1-1] // (0.2752 0.107 0.1682)[1-1-1] and 0.0612(0-1-1)[1-1 1] + 0.107(-1-2-1)[1-1 1] // (-0.107 -0.2752 -0.1682)[1-1 1]

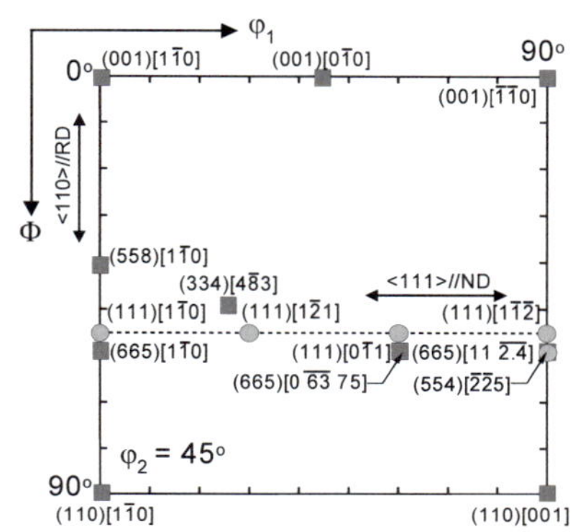

Figure 51. Section of φ_2 =45° in Euler space with locations of important orientations and fibers.

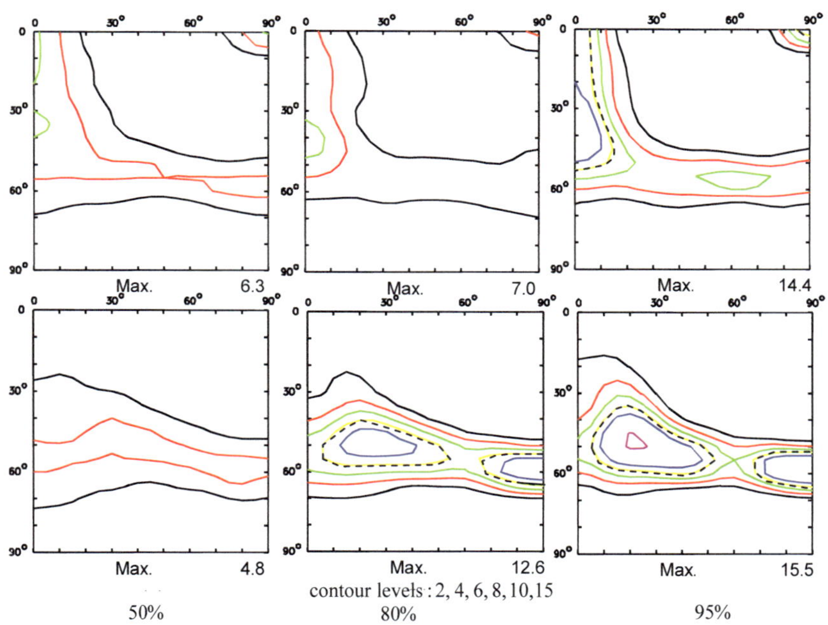

Figure 52. ODFs (φ_2= 45°) of 50, 80, and 95% rolled IF steel sheets (top) before and (bottom) after annealing at 695 °C for 1000 s [71]

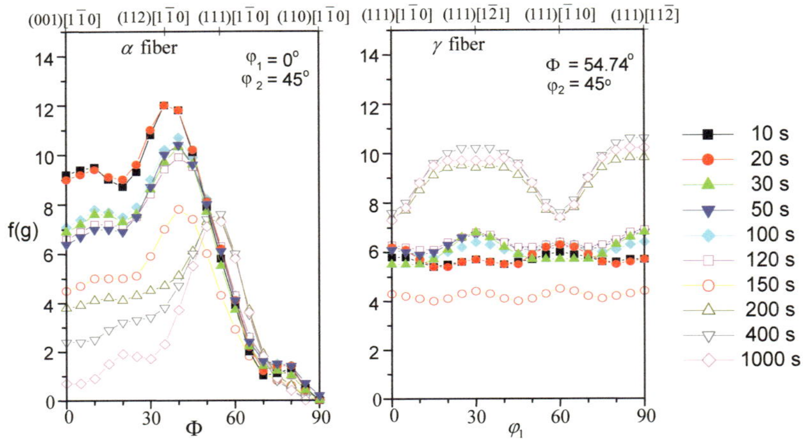

Figure 53. Orientation densities along α and γ fibers for IF steel sheets cold-rolled by 80% and subsequently annealed at 695°C for 10 to 1000 s [71].

These two slip systems are depicted as locating in the opposite sides of the rolling plane as shown in Figure 54a, and they are physically equivalent. They may not be activated homogeneously, even though they are equally activated macroscopically. In this case, AMSD is [1-1-1] or [1-1 1]. It should be mentioned that all active slip directions are not summed unlike fcc metals in which all slip directions are related to each other through associated slip planes. Figure 54b shows angular relationships among MYMD [100], ND [111], and RD [2-1-1] in the (111)[2-1-1] grains, whose orientation has been supposed to be the Rex texture of the (111)[1-1 0] rolling texture. It can be seen that [1-1 1] in Figure 54a is not parallel to [100] in Figure 54b. According to SERM, the {111}<110> rolling texture is not likely to link with the {111}<112> Rex texture.

Examining the experimental results more closely, the evolution of the {665}<1 1 2.4> Rex texture [(φ_1,Φ,φ_2)=(90°,59.4°,45°)] appears to be linked to the {665}<110> deformation texture [(φ_1,Φ,φ_2)=(0°,59.5°,45°) or (54.8°,58.7°,50°)]. Let the (665)[1-1 0] orientation be an ideal orientation representing the {665}<110> deformation texture. The (665)[1-1 0] orientation is calculated to be stable using the rate sensitive slip with pancake relaxations. Calculated active slip systems and their activities in the (665)[1-1 0] crystal are given in Table 5. Active slip directions are [1-1-1], [1-1 1], and [1 1-1]. It is noted that the [1-1-1] and [1-1 1] slip directions are chosen to be at acute angles with RD and physically equivalent. The [1 1-1] direction is normal to RD. The relationship between various directions is shown in Figure 55. AMSD is therefore (0.0536 + 0.0086 + 0.0843)[1-1-1] + (0.0368 + 0.0148 + 0.0148)[-1-1 1] = [0.0801 -0.2129 -0.0801] // [1 -2.658 -1] or (0.0536 + 0.0086 + 0.0843)[1-1 1] + (0.0368 + 0.0148 + 0.0148)[1 1-1] = [0.2129 - 0.0801 0.0801] // [2.658 -1 1]. The second term direction in AMSD is determined physically with reference to Figure 55a. For the first term direction of [1-1 1], the angle between the [1-1 1] direction and TD [-1-1 2.4] direction is less than 90°. Therefore, the angle between the [-1-1 2.4] direction and the second direction, [-1-1 1] or [1 1-1], should be larger than 90° and so the [1 1-1] direction is chosen.

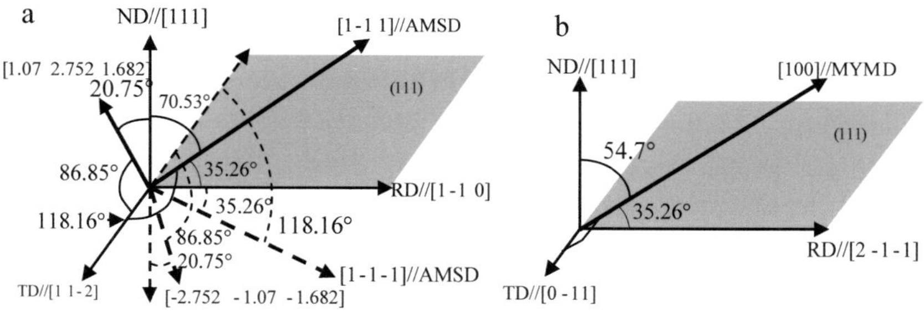

Figure 54. Angular relationships (a) among AMSD, ND, and RD in rolled (111)[1-1 0] crystal and. (b) among MYMD, ND, and RD in (111)[2-1-1] crystal.

Slip direction		1-1-1			1-1 1			1 1-1	
Slip plane	101	211	312	0-1-1	-1-2-1	-1-3-2	112	123	213
$\lvert d\gamma^{(k)}/d\varepsilon_{11}\rvert$, $d\varepsilon_{11}$=0.01	0.0536	0.0086	0.0843	0.0536	0.0086	0.0843	0.0368	0.0148	0.0148

Table 5. Shear strain rates on slip systems in plane strain compressed (665)[1-1 0] crystal calculated based on rate sensitive pancake model [71]

For the (665)[-1-1 2.4] orientation as an orientation representing the {665}<1 1 2.4> Rex texture, the angles among ND, TD, RD, and [001] are shown in Figure 55b. Comparison of Figures 55a and 55b shows that AMSD in the deformed specimen is almost parallel to [001], MYMD of iron, in the Rexed specimen. This is compatible with SERM. In other words, the transformation from the (665)[1-1 0] deformation orientation to the (665)[1-1 2.4] Rex orientation is compatible with SERM. The deformed matrix and Rexed grains share the [665] ND (2nd priority in Section 2). Taking symmetry into account, the {665}<110> rolling texture is calculated to transform to the {665}<1 1 2.4> Rex texture, in agreement with the experimental result. This transformation relationship may be approximated by the transformation from the {111}<110> deformation texture to the {111}<112> Rex texture.

The {111}<112> orientation is not stable with respect to plane-strain compression. However, if the orientation survived during rolling, grains with the orientation must have been plane-strain compressed. The plane-strain compressed (111)[1 1-2] crystal is calculated, by the full constrains model, to have slip systems of (110)[1-1-1] and (110) [-1 1-1], whose activities are the same, if we consider slip systems on one side of the rolling plane. It is noted that the slip directions are at acute angles to RD and on the same slip plane. AMSD is calculated to be [1-1-1] + [-1 1-1] = [0 0-2], which is parallel to a MYMD (Figure 56a). Therefore, the {111}<112> deformation texture is likely to remain unchanged during annealing (1st priority in Section 2). The {111}<112> grains may act as nuclei.

Yoshinaga et al. [74] observed that a {111}<112> nucleation texture was strongly formed in 65% rolled iron electrodeposit with a weak {111}<112> texture, resulting in the {111}<112> Rex texture, whereas a {111}<110> nucleation texture was formed in 80% rolled electrodeposit having a strong{111}<112> texture, resulting in the {111}<110> Rex texture. They noted the importance of the nucleation texture in the Rex texture formation and attributed to the {111}<110> Rex texturing in the 80% rolled sheet to higher mobility of grain boundaries between the {111}<110> grains and the{111}<112> deformed matrix. They did not account for the differences in nucleation texture between the 65% and 80% rolled sheets.

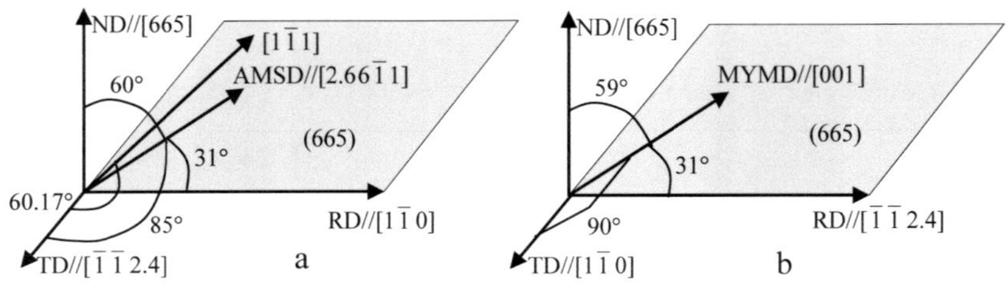

Figure 55. (a) AMSD in (665)[1-1 0] rolled crystal; (b) MYMD in (665)[-1-1 2.4] Rexed crystal [71].

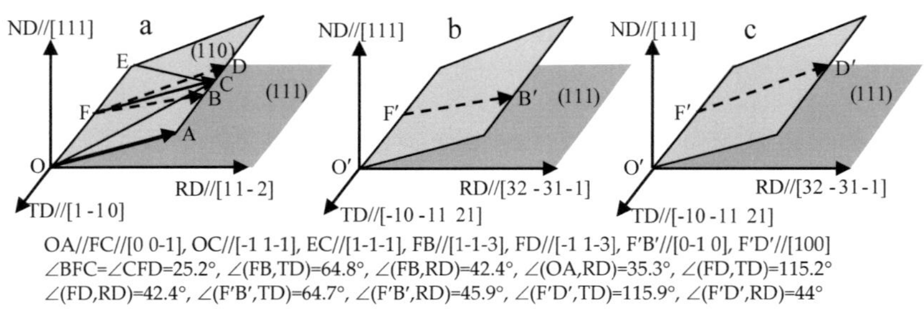

Figure 56. Explanation of {111}<112> rolling texture changing to {111}<112> or {111}<110> after Rex. F′B′ and F′D′ are MYMDs in Rexed state and are almost parallel to AMSDs, FB and FD, in deformed state, respectively [5].

According to SERM, the {111}<112> deformation texture is likely to remain unchanged after Rex because AMSD in the deformed state is parallel to MYMD, as mentioned above. If the activities of the slip systems of (110)[-1 1-1] and (110)[1-1-1] in Figure 56a are well balanced, MYMD becomes [0 0-1]. This may be the case in the 65% rolled sheet. As the rolling reduction increases, the balance can be broken. When the (110)[1-1-1] slip system is two times more active than the (110)[-1 1-1] system, AMSD is parallel to the [1-1-3] direction (2[1-1-1] + [-1 1-1] = [1-1 -3]). Similarly if the (110)[-1 1-1] system is two times more active than the (110)[1-1-1] slip system, AMSD is parallel to the [-1 1-3] direction. These directions are shown in Figure 56a. If one of the two slips takes place in one layer and another one does in another layer and so on, as in ε_{23} relaxation, the rolling texture macroscopically appears the same as in the balanced slip. When these AMSDs are made to be parallel to MYMD, one of the <100> directions, as shown in Figures 56b and 56c, we come to the result that the {111}<112> rolling texture is linked with the {111}<32 31 1> Rex texture that is approximated by the {111}<110> texture.

http://dx.doi.org/10.5772/54123

As the Rexed {665}<1 1 2.4> and {111}<112> grains grow, they are likely to meet the α fiber grains. If the Rexed grains are not in a favorable orientation relationship with the α fiber grains, they may not grow at the expense of the α fiber grains. This is discussed in the next section.

8.2. Recrystallization in α fiber grains

Park et al. [75,76] discussed orientation relationships between the rolling and Rex textures in rolled IF steel sheets based on both SERM and the conventional OG, in which the α-fiber rolling texture was assumed to transform to the γ-fiber Rex texture. The {001}<110> and {112}<110>rolling orientations, which are main components in the α-fiber texture, are calculated to be stable using the full constraints Taylor model. For the (001)[110] orientation as an orientation representing the {001}<110> orientation, active slip systems are calculated to be (1 1-2)[111] and (112)[1 1-1] from the full constraints Taylor model. Therefore, AMSD can be [111] or [1 1-1]. Figure 57a shows the angular relation between the [111] direction and the (001)[110] specimen axes. Figure 57b shows the angular relation between the [001] direction, which is a MYMD, and the axes of the specimen with the (111)[-1-1 2] Rex texture. It can be seen from Figure 57 that AMSD in the deformed state is parallel to MYMD in the Rexed state and TD is shared by the deformed and Rexed states (2nd priority in Section 2). Taking the symmetry into account, the {001}<110> deformation texture is calculated to transform into the {111}<112> Rex texture. This transformation was observed in the experimental results (Figures 52 and 53, [75], [77]). It is often addressed that the {001}<110> orientation is difficult to be Rexed. It may be attributed to the fact that the orientation has a low Taylor factor [66].

For the (558)[1-1 0] orientation as an orientation representing the {558}<110> orientation, active slip systems are calculated to be 2.283(101)[1-1-1], (101)[-1-1 1], 2.283(0-1-1)[1-1 1], and (0-1-1)[1 1-1] from the full constraints Taylor model, where the factor 2.283 in front of slip systems indicates that their activities are 2.283 times higher than other slip systems [66]. The slip systems reduce effectively to (101)[1-2.56-1] and (0-1-1)[2.56-1 1]. Therefore, AMSD becomes [1-2.56-1] or [2.56 -1 1]. Figure 58 shows that the [1-2.56-1] direction in the (558)[1-10] crystal is nearly parallel to MYMD in the Rexed state, and the [101] direction is shared by the deformed and Rexed states (2nd priority in Section 2). Taking the symmetry into account, the {558}<110> deformation texture is calculated to transform into the {334}<483> Rex texture. This transformation relation was observed in the experimental result. The {334}<483> orientation is away from the {111}<112> orientation. An exact correspondence between the (112)[-110] deformation and (-2.45 2 –2.45)[1 2.45 1] Rex orientations can be seen in Figure 59.

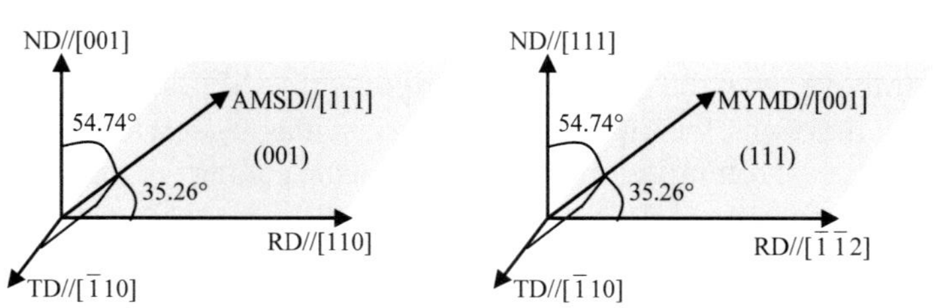

Figure 57. AMSD in (001)[110] rolled crystal and MYMD in (111)[-1-1 2] Rexed crystal [71].

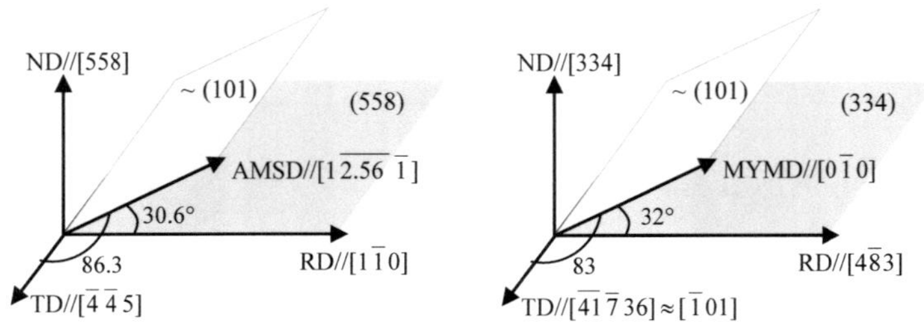

Figure 58. AMSD in (558)[1-1 0] rolled crystal and MYMD in (334)[4-8 3] Rexed crystal [71].

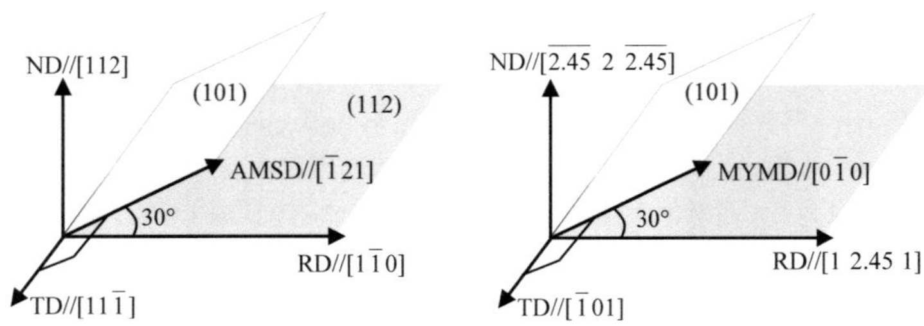

Figure 59. AMSD in (112)[-110] rolled crystal and MYMD in (-2.45 2 -2.45)[1 2.45 1] Rexed crystal [71].

Park et al. [75] studied relationships between rolling and Rex textures of IF steel. When the {112}<110>, {225}<110>, and {112}<110> components had the highest density in cold rolling texture, the {567}<943>, {223}<472>, and {554}<225> components had the highest density in Rex texture, respectively. Rolling and Rex textures of low carbon steel (C in solution), and Fe-16%Cr and Fe-3%Si steels indicate that the strong rolling texture components {001}<110> and {112}<110> have an effect on the evolution of a very strong Rex texture {111}<112> [77].

Park et al. [76] investigated the macrotexture changes in 75% cold-rolled IF steel with annealing time at 650ºC along the α-fiber. The cold rolling texture showed the development of the α fiber as typical in bcc steels. The orientation densities of the α-fiber increased slightly after annealing for 300 s. This is a well-known recovery phenomenon. A part of the α-fiber, near {114}<110>, substantially decreased after annealing for 1000 s. EBSD analysis indicated that the {556}<175> Rex component was formed at the expense of the {114}<110> deformation component. This texture transformation could be explained by SERM. Relationships between various rolling and Rex textures are summarized in Table 6.

These results can be explained based on SERM [4,75,76]. Figure 60 shows drawings relating the rolling texture components to the Rex texture components. AMSDs can be easily obtained by choosing the <111> directions, the slip directions in bcc metals, closest to 45° to the compression axis without calculation of rolling deformation. For the cold rolling texture (001)[110], TD is calculated by the vector product of [110] and [001] to be [-110]. The <111> directions closest to 45° with the [001] compression axis are [-111], [-1-1 1], and [111]. The [-111] direction is likely to contribute to spread of the width of sheets. Therefore, slip along the [-111] direction is unlikely. The effective slip planes are likely to be parallel to TD and contain the [111] and [-1-1 1] slip directions. The planes are those normal to the vector product of the [-110] TD and the [-1-1 1] and/or

[111] directions. They are calculated to be the (112) and/or (1 1-2) planes. The related slip systems are therefore (112)[-1-1 1] and (1 1-2)[111]. These systems are physically equivalent. Therefore, it is sufficient to choose one of them. Let us choose the [111] direction. The [111] direction and other related directions and planes are shown in Figure 60a. The [111] direction is on ND-RD plane. Therefore, it is likely to be AMSD.

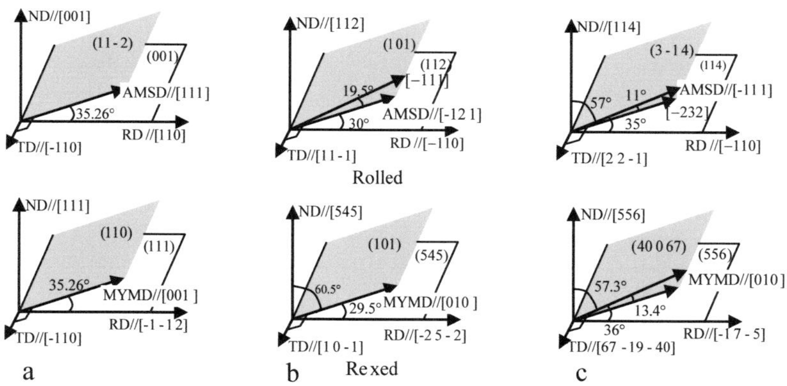

Figure 60. Correspondence between AMSD in rolled state (top), MYMD in Rexed state (bottom) in orientation relationships between (a) (001)[110] roll ↔ (111)[-1-1 2] Rex, (b) (112)[-110] roll ↔ (545)[-2 5-2] Rex, (c) (114)[-110] roll ↔ (556)[-1 7-5] Rex.

If the [111] direction in the deformed state is set to be parallel to MYMD [001] in the Rexed state, the (111) plane becomes parallel to the rolling plane and the [-1-1 2] direction becomes parallel to RD in the Rexed state, giving rise to the (111)[-1-1 2] Rex texture as shown in Figure 60a. This result is the same as that obtained based on the full-constraints Taylor model (Figure 57). Therefore, the {001}<110> may be responsible for the measured {111}<112> Rex texture. It is noted that the [-110] TD is shared by both the deformed and Rexed states (2nd priority in Section 2). It is also noted that the angle between AMSD and RD is about 30° which is the usually observed angle between the shear band and RD.

Other examples in Figure 60 are self-explainable. In all the examples except Figure 60d, the <110> directions are shared by the deformed and Rexed states. In fact, the Rex textures in Figure 60b and 60c are very similar. This is the reason why there exists an angular relation between the deformed and Rexed states about the <110> axes (Table 6). This has often been interpreted to be associated with CSL boundaries. However, there is no consistency in the CSL boundaries. Anyhow the high density orientations along the α fiber change to near {111}<112> orientations on Rex.

As the Rexed γ fiber grains grow, they are likely to meet the α fiber grains. Main components in α fiber including the {112}<110> orientation are predicted to tend to change to near {111}<112> orientations according to SERM. Therefore, the {111}<112> Rexed grains will grow at the expense of the α fiber grains with little disturbance of orientation. It is interesting to note that SERM can satisfy the relation between the deformation and Rex textures in the nucleation and growth stages. The two prominent components, (334)[4-8 3] and (554)[-2-2 5], in the Rex texture are related to the (558)[1-1 0] and (112)[1-1 0] components in the rolling texture, respectively.

Rolling texture Component	Rex texure Component	RD↔AMSD (°)	RD↔MYMD (°)	Observed OR	Coincidence site lattice relation
{112}<110>	{567}<943>	30	29	30°<110>	Σ19a(26.5°<110>)
{225}<110>	{223}<472>	35.3	32.6	25°<110>	Σ19a(26.5°<110>)
{112}<110>	{554}<225>	30	29.5	35°<110>	Σ9(38.9°<110>)
{001}<110>	{111}<112>	35.3	35.3	55°<110>	Σ11(50.5°<110>)
{114}<110>	{556}<175>	35	36		
{558}<110>	{334}<483>	30.6	32	29°<110>	Σ19a (26.5°<110>

Table 6. Orientation relationships (OR) between major components which dominate rolling and Rex textures.

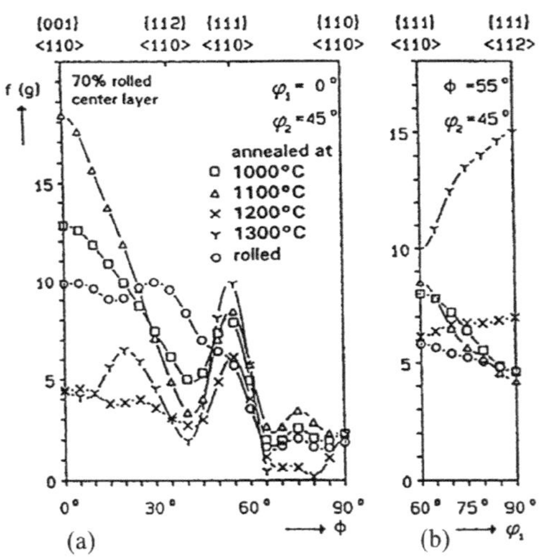

Figure 61. Orientation densities along α- and β- fibers for 70% rolled and annealed Ta [78].

8.3. Recrystallization in non-ferrous bcc metals

The texture evolution in Ta after 70% rolling and subsequent annealing at various temperatures is shown in Figure 61 [78]. The rolling texture of Ta is characterized by a partial α-fiber extending from {001}<110> to {111}<110> and a complete γ-fiber {111}<uvw>. The major deformation texture components are {112}<110> and {001}<110> as in steel. MYMD of Ta is <100> (A>0 in Table 6), the development of the Rex texture is expected to be similar to that in steel. It can be seen that an enhancement of {001}<110> due to recovery and a strong decrease in {001}<110> to {112}<110> accompanied by a strong increase in γ-fiber {111}<112> and/or {554}<225> due to Rex. The Rex behavior is readily understood from Figure 60 [4, 75].

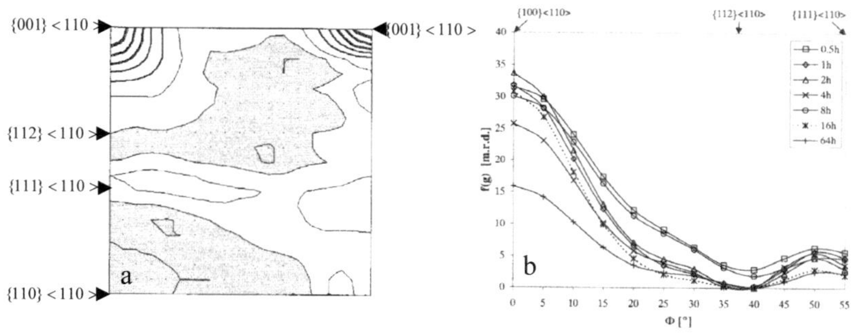

Figure 62. (a) φ_2 =45° ODF section of center of Mo sheet annealed for 1h at 925°C and (b) orientation density along α-fiber in central zone of Mo sheet annealed at 850°C [79].

The deformation texture of rolled Mo sheets was characterized by a weak γ-fiber and α-fiber with a strong {100}<110> component [79]. Full Rex does not change the rolling texture but reduces its intensity (Figure 62). This result is compatible with SERM considering that the <111> directions are not only slip directions, which is approximately AMSD (Figure 60a, top), but also MYMD of Mo (A<0 in Table 7) (1st priority in Section 2). The decrease in orientation density during annealing may be attributed to A being close to unity.

Since the slip systems of W are {112}<111> [80], it is predicted that the {001}<110> component dominates the rolling texture as shown in Figure 60a. Figure 63a shows the rolling texture which is dominated by the {100}<011> component as predicted. The deformation texture is approximately randomized after Rex (Figure 63b). This is compatible with SERM because W is almost isotropic in its elastic properties (A≈1 in Table 7).

Material	Temp., K	f S_{11}	f S_{44}	f S_{12}	A	Reference
Ta	1300	8.137	13.966	-3.150	1.6164	82
	1400	8.297	14.025	-3.224	1.6429	
	1500	8.408	14.184	-3.274	1.6472	
	1600	8.576	14.409	-3.357	1.6563	
Mo	273	2.607	9.158	-0.622	0.7052	83
	373	2.655	9.242	-0.682	0.7092	
	973	3.010	9.823	-0.833	0.7824	
W	100	2.398	6.158	-0.665	0.9948	84
	297	2.454	6.218	-0.690	1.0113	85
	973	2.711	6.553	-0.798	1.0710	85
	2073	3.509	7.375	-1.160	1.2662	85

Table 7. S_{ij} (GPa^{-1}) and $A=2(S_{11}-S_{12})/S_{44}$ for Ta, Mo, and W. f =1000

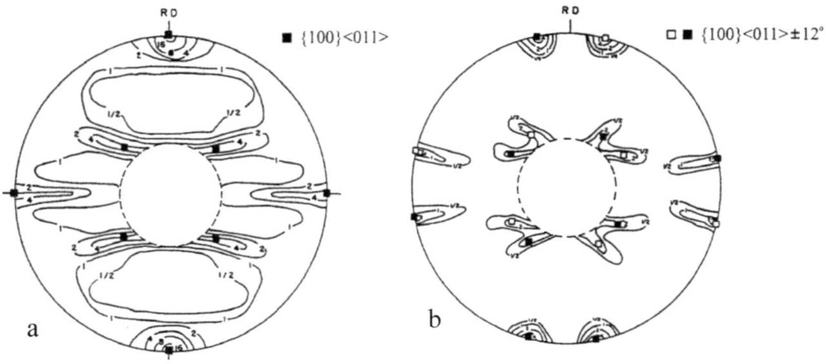

Figure 63. Pole figure of W sheet (a) after cold rolling by 96% and (b) subsequent annealing at 2000 °C for 30 min [81]. Max intensity: (a) >16 and (b) >4. Contour levels: 1/2, 1, 2, 4, 8, 16.

9. Conclusion

The Rex textures of freestanding electro- and vapor-deposits of metals and heavily deformed metals and alloys whose stored energies due to dislocations constitute the main driving forces for Rex can be determined such that AMSDs in the fabricated matrix can be along MYMDs in Rexed grains or nuclei, and by other conditions which can maximize the strain-energy release in the system. The strain-energy-release-maximization theory can explain the following results.

1. The <100>, <111> and <110> electro- and vapor-deposition textures of Cu, Ni, and Ag transform to the <100>, <100>, and <√3 1 0> textures, respectively, after Rex.

2. The <100> and <111> deposition textures of Cr remain unchanged after Rex.

3. The <111>+<100> drawing textures of uniaxially drawn Ag, Al, Cu, and Au wires change to the <100>textures after Rex.

4. Channel-die compressed {110}<001> Al single crystal keeps its {110}<001>deformation texture after Rex.

5. The {135}<2 1-1> Al sheet obtained by channel-die compression of Al crystals of {123}<412> orientations by 90% develops {-0.0062 0.2781 0.9606}<0.9907 0.1322 -0.0319> ≈ {0 1 3.5}<31 4 -1> after Rex.

6. An Al crystal of {112}<111> obtained by channel-die compression of a (001)[110] Al single crystal develops {001}<√610> after Rex.

7. The (123)[-6-3 4] + (321)[-436] + {112}<111> Cu sheet obtained after reversible rolling of a (123)[-6-3 4] Cu single crystal by 99.5% under oil lubrication develops the Rex texture of major {001} <100> + minor (0 3-1)[100] orientations. The {001}<100> and (0 3-1)[100]

components are calculated to result from the {112}<111> and (123)[-6-3 4] components in the deformation textures, respectively.

8. The {011}<211>, {112}<111>, and {123}<634> components in the rolling texture of cold-rolled polycrystalline fcc metals and alloys with medium to high stacking fault energy are respectively linked with the {011}<100>, {100}<001>, and {031}<100> component in the Rex texture.

9. The {111}<112> bcc crystal undergoing plane strain rolling can develop three different Rex textures of {441}<118>≈ {110}<001>, {111}<112>, and ~{111}<110> depending on local slip systems and their activities in the same gloval deformation.

10. The {665}<110>, {001}<110>, {558}<110>, {112}<110>, {114}<110> components in the rolling texture of steel are respectively linked with the {665}<1 1 2.4>, {111}<112>, {334}<483>, {545}<252>, {556}<175> components in the Rex texture.

11. The rolling and Rex textures of Ta are similar to those of steel.

12. Full Rex of Mo does not change the rolling texture but reduces its intensity.

13. The rolling texture of W transforms to a texure which can be approximated by random orientation distribution after Rex.

Acknowledgements

This study was supported by a grant (0592-20120019) from POSCO and a grant(0417-20110114) from Center for Iron & Steel Research of RIAM in Seoul National University.

Author details

Dong Nyung Lee and Heung Nam Han

Department of Materials Science and Engineering, Seoul National University, Seoul, Republic of Korea

References

[1] Burgers WG, Louwerse PC. Über den Zusammenhang zwischen Deformationsvorgang und Rekristallisationstextur bei Aluminium. Zeitschrift für Physik 1931; 61 605-678.

[2] Barrett CS. Recrystallization texture in aluminum after compression. Trans AIME 1940; 137 128-145.

[3] Gottstein G. Evolution of recrystallization texture- classical approaches and recent advances. Materials Science Forum 2002; 408-412 1-24.

[4] Lee DN. The evolution of recrystallization textures from deformation textures. Scripta Metallurgica et Materialia 1995;32 1689-1694.

[5] Lee DN. Relationship between deformation and recrystallization textures. Philosophical Magazine 2005;85 297-322.

[6] Lee DN. Currrent understanding of annealing texture evolution in thin films and interconnects. Zeitschrift für Metallkunde 2005;96 259-268.

[7] Lee DN. Recrystallization-texture theories in light of strain-energy-release-maximization. Materials Science Forum 2010;638-642 182-189.

[8] Lee DN. Strain energy release maximization model for evolution of recrystallization textures. International Journal of Mechanical Sciences 2000;42 1645-1678.

[9] Zehetbauer M. Cold work hardening in stages IV and V of F.C.C. metals—II. Model fits and physical results. Acta Metallurgica et Materialia 1993;41 589-599.

[10] Lee DN. A stability criterion for deformation and deposition textures of metals during annealing. Journal of Materials Processing Technology 2001;117 307-310.

[11] Lee DN. Maximum energy release theory for recrystallization textures. Metals and Materials 1996; 2 121-131.

[12] Lee Y-S, Lee DN. Characterization of dislocations in copper electrodeposits. Journal of Materials Science 2000;35 6161-6168.

[13] Lee DN, Kang S, Yang J. Relationship between initial and recrystallization textures of copper electrodeposits.Plating and Surface Finishing 1995;82(3) 76-79.

[14] Choi J-H, Kang S, Lee DN. Relationship between deposition and recrystallization textures of copper and chromium electrodeposits. Journal of Materials Science 2000;35 4055-4066.

[15] Yang JS, Lee DN. The deposition and recrystallization textures of copper electrodeposits obtained from a copper cyanide bath. Metals and Materials 1999;5 465-470.

[16] Lee DN, Han HN. Orientation relationship between precipitates and their parent phases in steels at low transformation temperatures. Journal of Solid Mechanics and Materials Engineering 2012;6(5) 323-338.

[17] Chang YA, Himmel L. Temperature dependence of the elastic constants of Cu, Ag, and Au above room temperature. Journal of Applied Physics 1966;37 3567-3572.

[18] Kang SY, Lee DN. Recrstallization texture of a copper electrdeposit with the <111> and <110> duplex orientation. Materials Science Forum 2002;408-412 895-900.

[19] Lee DN, Kim YK. In: Kanematsu H. (ed) Variations of texture, surface morphology and mechanical property of copper foils with electrolysis conditions. Proceedings of the 2nd

Asian Metal Finishing Forum, 1-3 June 1985, Tokyo, Japan: The Metal Finishing Society of Japan; 1985. pp. 130-133.

[20] Alers GA, Karbon JA. Elastic moduli of the lead-thallium alloys. Journal of Applied Physics 1966;37 4252-4255.

[21] Kim I, Lee SK. Initial and recrystallization textures of Ni electrodeposits. Textures and Microstructures 2000;34 159-169.

[22] Nam H-S, Lee DN. Recrystallization textures of silver electrodeposits. Journal of the Electrochemical Society 1999;146 (9), 3300-3308.

[23] Bolef DI, de Klerk J. Anomalies in the elastic constants and thermal expansion of chromium single crystals. Physical Review 1963;129 1063-1067.

[24] Patten JW, McClanahan ED, Johnston JW. Room-temperature recrystallization in thick bias-sputtered copper deposits. Journal of Applied Physics 1971;42(11) 4371-4377.

[25] Greiser J, Mullner P, Arzt E. Abnormal growth of giant grains in silver thin films. Acta materialia 2001;49 1041-1050.

[26] Lee DN. Textures and microstructures of vapor deposits. Journal of Materials Science 1999;34 2575-2582.

[27] Carel R, Thompson CV, Frost HJ. Computer simulation of strain energy effects vs surface and interface energy effects on grain growth in thin films. Acta materialia 1996;44 2479-2494.

[28] Lee DN. Texture development in thin films. Materials Science Forum 2002;408-412 75-94.

[29] Hibbard Jr WR. Deformation texture of drawn face centered cubic metal wires. Transactions of AIME 1950;77 581-585

[30] Inoue H, Nakazu N, Yamamoto H. Development of recrystallization texture in drawn aluminum wire. In: Nagashima S. (ed) Proceedings ICOTOM 6, 28 September-3 October 1981, Tokyo, Japan: The Iron and Steel Institute of Japan; 1981. P. 591.

[31] Park H, Lee DN. Effects of shear strain and drawing pass on the texture development in copper wire. Materials Science Forum 2002;408-412 637-642.

[32] Shin HJ, Jeong H-T, Lee DN. Deformation and annealing textures of silver wire. Materials Science and Engineering 2000;A279 244-253.

[33] Ahlborn H, Wassermann G. Einfluss von Verformungsgrad und Temperatur auf die Textur von Silberdrahten. Zeitschrift für Metallkunde 1963;54 1-6.

[34] Lee DN, Chung YH, Shin MC. Preferred orientation in extruded aluminum alloy rod. Scripta Metallurgica 1983;17 339-343.

[35] Park H, Lee DN. The evolution of annealing textures in 90 pct drawn copper wire, Metallurgical and Materials Transactions 2003;34A 531-541.

[36] Cho J-H, Cho J-S, Moon J-T, Lee J, Cho YH, Kim YW, Rollett AD, Oh KH. Recrystallization and grain growth of cold-drawn gold bonding wire. Metallurgical and Materials Transactions A 2003;34A 1113-1125.

[37] Ferry M, Humphreys FJ. Discontinuous subgrain growth in deformed and annealed {110}<001> aluminium single crystals. Acta Materialia 1996;44 1293-1308.

[38] Lee DN. A stability criterion for deformation and deposition textures of metals during annealing. Journal of Materials Processing Technology 2001;117 307-310.

[39] Sutton PM. The Variation of the elastic constants of crystalline aluminum with temperature between 63°K and 773°K. Physical Review 1953;91 816-821.

[40] Blicharski M, Liu J, Hu H. Annealing of aluminum bicrystals with S orientations deformed by channel die compression. Acta Materialia 1995;43 3125-3138.

[41] Lee DN, Jeong H-T. Recrystallization textures of aluminum bycrystals with S orieintation deformed by channel die compression. Materials Science and Engineering 1999;A269 49-58

[42] Butler Jr JF, Blickarski M, Hu H. The formation of dislocation structure and nucleation of recrystallized grains in an aluminum single crystal.Textures and Microstructures 1991;14-18 611-616.

[43] Lee DN (1996). Recrystallization texture of plane strain compressed aluminum single crystal. Texturures and Microstructures 1996;26-37 361-367

[44] Kamijo T, Fujiwara A, Yoneda Y, Fukutomi H. Formation of cube texture in copper single crystals.Acta Metallurgica and Materialia 1991;39 1947-1952.

[45] Lee DN, Shin H-J. Recrystallization texture of (123)[-6-3 4] copper single crystal cold rolled copper sheet. Materials Science Forum 2003;426-432 83-90.

[46] Necker CT, Doherty RD, Rollett AD. Quantitative measurement of the development of recrystallization texture in OFE copper. Textures and Microstructures 1991:14-18 635-640.

[47] Engler O. Recrystallization textures in copper-manganese alloys. Acta Materialia 2001;49 1237-1247.

[48] Koizumi M, Saitou T, Inagaki H. Effect of Mg content of the development of rolling and recrystallization textures in Al-Mg alloys. In Szpunar JA. (ed.) Proceedings of the Twelfth International Conference on Textures of Materials Vol.2, 848-853, Montreal, 9-13 August 1999, Ottawa: NRC Research Press; 1999.

[49] Huh MY, Cho YS, Engler O. Effect of lubrication on the evolution of microstructure and texture during rolling and recrystallization of copper. Materials Science and Engineering 1998;A247 152-164.

[50] Ryu JH, Lee DN. The effect of precipitation on the evolution of recrystallization texture in AA8011 aluminum alloy sheet. Materials Science and Engineering 2002;A336 225-232.

[51] Caleyo F, Baudin T, Penelle R, Venegas V. EBSD study the development of cube recrystallization texture in Fe-50%Ni. Scripta Materialia 2001;45 413-420.

[52] Lee DN, Hong S-H. (2001). Recrystallization textures of plane strain rolled polycrystalline aluminum alloys and copper, In: Recrystallization and Grain Growth, G. Gottstein, D.A. Molodov, (Ed.), 1349-1354, Springer, Berlin.

[53] Godfrey A, Juul Jensen D, Hansen N. Measurement of orientation dependent stored energy of deformation on a local scale. In Gottstein G, Molodov DA (eds.) Recrystallization and Grain Growth, Aachen, 27-31 August 2001, Springer, Berlin, pp843-848.

[54] Hong S-H, Lee DN (2003). The evolution of the cube recrystallization texture in cold rolled copper sheets. Materials Science and Engineering 2003;A351 133-147.

[55] Bauer RE, Mecking H, Lücke K. Textures of copper single crystals after rolling at room temperature. Materials Science and Engineering 1977;27 163-180.

[56] Honeff H, Mecking H. A method for the determination of the active slip systems and orientation changes during single crystal deformation. In: Gottstein G, Lücke K. (eds.) Textures of Materials Vol. 1: proceedings of ICOTOM5, Aachen, Springer-Verlag, Berlin; 1978. P.265.

[57] Van Houtte P. Adaptation of the Taylor theory to the typical substructure of some cold rolled fcc metals. In: Nagashima S. (ed.) Textures of Materials Vol. 1: proceedings of ICOTOM6, The Iron and Steel Institute of Japan, Tokyo; 1981. p. 248.

[58] Steffens T, Schwink C, Korner A, Karnthaler HP. Transmission electron microscopy sudy of the stacking-fault energy and dislocation structure in copper-manganese alloys. Philosophical Magazine 1987;A56(2) 161-173.

[59] Engler O. Deformation and Texture of copper-manganese alloys. Acta materialia 2000;48 4827-4840.

[60] Engler O. Recrystallization textures in copper-manganese alloys. Acta materialia 2001;49 1237-1247.

[61] Lee DN, Shin H-J, Hong S-H. The evolution of the cube, rotated cube and Goss recrystallization textures in rolled copper and Cu-Mn alloys. Key Engineering Materials 2003;233-236 515-520.

[62] Gottstein G. Physicalische Grundlagen der Materialkunde, Springer-Verlag; 1998.

[63] Lee DN, Jeong H-T. The evolution of the goss texture in silicon steel. In: Imam AM, DeNale R, Hanada S, Zhong Z, Lee DN. (eds.) Advanced Materials and Processing: proceedings of the third pacific rim international conference on advanced materials and processing, 12-16 July 1998, Hawaii, Pennsylvania: TMS; 1998.

[64] Lee SH, Lee DN. Shear rolling and recrystallization textures of interstitial-free steel sheet. Materials Science and Engineering 1998;A249 84-90.

[65] Lee DN, Kim K-H. Effects of asymmetric rolling parameters on texture development in aluminum sheets. In: Demer MY. (ed.) Innovations in Processing and Manufacturing of Sheet Materials: proceedings of the second gloval symposium on innovation in materials processing and manufacturing: sheet materials, 11-15 February 2001, New Orleans, Louisiana. Pennsylvania: TMS; 2001.

[66] Dunn CG. Cold-rolled and primary recrystallization textures in cold-rolled single crystals of silicon iron. Acta Metallurgica 1954;2 173-183.

[67] Raabe D, Lücke K. Rolling and annealing textures of bcc metals. Materials Science Forum 1994;157-162 597-610.

[68] Dorner D, Lahn L, Zaefferer S. Investigation of the primary recrystallization microstructure of cold rolled and annealed Fe 3% Si single crystals with Goss orientation, Materials Science Forum 2004;467-470 129-134.

[69] Dorner D, Lahn L, Zaefferer S. Survival of Goss grains during cold rolling of a silicon steel single crystal. Materials Science Forum 2005;495-497 1061-1066.

[70] Lee DN, Jeong H-T. The evolution of the Goss texture in silicon steel. Scripta Materialia 1998;38 1219-1223.

[71] Hong S-H, Lee DN. Recrystallization textures in cold-rolled Ti bearing IF steel sheets. ISIJ International 2002;42 1278-1287.

[72] Seidal L, Hölscher M, Lücke K. Rolling and recrystallization textures in iron–3% silicon. Textures and Microstructures 1989;11 171-185.

[73] Von Schlippernbach U, Emren F, Lücke K. Investigation of the development of the cold rolling texture in deep drawing steels by ODF-analysis. Acta metallurgica 1986;34 1289-1301.

[74] Yoshinaga N, Vanderschueren D, Kestens L, Ushioda K, Dilewijns J. Cold-rolling and recrystallization texture formation in electro-deposited pure iron with a sharp and homogeneous gamma-fiber. ISIJ International 1998;38 610-616.

[75] Park YB, Lee DN, Gottstein G. Evolution of recrystallization textures from cold rolling textures in interstitial free steel. Materials Science and Technology 1997;13 289-298.

[76] Park YB, Lee DN, Gottstein G. The evolution of recrystallization textures in body centered cubic metals. Acta Materialia 1998;46 3371-3379.

[77] Lücke K, Holscher M. Rolling and recrystallization textures of BCC steels. Textures and Microstructures 1991;14-18 585-596.

[78] Raabe D, Lücke K. Rolling and annealing textures of bcc metals. Materials Science Forum 1994;157-162 597-610.

[79] Hünsche I, Oertel C-G, Tamm R, Skrotzki W, Knabl W. Microstructure and texture development during recrystallization of rolled molybdenum sheets. Materials Science Forum 2004;467-470 495-500.

[80] Wassermann G, Grewen J. Texturen metallischer Werkstoffe. Berlin: Springer-Verlag; 1962.

[81] Pugh JW. The temperature dependence of preferred orientations in rolled tungsten.Transactions AIME 1958;212 637.

[82] Walker E, Bujard P. Anomalous temperature behavior of the shear elastic constant c 44 in tantalum. Solid State Communications 1980;34 691-693.

[83] Dickinson JM, Armstrong PE. Temperature Dependence of the Elastic Constants of Molybdenum. Journal of applied Physics 1967;38 602-606.

[84] Bolef DI, de Klerk J. Elastic Constants of Single-Crystal Mo and W between 77° and 500°K. Journal of applied Physics 1962;33 2311-2314.

[85] Lowrie R, Gonas AM, Single-Crystal Elastic Properties of Tungsten from 24° to 1800°C. Journal of applied Physics 1967;38 4505-4509.

Chapter 2

Characterization for Dynamic Recrystallization Kinetics Based on Stress-Strain Curves

Quan Guo-Zheng

Additional information is available at the end of the chapter

http://dx.doi.org/10.5772/54285

1. Introduction

Most metals and alloys have become increasingly important in a variety of applications. The most important of these properties are ease of high strength, relatively good ductility, and good corrosion resistance. One of the ways of acquiring the best combination of these properties is to select the microstructure, which in turn depends on thermo-mechanical history as well as on chemical composition [1]. An optimization of the thermo-mechanical process can be achieved through an understanding of the entire forming process and the metallurgical variables affecting the micro-structural features occurring during deformation operations carried out during deformation operations carried out at high temperatures. Most applications of these alloys are in the chemical process equipment, petrochemical, aerospace industry, and medical tools. The understanding of metals and alloys behavior at hot deformation condition has a great importance for designers of hot metal forming processes (hot rolling, forging and extrusion) because of its effective role on metal flow pattern, and the constitutive relationships are often used to describe the plastic flow properties of metals and alloys in a form that can be used in computer code to model the forging response of mechanical part members under the prevailing loading conditions [2].

During hot forming process alloy is liable to undergo work hardening (WH), dynamic recovery (DRV) and dynamic recrystallization (DRX), three metallurgical phenomena for controlling microstructure and mechanical properties [3]. The relay of softening mechanism from strain-hardening and dynamic recovery to DRX is the reason the term discontinuous has been earned. At a microstructural level DRX begins when strain hardening plus recovery can no longer store more immobile dislocations. When the critical strain is reached, on face centered cubic (fcc) metals of medium to low stacking fault energy, strain-hardening and dynamic recovery cease to be the principle mechanisms responsible of the stress-strain response, DRX accompanies the process. However, DRX is not a phenomenon restricted to fcc metals, it has been described on ice, some minerals, and even high purity α–Fe (bcc metal) [4-6]. In the deformed material DRX would affect the crystallographic texture and thus, material anisotropy. For example, DRX would eliminate some crystal defects, such as part of dislocations resulting from work hardening, which will improve hot plasticity, refine microstructure, and reduce the deformation resistance [7]. High stacking fault energy (SFE) metals, such as aluminium alloys, alpha titanium alloys, and ferritic steels, undergo continuous dynamic recrystallization (CDRX) rather than discontinuous dynamic recrystallization (DDRX) during high temperature deformation. In particular, due to the high efficiency of dynamic recovery, new grains are not formed by a classical nucleation mechanism; the recrystallized microstructure develops instead by the progressive transformation of subgrains into new grains, within the deformed original grains. Dislocations produced by strain hardening accumulate progressively in low-angle (subgrain) boundaries (LABs), leading to the increase of their misorientation angle and the formation of high-angle (grain) boundaries (HABs), when a critical value of the misorientation angle is reached. The microstructure is thus intermediate between a subgrain and a grain structure: while grains and subgrains are entirely delimited by HABs and LABs, respectively, it will be referred to as an aggregate of crystallites, which are bounded partly by LABs and partly by HABs. On the contrary, low stacking fault energy (SFE) metals, such as magnesium alloys, austenitic steels, and beta titanium alloys, undergo discontinuous dynamic recrystallization (DDRX) rather than continuous dynamic recrystallization (CDRX) during high temperature deformation [8].

2. Description of softening flow behavior coupling with DRX

Hot working behavior of alloys is generally reflected on flow curves which are a direct consequence of microstructural changes: the generation of dislocations, work hardening, WH, the rearrangement of dislocations, their self-annihilation, and their absorption by grain boundaries, DRV, the nucleation and growth of new grains, DRX. The latter is one of the most important softening mechanisms at high temperatures. This is a characteristic of low and medium stacking fault energy, SFE, materials e.g., γ-iron, the austenitic stainless steels, and copper. The most significant changes in the structure-sensitive properties occur during the primary recrystallization stage. In this stage the deformed lattice is completely replaced by a new unstrained one by means of a nucleation and growth process, in which practically stress-free grains grow from nuclei formed in the deformed matrix. The orientation of the new grains differs considerably from that of the crystals they consume, so that the growth process must be regarded as incoherent, i.e. it takes place by the advance of large-angle boundaries separating the new crystals from the strained matrix [1].

DRX occurs during straining of metals at high temperature, characterized by a nucleation rate of low dislocation density grains and a posterior growth rate that can produce a homogeneous grain size when equilibrium is reached. The process of recrystallization may be pictured as follows. After deformation, polygonization of the bent lattice regions on a fine scale occurs and this results in the formation of several regions in the lattice where the strain energy is lower than in the surrounding matrix; this is a necessary primary condition for nucleation. During this initial period when the angles between the sub-grains are small and less than one degree, the sub-grains form and grow quite rapidly. However, as the sub-grains grow to such a size that the angles between them become of the order of a few degrees, the growth of any given sub-grain at the expense of the others is very slow. Eventually one of the sub-grains will grow to such a size that the boundary mobility begins to increase with increasing angle. A large angle boundary, $\theta \approx 30\text{~}40°$, has a high mobility because of the large lattice irregularities or 'gaps' which exist in the boundary transition layer. The atoms on such a boundary can easily transfer their allegiance from one crystal to the other. This sub-grain is then able to grow at a much faster rate than the other sub-grains which surround it and so acts as the nucleus of a recrystallized grain. The further it grows, the greater will be the difference in orientation between the nucleus and the matrix it meets and consumes, until it finally becomes recognizable as a new strain-free crystal separated from its surroundings by a large-angle boundary [9].

Fig.1 shows typical flow curves during cold and hot deformation. During hot deformation, the shape of the flow curve can be 'restricted', or work hardening rates counterbalanced, by dynamic recovery or by dynamic recrystallization (i.e. discontinuous dynamic recrystallization). Dynamic recovery is typical of high-SFE metals (e.g. aluminium, low-carbon ferritic steel, etc.), where the flow stress saturates after an initial period of work hardening. This saturation value depends on temperature, strain rate and composition. On the other hand, as shown in Fig.1, a broad peak (or multiple peaks) typically accompany dynamic recrystallization. Fig.2 illustrates schematically the microstructure developments during dynamic recovery and dynamic recrystallization. During dynamic recovery, the original grains get increasingly strained, but the sub-boundaries remain more or less equiaxed. This implies that the substructure is 'dynamic' and re-adapts continuously to the increasing strain. In low-SFE metals (e.g. austenitic stainless steel, copper, etc.), the process of recovery is slower and this, in turn, may allow sufficient stored energy build-up. At a critical strain, and correspondingly at a value/variation in driving force, dynamically recrystallized grains appear at the original grain boundaries – resulting in the so-called 'necklace structure'. With further deformation, more and more potential nuclei are activated and new recrystallized grains appear. At the same time, the grains, which had already recrystallized in a previous stage, are deformed again. After a certain amount of strain, saturation/equilibrium sets in (see Fig.2b). Typically equilibrium is reached between the hardening due to dislocation accumulation and the softening due to dynamic recrystallization. At this stage, the flow curve reaches a plateau and the microstructure consist of a dynamic mixture of grains with various dislocation densities. It is important, at this stage, to bring out further the structural developments and structure–property correlation accompanying dynamic recovery and dynamic recrystallization respectively [10].

The true compressive stress-strain curves for as-extruded 7075 aluminum alloy under different temperatures and strain rates are illustrated in Fig. 3a~d. The flow stress as well as the shape of the flow curves is sensitively dependent on temperature and strain rate. Comparing these curves with one another, it is found that increasing strain rate or decreasing deformation temperature makes the flow stress level increase, in other words, it prevents the occurrence of

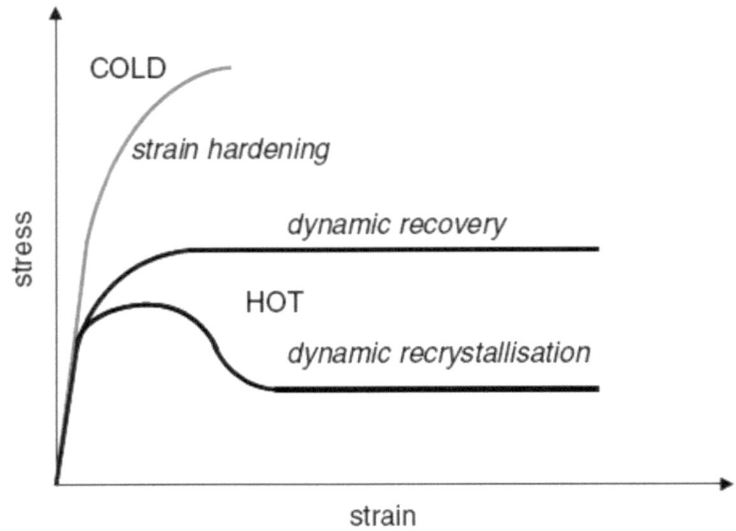

Figure 1. Typical flow curves during cold and hot deformation

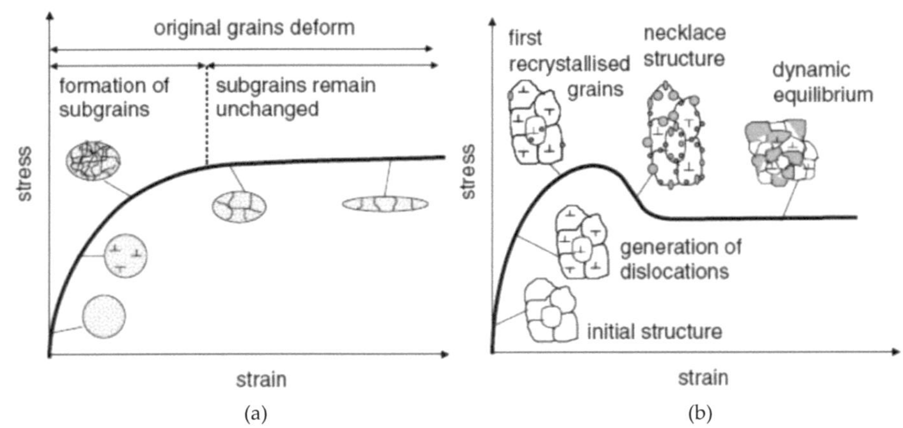

Figure 2. Evolution of the microstructure during (a) hot deformation of a material showing recovery and (b) continuous dynamic recrystallization (CDRX).

softening due to DRX and dynamic recovery (DRV) and makes the deformed metals exhibit work hardening (WH). The cause lies in the fact that higher strain rate and lower temperature provide shorter time for the energy accumulation and lower mobilities at boundaries which result in the nucleation and growth of dynamically recrystallized grains and dislocation annihilation. For every curve, after a rapid increase in the stress to a peak value, the flow stress decreases monotonically towards a steady state regime with a varying softening rate which typically indicates the onset of DRX. In further, from all the true stress-strain curves, it can be summarized that the stress evolution with strain exhibits three distinct stages. At the first stage where work hardening (WH) predominates, flow stress exhibits a rapid increase to a critical value. At the second stage, flow stress exhibits a smaller and smaller increase until a peak value or an inflection of work-hardening rate, which shows that the thermal softening due to DRX and dynamic recovery (DRV) becomes more and more predominant, then it exceeds WH. At the third stage, two types of curve variation tendency can be generalized as following: decreasing gradually to a steady state with DRX softening (573~723 K & 0.01 s^{-1}, 623~723 K & 0.1 s^{-1}, 623~723 K & 1 s^{-1}, 723 K & 10 s^{-1}), decreasing continuously with significant DRX softening (573 K & 0.1 s^{-1}, 573 K & 1 s^{-1}, 573~623 K & 10 s^{-1}). Thus, it can be concluded that the typical form of flow curve with DRX softening, including a single peak followed by a steady state flow as a plateau, is more recognizable at higher temperatures and lower strain rates. That is because at lower strain rates and higher temperatures, the higher DRX softening rate slows down the rate of work-hardening, and both the peak stress and the onset of steady state flow are therefore shifted to lower strain levels [11].

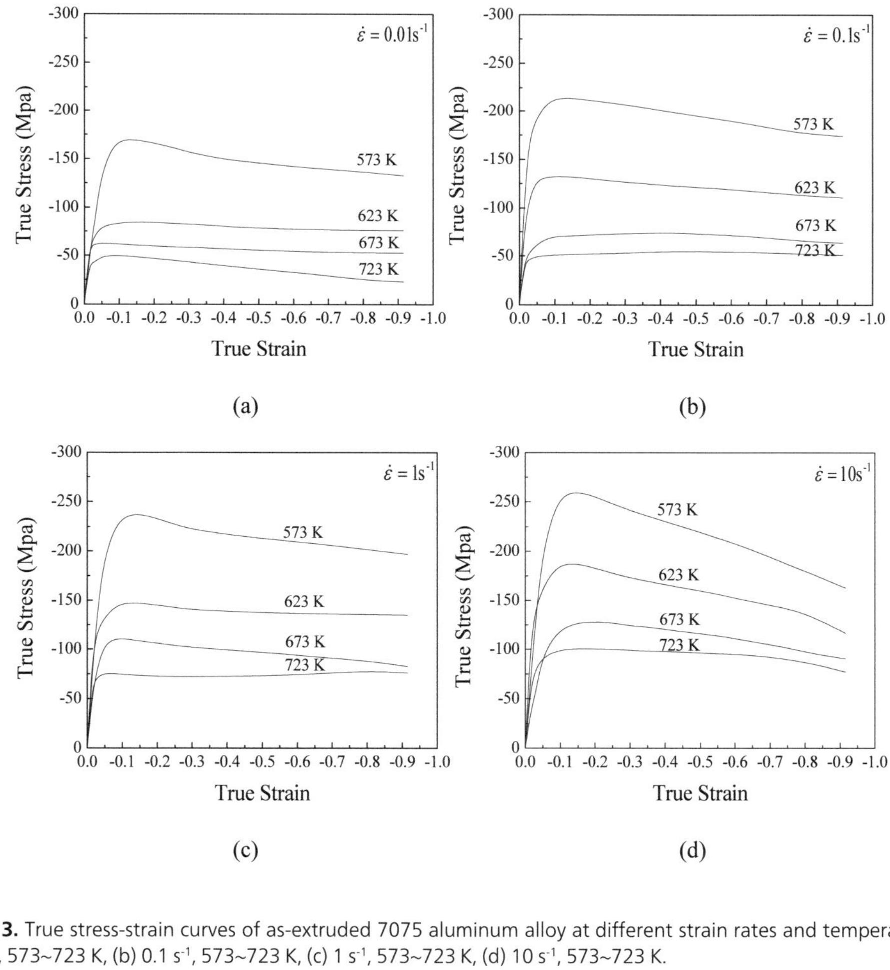

Figure 3. True stress-strain curves of as-extruded 7075 aluminum alloy at different strain rates and temperatures. (a) 0.01 s^{-1}, 573~723 K, (b) 0.1 s^{-1}, 573~723 K, (c) 1 s^{-1}, 573~723 K, (d) 10 s^{-1}, 573~723 K.

The similar flow behavior of as-cast AZ80 magnesium alloy with as-extruded 7075 aluminum alloy is illustrated in Fig. 4a~d. Both deformation temperature and strain rate have considerable influence on the flow stress of AZ80 magnesium alloy. From the true stress-strain curves in Fig. 4a~d, it also can be seen that the stress evolution with strain exhibits three distinct stages. At the first stage where work hardening (WH) predominates and cause dislocations to polygonize into stable subgrains, flow stress exhibits a rapid increase to a critical value with increasing strain, meanwhile the stored energy in the grain boundaries originates from a large difference in dislocation density within subgrains or grains and grows rapidly to DRX activation energy. When the critical driving force is attained, new grains are nucleated along the grain boundaries, deformation bands and dislocations, resulting in equiaxed DRX grains. At the second stage, flow stress exhibits a smaller and smaller increase until a peak value or an inflection of work-hardening rate, which shows that the thermal softening due to DRX and dynamic recovery (DRV) becomes more and more predominant, then it exceeds WH. At the third stage, two types of curve variation tendency can be generalized as following: decreasing gradually to a steady state with DRX softening (573~673 K & 0.01 s^{-1}, 623~673 K & 0.1 s^{-1}, 573 K & 1 s^{-1},, 673 K & 1 s^{-1}), and decreasing continuously with significant DRX softening (523 K & 0.01 s^{-1}, 523~573 K & 0.1 s^{-1}, 523 K & 1 s^{-1},, 623 K & 1 s^{-1}, 523~673 K & 10 s^{-1}) [12-14].

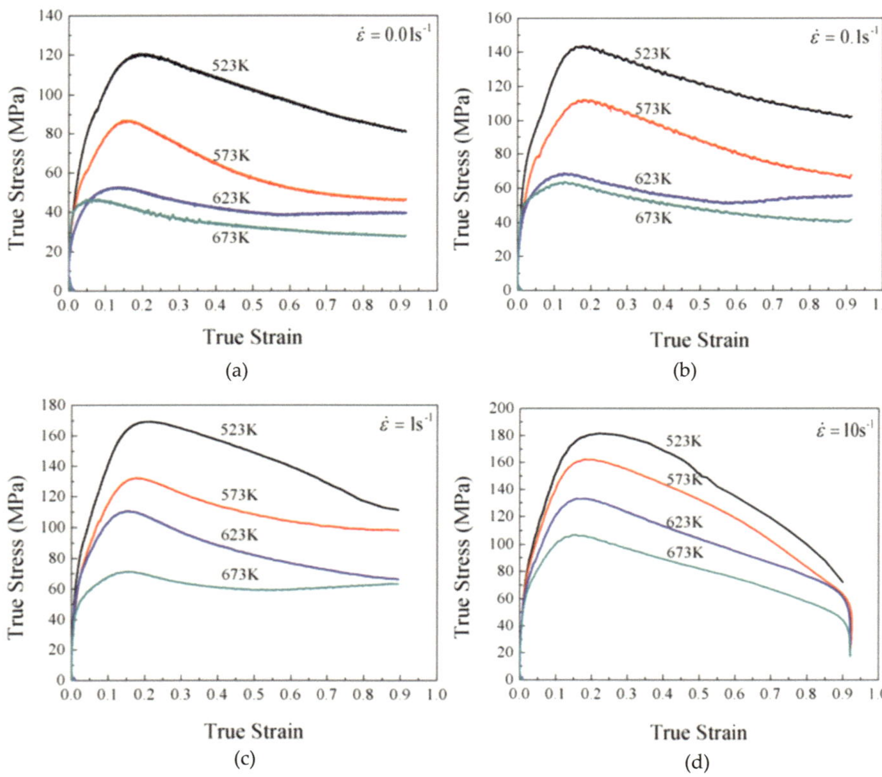

Figure 4. True stress-strain curves of as-cast AZ80 magnesium alloy obtained by Gleeble 1500 under different deformation temperatures with strain rates (a) 0.01 s^{-1}, (b) 0.1 s^{-1}, (c) 1 s^{-1}, (d) 10 s^{-1}.

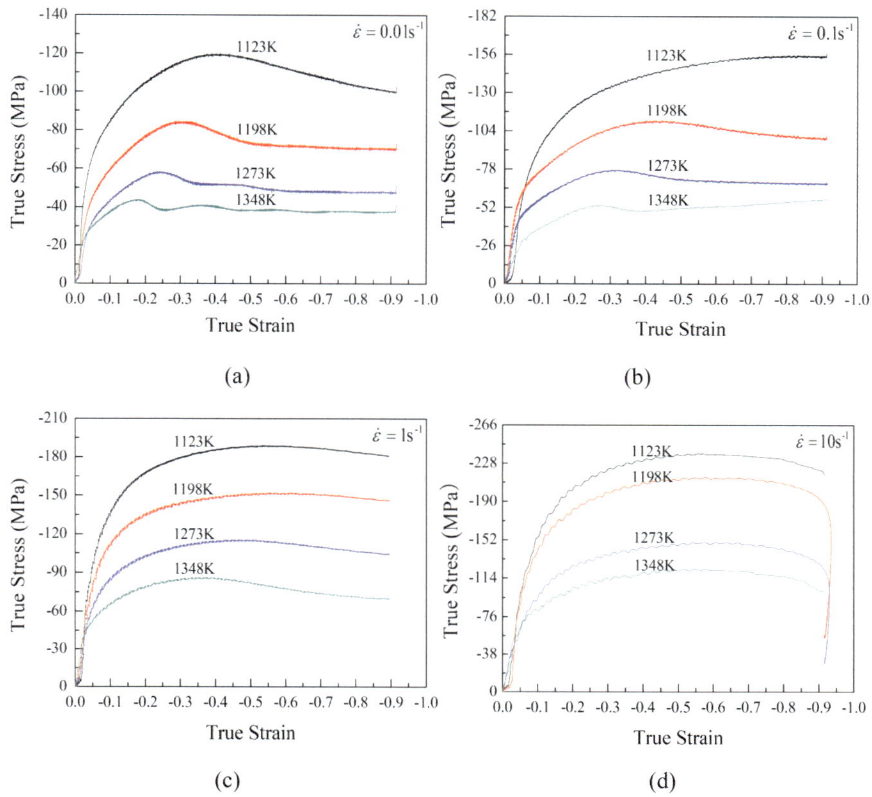

Figure 5. True stress-strain curves of as-extruded 42CrMo high-strength steel obtained by Gleeble 1500 under the different deformation temperatures with strain rates. (a) 0.01 s^{-1}, (b) 0.1 s^{-1}, (c) 1 s^{-1}, (d) 10 s^{-1}.

A little different flow behavior of as-cast 42CrMo high-strength steel from as-extruded 7075 aluminum alloy and as-cast AZ80 magnesium alloy is illustrated in Fig. 5a~d. From the true stress-strain curves in Fig. 5a~d, it also can be seen that the stress evolution with strain exhibits three distinct stages. But the difference is as follows: at the third stage, three types of curve variation tendency can be generalized as following: decreasing gradually to a steady state with DRX softening (1123~1348 K & 0.01 s^{-1}, 1198~1348 K & 0.1 s^{-1}, 1273~1348 K & 1 s^{-1}), maintaining higher stress level without significant softening and work-hardening (1123~1198 K & 1 s^{-1}, 1123~1348 K & 10 s^{-1}), and increasing continuously with significant work-hardening (1123 K & 0.1 s^{-1}) [15-17].

3. DRX critical strain and DRX kinetic model

3.1. The initiation of DRX

From the true compressive stress-strain data of as-extruded 42CrMo high-strength steel shown in Fig. 5a~d, the values of the strain hardening rate ($\theta = d\sigma / d\varepsilon$) were calculated. The critical conditions for the onset of DRX can be attained when the value of $|-d\theta / d\sigma|$, where strain hardening rate $\theta = d\sigma / d\varepsilon$, reaches the minimum which corresponds to an inflection of $d\sigma / d\varepsilon$ versus σ curve. In this study, analysis of inflections in the plot of $d\sigma / d\varepsilon$ versus σ up to the peak point of the true stress-strain curve has been performed to reveal whether DRX occurs. Results confirm that the $d\sigma / d\varepsilon$ versus σ curves have characteristic inflections as shown in Fig. 6a~d, which indicates that DRX is initiated at corresponding deformation conditions. The critical stress to initiation can be identified, and hence the corresponding critical strain to initiation can be obtained from true stress-strain curve. As a result, the values of critical strain and peak stress at different deformation conditions were shown in Table.1, from which it can be seen that the critical strain and critical stress depend on temperature and strain rate nonlinearly, and it is summarized that $\varepsilon_c / \varepsilon_p$=0.165~0.572 , σ_c / σ_p=0.645~0.956 [15].

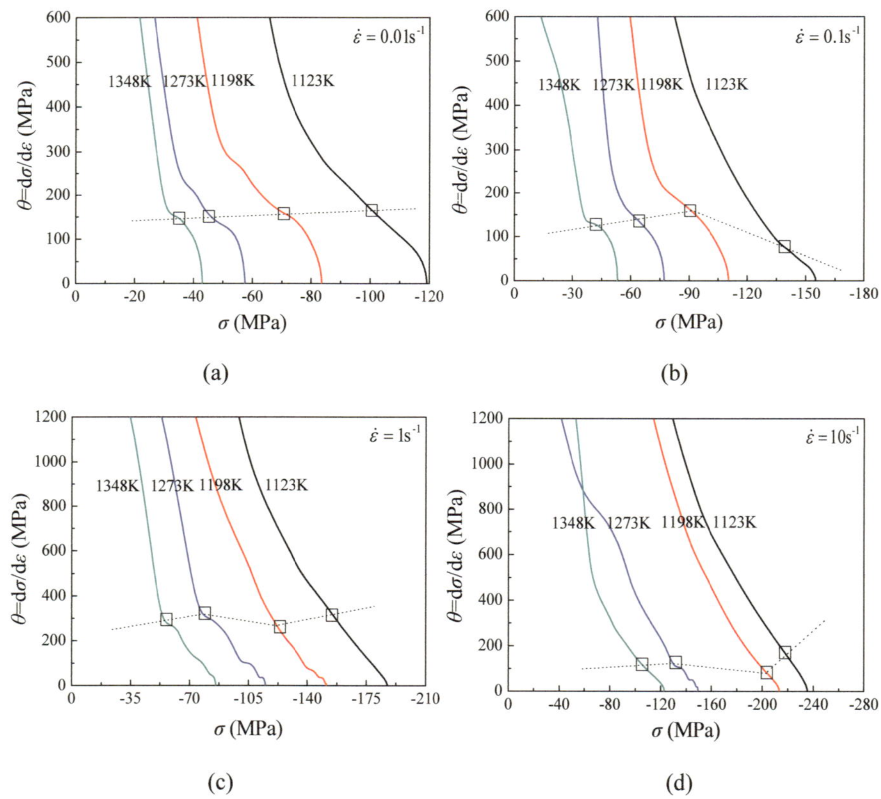

Figure 6. Formula: Eqn012.wmf>versus dσ / dε plots up to the peak points of the true stress-strain curves under different deformation temperatures with strain rates (a) 0.01 s^{-1}, (b) 0.1 s^{-1}, (c) 1 s^{-1}, (d) 10 s^{-1}.

http://dx.doi.org/10.5772/54285

		Strain rate (s^{-1})	Temperature (K)			
			1123	**1198**	**1273**	**1348**
True strain	σ	0.01	-0.177	-0.160	-0.113	-0.083
		0.1	-0.336	-0.172	-0.140	-0.142
		1	-0.146	-0.135	-0.077	-0.059
		10	-0.259	-0.359	-0.247	-0.218
	ε_c	0.01	-0.400	-0.309	-0.236	-0.355
		0.1	-0.818	-0.436	-0.318	-0.264
		1	-0.545	-0.591	-0.455	-0.355
		10	-0.564	-0.627	-0.600	-0.573
True stress (MPa)	ε_p	0.01	-100.875	-70.523	-45.244	-35.122
		0.1	-137.038	-89.748	-63.469	-42.362
		1	-153.988	-122.629	-76.918	-55.127
		10	-215.180	-203.734	-131.429	-105.296
	σ_c	0.01	-119.111	-83.558	-57.221	-40.714
		0.1	-155.026	-110.287	-77.083	-53.280
		1	-187.522	-151.374	-114.790	-85.495
		10	-235.405	-213.198	-149.191	-122.623
σ_p		0.01	0.441	0.519	0.476	0.234
		0.1	0.411	0.394	0.439	0.540
		1	0.268	0.229	0.170	0.165
		10	0.459	0.572	0.412	0.381
$\varepsilon_c / \varepsilon_p$		0.01	0.847	0.844	0.791	0.863
		0.1	0.884	0.814	0.823	0.795
		1	0.821	0.810	0.670	0.645
		10	0.914	0.956	0.881	0.859

Table 1. Values of σ_c / σ_p , ε_c , σ_c and ε_p at different deformation conditions.

3.2. Arrhenius equation for flow behavior with DRX

It is known that the thermally activated stored energy developed during deformation controls softening mechanisms which induce different DRX softening and work-hardening. The activation energy of DRX, an important material parameter, determines the critical conditions for DRX initiation. So far, several empirical equations have been proposed to determine the deformation activation energy and hot deformation behavior of metals. The most frequently used one is Arrhenius equation which designs a famous Zener-Hollomon parameter, σ_p , to represent the effects of the temperatures and strain rate on the deformation behaviors, and then uncovers the approximative hyperbolic law between Z parameter and flow stress [15].

$$Z = \dot{\varepsilon}\exp(Q/RT) \tag{1}$$

$$\dot{\varepsilon} = AF(\sigma)\exp(-Q/RT) \tag{2}$$

Where,

where $F(\sigma)=\begin{cases} |\sigma|^n & \alpha|\sigma|<0.8 \\ \exp(\beta|\sigma|) & \alpha|\sigma|>1.2 \\ [\sinh(\alpha|\sigma|)]^n & \text{for all } \sigma \end{cases}$ is the strain rate (s^{-1}), $\dot{\varepsilon}$ is the universal gas constant (8.31 J mol^{-1} K^{-1}), R is the absolute temperature (K), T is the activation energy of DRX (kJ mol^{-1}), Q is the flow stress (MPa) for a given stain, σ , A and α are the material constants (n).

3.2.1. Calculation of material constant $\alpha=\beta/n$

For the low stress level (n), substituting the power law of $\alpha\sigma<0.8$ into Eq. (2) and taking natural logarithms on both sides of Eq. (2) give

$$\ln\dot{\varepsilon} = \ln A + n\ln|\sigma| - Q/RT \tag{3}$$

Then, $\ln\dot{\varepsilon}=\ln A + n\ln|\sigma| - Q/RT$. In 2010, Quan et al. [17] plotted the relationships between the true stress and true strain of 42CrMo high-strength steel in ln-ln scale under different temperatures and strain rates, and hence found a true strain range of -0.08~-0.18 including part of the first stage and the second stage described in the previous, in which all the stresses increase gradually with almost the same ratios. Therefore, this true strain range was accepted as a steady WH stage corresponding to low stress level. In further, Quan et al. [17] fitted the relationships between the stress and the strain rate as the true strain was -0.14, and then found almost equally linear relationships which revealed that the influence of temperature was very small. Thus, it can be deduced that to evaluate the material constant $n=d\ln\dot{\varepsilon}/d\ln|\sigma|$ of Arrhenius equation, the stress-strain data in the true strain range of -0.08~-0.18 contribute to the minimum calculation tolerance. Here true strain n was chose. Fig. 7 shows the relationships between $\varepsilon=-0.1$ and $\ln|\sigma|$ for $\ln\dot{\varepsilon}$ under different temperatures. The linear relationship is observed for each temperature and the slope rates are almost similar with each other. The mean value of all the slope rates is accepted as the inverse of material constant $\varepsilon=-0.1$, thus n value is obtained as 8.27780.

3.2.2. Calculation of material constant $\ln\dot{\varepsilon}$

For the high stress level (β), substituting the exponential law of $\alpha|\sigma|>1.2$ into Eq. (2) and taking natural logarithms on both sides of Eq. (2) give

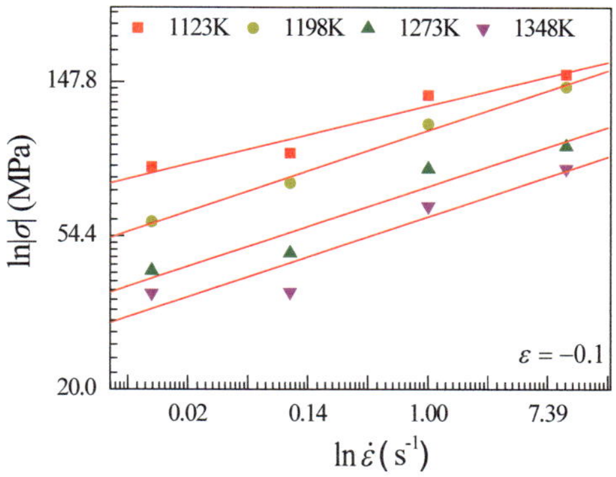

Figure 7. The relationships between *n* and ln*σ* .

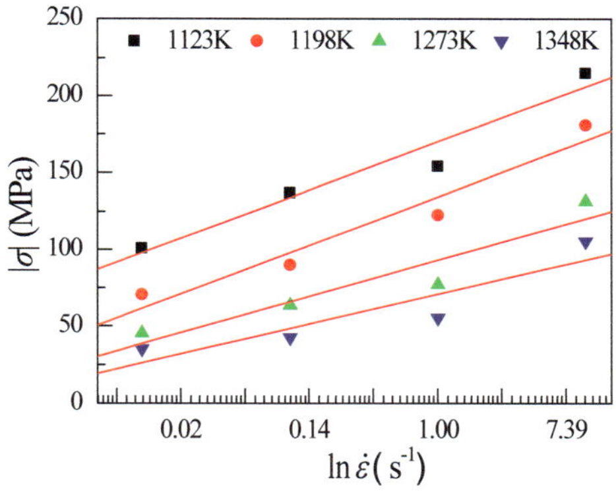

Figure 8. The relationships between $\alpha = \beta / n = 0.00913$ and $|\sigma|$.

$$\ln \dot{\varepsilon} = \ln A + \beta|\sigma| - Q/RT \tag{4}$$

Then, $\ln\dot{\varepsilon} = \ln A + \beta|\sigma| - Q/RT$. The peak stresses at different temperatures and strain rates can be identified for the target stresses with high level. The linear relationships between $\beta = d\ln\dot{\varepsilon} / |\sigma|$ and $|\sigma|$ at different temperatures were fitted out as Fig. 8. The mean value of all the slope rates is accepted as the inverse of material constant $\ln\dot{\varepsilon}$, thus β value is obtained as 0.07558 MPa^{-1}. Thus, another material constant β MPa^{-1}.

3.2.3. *Calculation of DRX activation energy* $\ln\dot{\varepsilon}$

For all the stress level (including low and high stress levels), Eq.(2) can be represented as the following

$$\ln\dot{\varepsilon} = \ln A + n\left[\ln\sinh(\alpha|\sigma|)\right] - Q/RT \tag{5}$$

If $\ln\dot{\varepsilon} = \ln A + n[\ln\sinh(\alpha \mid \sigma \mid)] - Q/RT$ is constant, there is a linear relationship between $\dot{\varepsilon}$ and $\ln\sinh(\alpha \mid \sigma \mid)$, and Eq. (5) can be rewritten as

$$Q = Rn\left\{d\left[\ln\sinh(\alpha|\sigma|)\right]/d(1/T)\right\} \tag{6}$$

The peak stresses at different temperatures and strain rates can be identified for the present target stresses. The linear relationships between $Q = Rn\{d[\ln\sinh(\alpha \mid \sigma \mid)]/d(1/T)\}$ and $\ln\sinh(\alpha\sigma)$ at different strain rates were fitted out as Fig. 9. The mean value of all the slope rates is accepted as $1/T$ value, then Q/Rn is calculated as 599.73210 kJ mol^{-1}. The activation energy of DRX is a term defined as the energy that must be overcome in order for the nucleation and growth of new surface or grain boundary to occur. In 2008, Lin et al. found that the activation energy of as-cast 42CrMo steel is not a constant but a variable 392~460 kJ mol^{-1} as a function of strain, and the peak value of DRX energy corresponds to the peak stress [18, 19]. In this investigate, the influence of strain on the variable activation energy was ignored to simplify the following calculations, and only the peak value of DRX energy was accepted as the activation energy of DRX. This simplification ensures the predicted occurrence of DRX by the derived equations. Lin et al. also pointed that the average value of the activation energy of as-cast 42CrMo steel is 438.865 kJ mol^{-1} [18, 19]. The average Q value of extruded 42CrMo steel, 599.73210 kJ mol^{-1}, is a little higher than that of as-cast 42CrMo steel adopted by Lin et al. The difference of two average Q values results from the different as-received statuses. In common, the higher deformation activation energy will be found in hot deformation of as-received steels with higher yield strength. It is obvious that the true stress data of extruded rods in this work are higher than that of as-cast billets in the work of Lin et al. In addition, the difference of experiment projects involving strain rate between this work and the work of Lin et al is another important reason for the difference of calculation result.

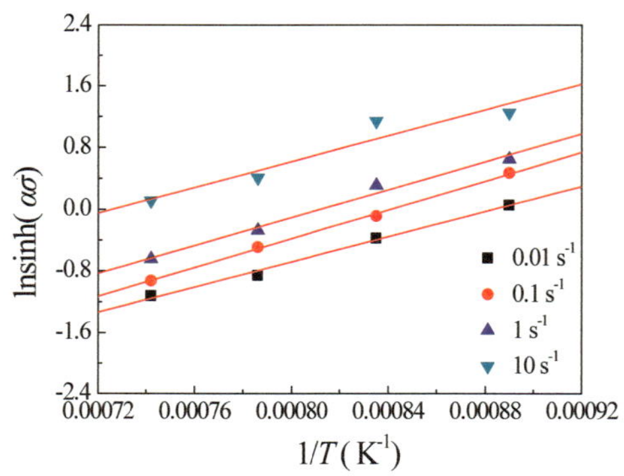

Figure 9. The relationships between Q and $\ln\sinh(\alpha|\sigma|)$.

3.2.4. Construction of constitutive equation

Substituting $1/T$, α , n and four sets of Q , $\dot{\varepsilon}$ and T into Eq. (5), the mean value of material constant σ is obtained as 2.44154×10^{25} s^{-1}. Thus, the relationship between A , $\dot{\varepsilon}$ and T can be expressed as

$$\dot{\varepsilon} = 2.44154 \times 10^{25}\left[\sinh(0.00913|\sigma|)^{8.27780}\right]\exp\left[-(599.73210\times10^{3})/8.31T\right] \quad (7)$$

Substituting $\dot{\varepsilon} = 2.44154\times10^{25}[\sinh(0.00913\,|\,\sigma\,|\,)^{8.27780}]\exp[-(599.73210\times10^{3})/8.31T]$ into Eq. (7), thus, the the flow stress can be expressed as

$$|\sigma| = 109.52903\ln\left\{[Z/(2.44154\times10^{25})]^{1/8.27780} + \{[Z/(2.44154\times10^{25})]^{2/8.27780} + 1\}^{1/2}\right\} \quad (8)$$

3.3. DRX kinetic model

During thermoplastic deformation process, dislocations continually increase and accumulate to such an extent that at a critical strain, DRX nucleus would form and grow up near grain boundaries, twin boundaries and deformation bands. It is well known that the conflicting effects coexist between the multiplication of dislocation due to continual hot deformation and the annihilation of dislocation due to DRX. When work-hardening corresponding to the former and DRX softening corresponding to the later are in dynamic balance, flow stress will keep constant with increasing strain, meanwhile deformation comes to a steady stage in which complete DRX grains have equiaxed shape and keep constant size. In common, the kinetics of DRX can be described in terms of normal S-curves of the recrystallized volume expressed as a function of time. In a constant strain rate, time can be replaced by strain and recrystallized volume fraction can be expressed by modified Avrami equation. Thus, the kinetics of DRX evolution can be predicted by the following equation [20].

$$X_{DRX} = 1 - \exp\left\{-\left[\left(\varepsilon - \varepsilon_c\right)/\varepsilon^*\right]^m\right\} \quad (9)$$

where $X_{DRX} = 1 - \exp\{-[(\varepsilon - \varepsilon_c)/\varepsilon^*]^m\}$ is the volume fraction of dynamic recrystallized grain and X_{DRX} is Avrami's constant. This expression, which is modified from the Avrami's equation, means that m depends on strain, strain rate and temperature.

The true stress-strain curve data after the peak stress point were adopted to calculate DRX softening rate (X_{DRX} versus $\theta = d\sigma/d\varepsilon$) plots, and the results were shown as Fig. 10a~d. The maximum softening rate corresponds to the negative peak of such plot. The strain for maximum softening rate, σ , identified from Fig. 10a~d, and the critical strain, ε^* , identified from Fig. 10a~d can be considered with a power function of dimensionless parameter, ε_c (Fig. 11a~b).

The function expressions linearly fitted by the method of least squares are Z/A and

$|\varepsilon^*| = 0.61822(Z/A)^{0.08207}$.

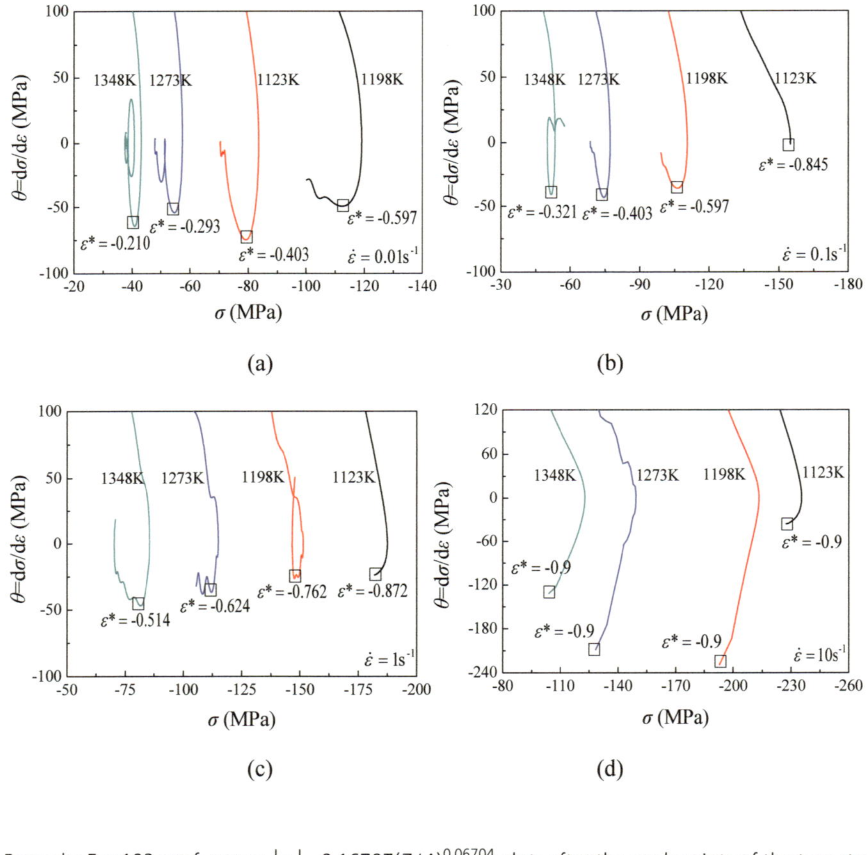

Figure 10. Formula: Eqn102.wmf>versus $|\varepsilon_c| = 0.16707(Z/A)^{0.06704}$ plots after the peak points of the true stress-strain curves under different deformation temperatures with strain rates (a) 0.01 s^{-1}, (b) 0.1 s^{-1}, (c) 1 s^{-1}, (d) 10 s^{-1}.

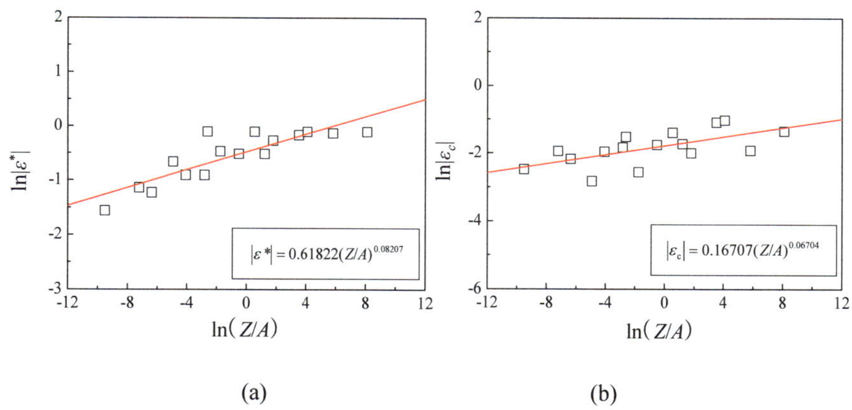

Figure 11. Relationships between the dimensionless parameter, Z/A, and (a) $d\sigma / d\varepsilon$, (b) σ .

In order to solve the Avrami's constant, ε^* , it is essential to identify the deformation conditions corresponding to ε_c meaning that the flow stress reaches a steady state in which complete DRX grains have equiaxed shape and keep constant size. From the true compressive stress-strain curves in Fig. 5a~d, and *m* versus $X_{\text{DRX}}=1$ plots in Fig. 10a~d, such the deformation conditions can be identified as shown in Table.2. Substituting these deformation conditions corresponding to $d\sigma / d\varepsilon$ into Eq. (9), the mean value of the Avrami's constant σ can be obtained as 3.85582. Thus, the kinetic model of DRX calculated from true compressive stress-strain curves can be expressed as Table.3.

True strain	Temperature (K)	Strain rate (s^{-1})
-0.5~-0.9	1198	0.01
-0.4~-0.9	1273	0.01
-0.3~-0.9	1348	0.01
-0.6~-0.9	1273	0.1
-0.4~-0.9	1348	0.1

Table 2. The deformation strain corresponding to $X_{\text{DRX}}=1$.

Volume fractions of dynamic recrystallization	Exponents
m	$X_{\text{DRX}}=1$ $X_{\text{DRX}}=1-\exp\{-[(\varepsilon-\varepsilon_c)/\varepsilon^*]^m\}$ $|\varepsilon^*| =0.61822(Z/A)^{0.08207}$ $|\varepsilon_c| =0.16707(Z/A)^{0.06704}s^{-1}$

Volume fractions of dynamic recrystallization	Exponents
	$Z = \dot{\varepsilon}\exp\left[(599.73210 \times 10^3) / 8.31T\right]$

Table 3. The kinetic model of DRX calculated from true compressive stress-strain curves.

Based on the calculation results of this model, the effect of deformation temperature, strain and strain rate on the recrystallized volume fraction is shown in Fig. 12a~d. These figures show that as the strain' absolute value increases, the DRX volume fraction increases and reaches a constant value of 1 meaning the completion of DRX process. Comparing these curves with one another, it is found that, for a specific strain rate, the deformation strain required for the same amount of DRX volume fraction increases with decreasing deformation temperature, which means that DRX is delayed to a longer time. In contrast, for a fixed temperature, the deformation strain required for the same amount of DRX volume fraction increases with increasing strain rate, which also means that DRX is delayed to a longer time. This effect can be attributed to decreased mobility of grain boundaries (growth kinetics) with increasing strain rate and decreasing temperature. Thus, under higher strain rates and lower temperatures, the deformed metal tends to incomplete DRX, that is to say, the DRX volume fraction tends to be less than 1.

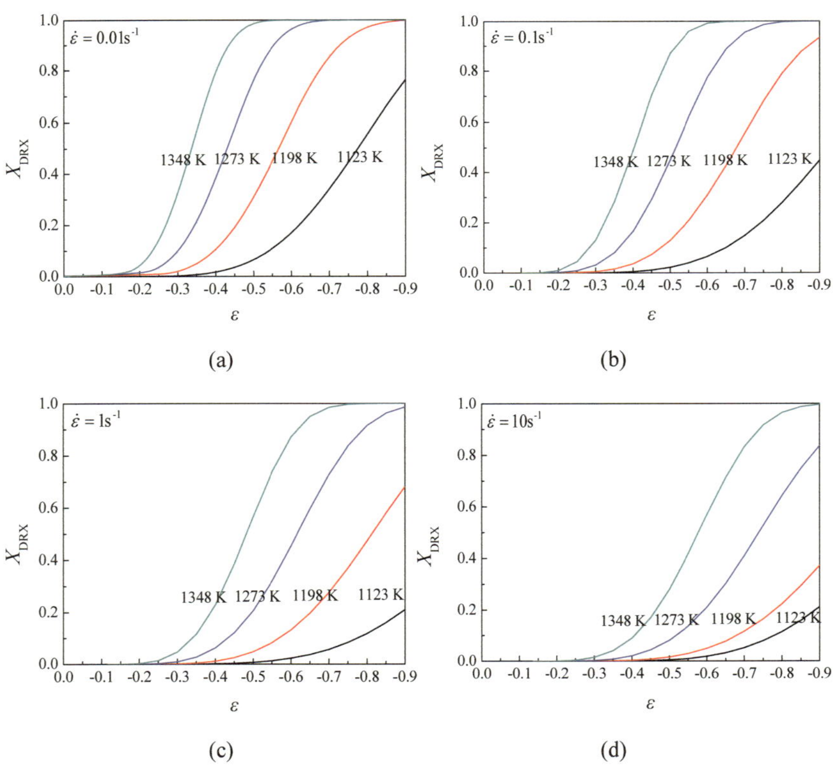

http://dx.doi.org/10.5772/54285

Figure 12. Predicted volume fractions of dynamic recrystallization obtained under different deformation temperatures with strain rates (a) 0.01 s^{-1}, (b) 0.1 s^{-1}, (c) 1 s^{-1}, (d) 10 s^{-1}.

4. Observation for size and fraction of DRX grains

The microstructures on the section plane of specimen deformed to the true strain of -0.9 were examined and analyzed under the optical microscope. Fig. 13 shows the as-received microstructure of as-extruded 42CrMo high-strength steel specimen with a single-phase FCC structure and a homogeneous aggregate of rough equiaxed polygonal grains, while with negligible volume fraction of inclusions or second-phase precipitates. The grain boundaries are straight to gently curved and often intersect at ~120° triple junctions. Fig. 14a~d show the typical microstructures of the specimens of as-extruded 42CrMo high-strength steel deformed to a strain of -0.9 at the temperature of 1123 K and at the strain rates of 0.01 s^{-1}, 0.1 s^{-1}, 1 s^{-1} and 10 s^{-1}, respectively. Fig. 15a~d show the typical microstructures of the specimens of as-extruded 42CrMo high-strength steel deformed to a strain of -0.9 at the temperature of 1198 K and at the strain rates of 0.01 s^{-1}, 0.1 s^{-1}, 1 s^{-1} and 10 s^{-1}, respectively. Fig. 16a~d show the typical microstructures of the specimens of as-extruded 42CrMo high-strength steel deformed to a strain of -0.9 at the temperature of 1273 K and at the strain rates of 0.01 s^{-1}, 0.1 s^{-1}, 1 s^{-1} and 10 s^{-1}, respectively. Fig. 17a~d show the typical microstructures of the specimens of as-extruded 42CrMo high-strength steel deformed to a strain of -0.9 at the temperature of 1348 K and at the strain rates of 0.01 s^{-1}, 0.1 s^{-1}, 1 s^{-1} and 10 s^{-1}, respectively. At such deformation conditions the recrystallized grains with wavy or corrugated grain boundaries can be easily identified from subgrains by the misorientation between adjacent grains, i.e. subgrains are surrounded by low angle boundaries while recrystallized grains have high angle boundaries. The deformed metal completely or partially transforms to a microstructure of approximately equiaxed defect-free grains which are predominantly bounded by high angle boundaries (i.e. a recrystallized microstructure) by relatively localized boundary migration.

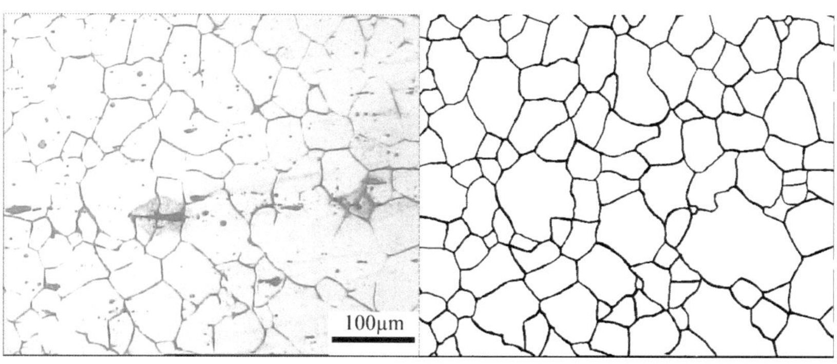

Figure 13. Optical microstructures and average grain size of as-extruded 42CrMo high-strength steel undeformed (starting material)

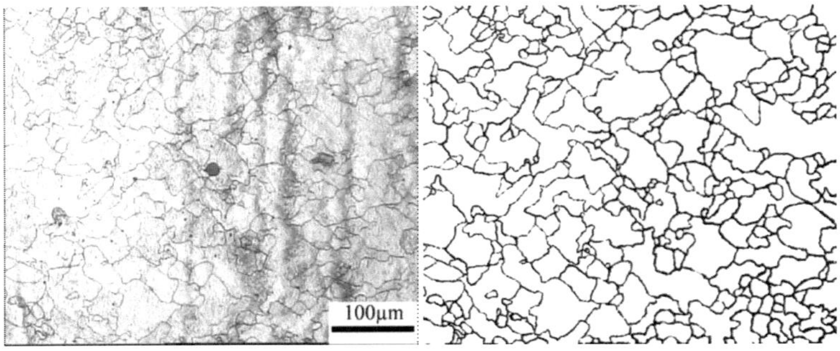

(a)

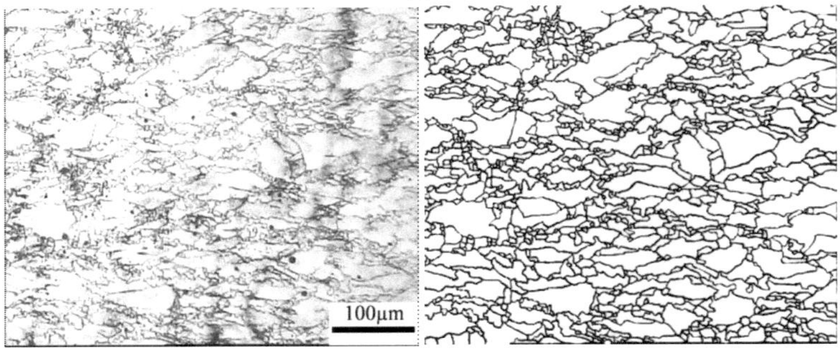

(b)

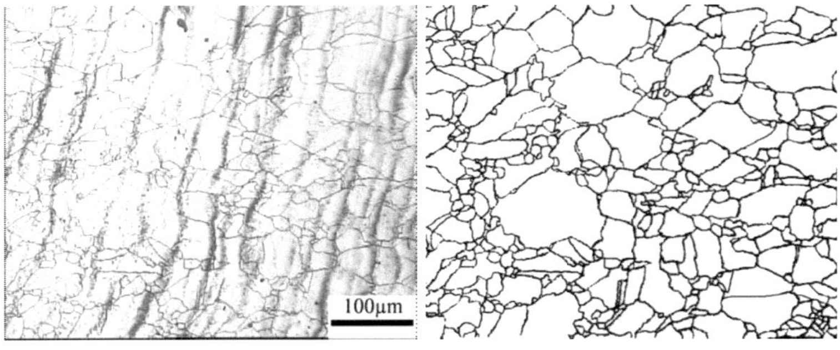

(c)

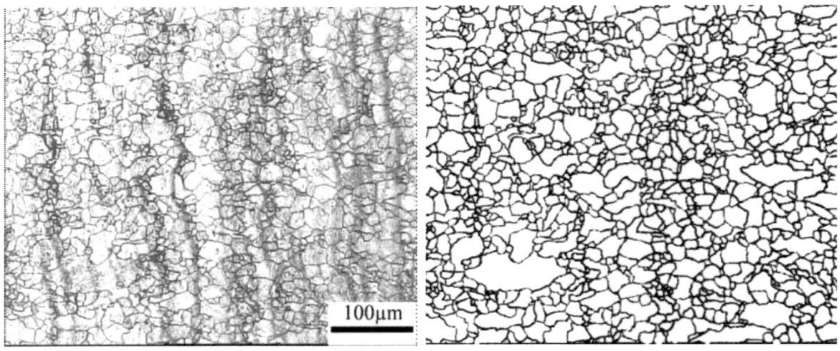

(d)

Figure 14. Optical microstructures of 42CrMo high-strength steel at a fix true strain of 0.9, a fix temperature of 1123 K and different strain rates: (a) 0.01 s^{-1}, (b) 0.1 s^{-1}, (c) 1 s^{-1}, (d) 10 s^{-1}.

http://dx.doi.org/10.5772/54285

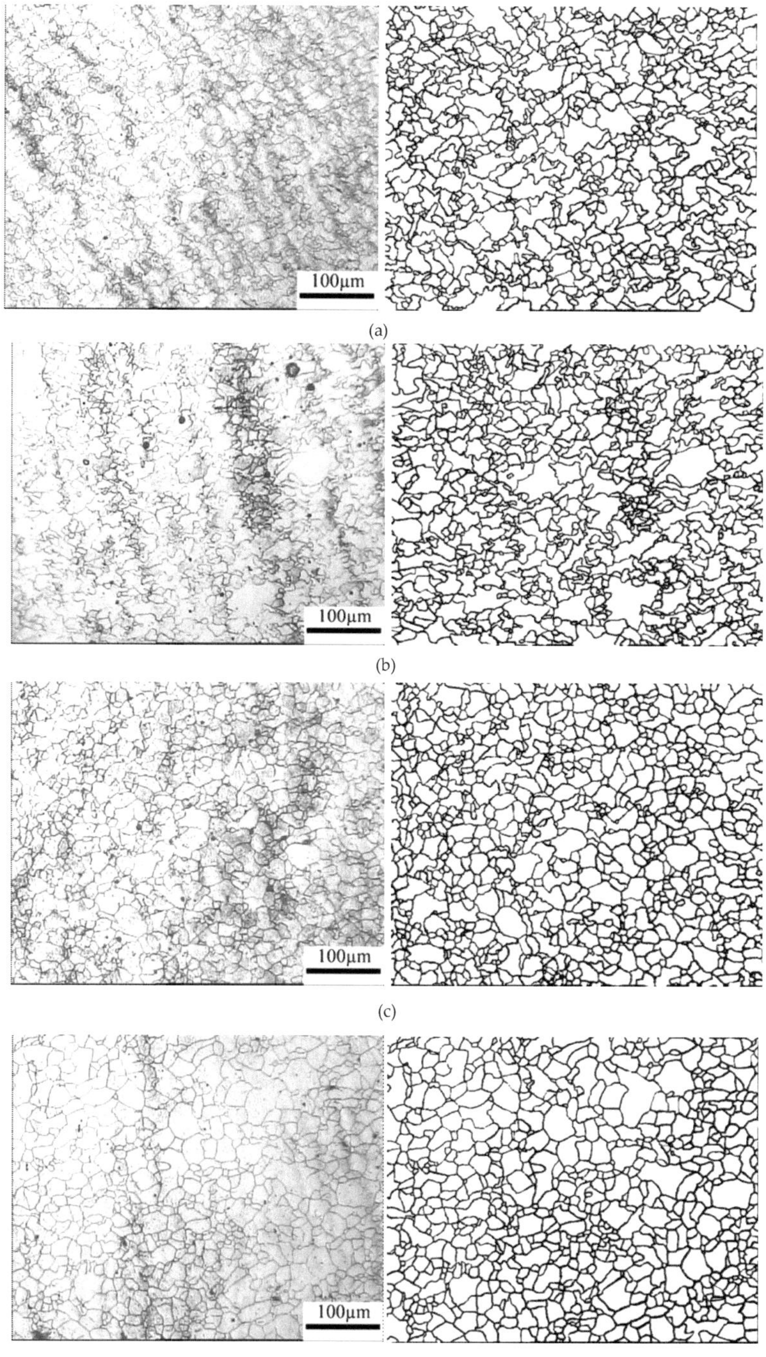

Figure 15. Optical microstructures of 42CrMo high-strength steel at a fix true strain of 0.9, a fix temperature of 1198 K and different strain rates: (a) 0.01 s^{-1}, (b) 0.1 s^{-1}, (c) 1 s^{-1}, (d) 10 s^{-1}.

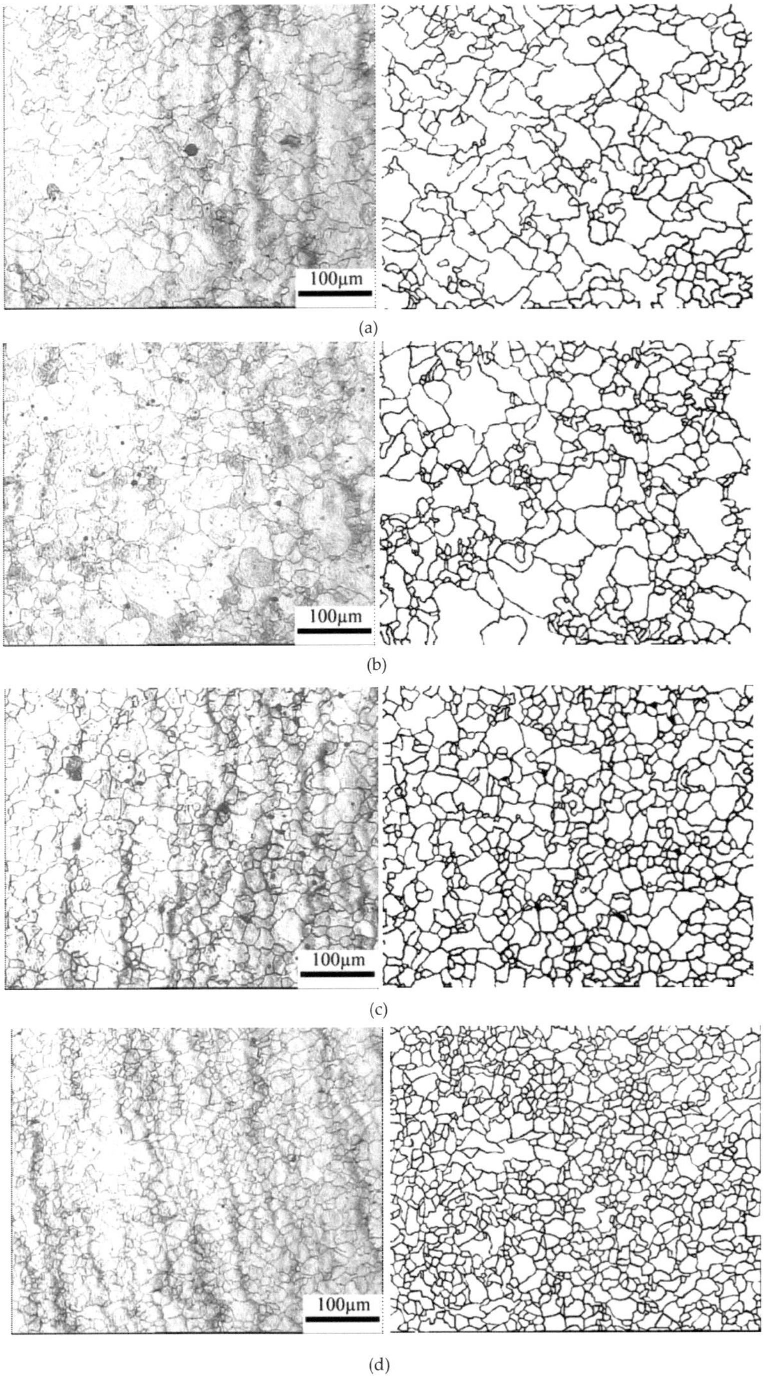

Figure 16. Optical microstructures of 42CrMo high-strength steel at a fix true strain of 0.9, a fix temperature of 1273 K and different strain rates: (a) 0.01 s^{-1}, (b) 0.1 s^{-1}, (c) 1 s^{-1}, (d) 10 s^{-1}.

http://dx.doi.org/10.5772/54285

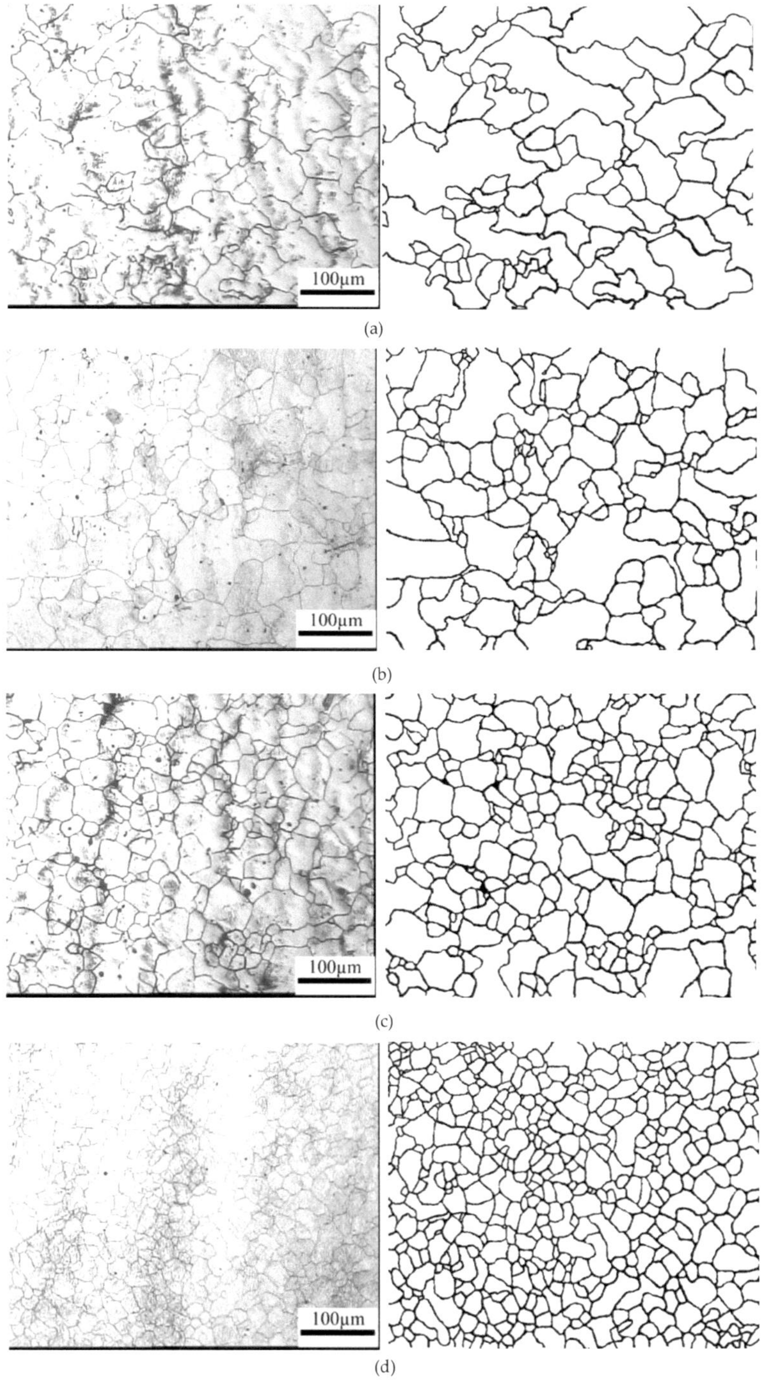

Figure 17. Optical microstructures of 42CrMo high-strength steel at a fix true strain of 0.9, a fix temperature of 1348 K and different strain rates: (a) 0.01 s^{-1}, (b) 0.1 s^{-1}, (c) 1 s^{-1}, (d) 10 s^{-1}.

Fig. 18 shows the grain size distribution of as-extruded 42CrMo high-strength steel undeformed (starting material). Fig. 19 shows the grain size distribution of 42CrMo high-strength steel at a fix true strain of 0.9, a fix temperature of 1123 K and different strain rates: (a) 0.01 s^{-1}, (b) 0.1 s^{-1}, (c) 1 s^{-1}, (d) 10 s^{-1}. Fig. 20 shows the grain size distribution of 42CrMo high-strength steel at a fix true strain of 0.9, a fix temperature of 1198 K and different strain rates: (a) 0.01 s^{-1}, (b) 0.1 s^{-1}, (c) 1 s^{-1}, (d) 10 s^{-1}. Fig. 21 shows the grain size distribution of 42CrMo high-strength steel at a fix true strain of 0.9, a fix temperature of 1273 K and different strain rates: (a) 0.01 s^{-1}, (b) 0.1 s^{-1}, (c) 1 s^{-1}, (d) 10 s^{-1}. Fig. 22 shows the grain size distribution of 42CrMo high-strength steel at a fix true strain of 0.9, a fix temperature of 1348 K and different strain rates: (a) 0.01 s^{-1}, (b) 0.1 s^{-1}, (c) 1 s^{-1}, (d) 10 s^{-1}. As depicted, under a fix temperature of 1123 K the microstructure of the as-cast billet with grain size of 53.1 μm became refined up to about 30.1 μm after upsetting under strain rate 0.01 s^{-1}, to about 25.4 μm under strain rate 0.1 s^{-1}, to about 20.4 μm under strain rate 1 s^{-1}, to about 15.6 μm under strain rate 10 s^{-1}. Under a fix temperature of 1198 K the microstructure of the as-cast billet with grain size of 53.1 μm became refined up to about 33.5 μm after upsetting under strain rate 0.01 s^{-1}, to about 26.9 μm under strain rate 0.1 s^{-1}, to about 21.0 μm under strain rate 1 s^{-1}, to about 18.5 μm under strain rate 10 s^{-1}. Under a fix temperature of 1273 K the microstructure of the as-cast billet with grain size of 53.1 μm became refined up to about 33.5 μm after upsetting under strain rate 0.01 s^{-1}, to about 27.3 μm under strain rate 0.1 s^{-1}, to about 19.7 μm under strain rate 1 s^{-1}, to about 15.7 μm under strain rate 10 s^{-1}. Under a fix temperature of 1348 K the microstructure of the as-cast billet with grain size of 53.1 μm became refined up to about 49.8 μm after upsetting under strain rate 0.01 s^{-1}, to about 38.2 μm under strain rate 0.1 s^{-1}, to about 32.2 μm under strain rate 1 s^{-1}, to about 24.4 μm under strain rate 10 s^{-1}. It can be summarized that under a fix temperature, as deformation strain rate increases, the microstructure of the as-received billet becomes more and more refined due to increasing migration energy stored in grain boundaries and decreasing grain growth time.

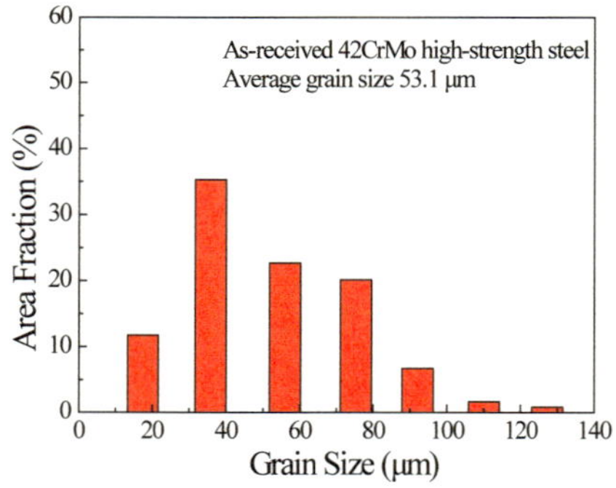

Figure 18. Grain size distribution of as-extruded 42CrMo high-strength steel undeformed (starting material).

http://dx.doi.org/10.5772/54285

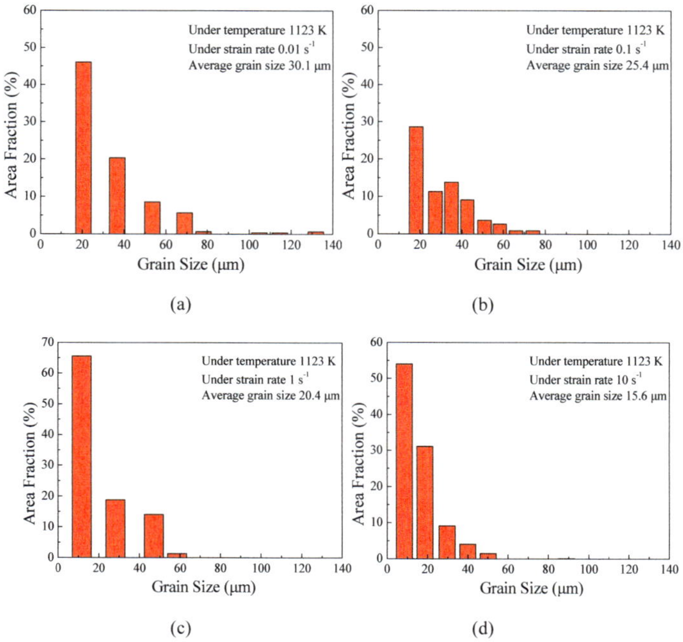

Figure 19. Grain size distribution of 42CrMo high-strength steel at a fix true strain of 0.9, a fix temperature of 1123 K and different strain rates: (a) 0.01 s^{-1}, (b) 0.1 s^{-1}, (c) 1 s^{-1}, (d) 10 s^{-1}.

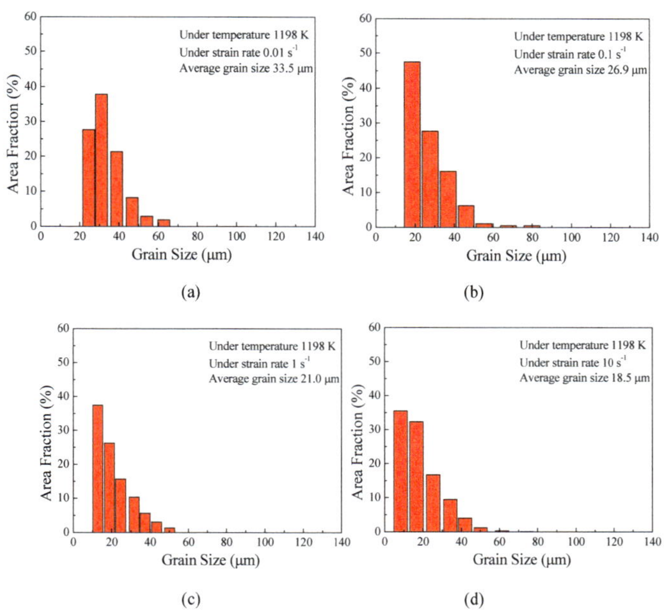

Figure 20. Grain size distribution of 42CrMo high-strength steel at a fix true strain of 0.9, a fix temperature of 1198 K and different strain rates: (a) 0.01 s⁻¹, (b) 0.1 s⁻¹, (c) 1 s⁻¹, (d) 10 s⁻¹.

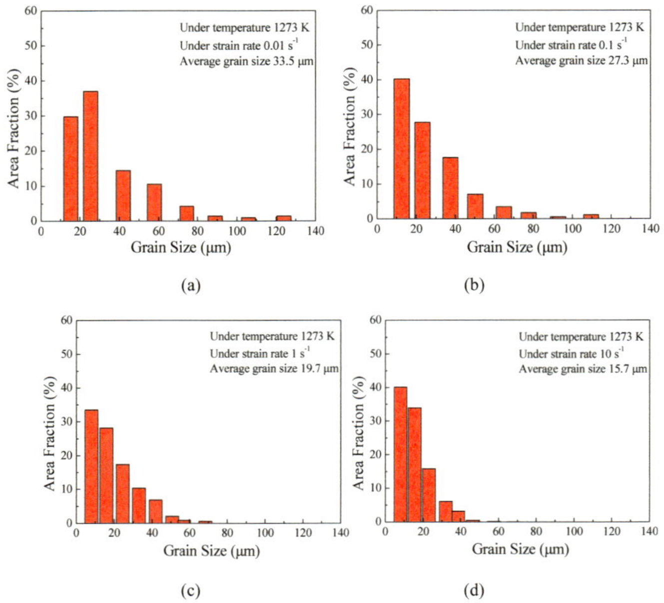

Figure 21. Grain size distribution of 42CrMo high-strength steel at a fix true strain of 0.9, a fix temperature of 1273 K and different strain rates: (a) 0.01 s^{-1}, (b) 0.1 s^{-1}, (c) 1 s^{-1}, (d) 10 s^{-1}.

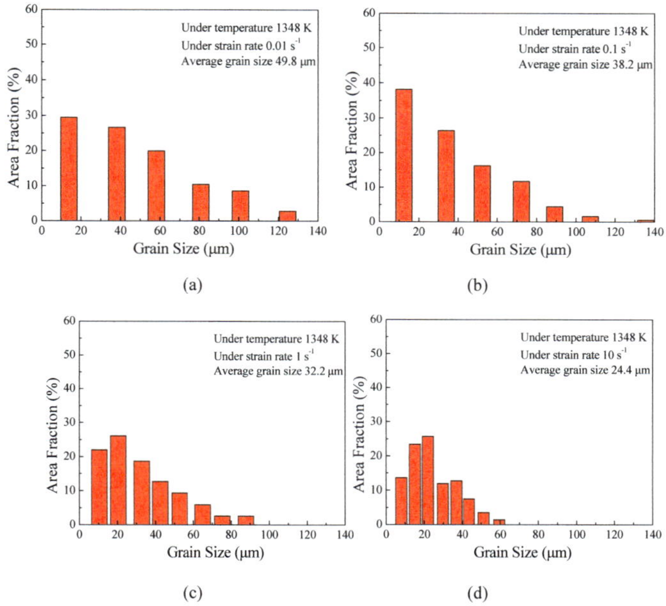

Figure 22. Grain size distribution of 42CrMo high-strength steel at a fix true strain of 0.9, a fix temperature of 1348 K and different strain rates: (a) 0.01 s^{-1}, (b) 0.1 s^{-1}, (c) 1 s^{-1}, (d) 10 s^{-1}.

5. Conclusions

In the deformed material DRX is one of the most important softening mechanisms at high temperatures. DRX occurs during straining of metals at high temperature, characterized by a nucleation rate of low dislocation density grains and a posterior growth rate that can produce a homogeneous grain size when equilibrium is reached. This is a characteristic of low and medium stacking fault energy, SFE, materials e.g., γ-iron, the austenitic stainless steels, and copper. Hot working behavior of alloys is generally reflected on flow curves which are a direct consequence of microstructural changes: the nucleation and growth of new grains, DRX, the generation of dislocations, work hardening, WH, the rearrangement of dislocations, their self-annihilation, and their absorption by grain boundaries, DRV. By the in-depth analysis of the coupling effect in DRX behavior and flow behavior the prediction of DRX evolution can be performed.

The characteristics of softening flow behavior coupling with DRX for as-extruded 42CrMo high-strength steel, as-cast AZ80 magnesium alloy and as-extruded 7075 aluminum alloy have been discussed and summarized as follows: (1) increasing strain rate or decreasing deformation temperature makes the flow stress level increase, in other words, it prevents the occurrence of softening due to DRX and dynamic recovery (DRV) and makes the deformed metals exhibit work hardening (WH); (2) for every curve, after a rapid increase in the stress to a peak value, the flow stress decreases monotonically towards a steady state regime (a steady state flow as a plateau due to DRX softening is more recognizable at higher temperatures and lower strain rates) with a varying softening rate which typically indicates the onset of DRX, and the stress evolution with strain exhibits three distinct stages; (3) at lower strain rates and higher temperatures, the higher DRX softening rate slows down the rate of work-hardening, and both the peak stress and the onset of steady state flow are therefore shifted to lower strain levels.

Three characteristic points (the critical strain for DRX initiation ($A=2.44154\times10^{25}$), the strain for peak stress ($m=3.85582$), and the strain for maximum softening rate (ε_c)) which indicate whether the evolution of DRX can be characterized by the process variables need to be identified from the conventional strain hardening rate curves. A modified Avrami equation, ε_p, has been introduced into this work to describe the kinetics of DRX, and then an integrated calculation process has been presented as an example of as-extruded 42CrMo high-strength steel. By the regression analysis for conventional hyperbolic sine equation, the dependence of flow stress on temperature and strain rate was described, and what's more, the activation energy of DRX (ε^*) and a dimensionless parameter controlling the stored energy ($X_{DRX}=1-\exp\{-[(\varepsilon-\varepsilon_c)/\varepsilon^*]^m\}$) were determined. In further, the strain for maximum softening rate, Q, and the critical strain, Z/A were described by the functions of ε^*. Thus, the evolution of DRX volume fraction was characterized by the modified Avrami type equation including the above parameters. Based on the calculation results of this model, the effect of deformation temperature, strain and strain rate on the recrystallized volume fraction is as follows: as the strain increases, the DRX volume fraction increases and reaches a constant value of 1 meaning the completion of DRX process; for a specific strain rate, the deformation strain required for the same amount of DRX volume fraction increases with decreasing deformation temperature, which means that DRX is delayed to a longer time; for a fixed temperature, the deformation strain required for the same amount of DRX volume fraction increases with increasing strain rate, which also means that DRX is delayed to a longer time.

The microstructures on the section planes of specimens deformed under different strain rates and temperatures were examined and analyzed under the optical microscope. The evolution of grain boundaries and grain size were presented as an example of as-extruded 42CrMo high-strength steel. It can be summarized that under a fix temperature, as deformation strain rate increases, the microstructure of the as-received billet becomes more and more refined due to increasing migration energy stored in grain boundaries and decreasing grain growth time.

Acknowledgements

This work was supported by National Key Technologies R & D Program of China (ZDZX-DFJGJ-08), Science and Technology Committee of Chongqing (cstc2009aa3012-1), Fundamental Research Funds for the Central Universities (Project No. CDJZR11130009).

Author details

Quan Guo-Zheng

Department of Material Processing & Control Engineering, School of Material Science and Engineering, Chongqing University, P.R., China

References

[1] Shokuhfar, A, Abbasi, S. M, & Ehsani, N. Dynamic recrystallization under hot deformation of a PH stainless steel. International Journal of ISSI, (2006).

[2] Quan Guo-zhengTong Ying, Luo Gang, Zhou Jie. A characterization for the flow behavior of 42CrMo steel. Computational Materials Science, (2010).

[3] Kentaro IharaYasuhiro Miura. Dynamic recrystallization in Al-Mg-Sc alloys. Materials Science and Engineering: A, (2004).

[4] Tsuji, N, Matsubara, Y, & Saito, Y. Dyanamic recrystallization of ferrite in interstitial free steel. Scripta Materialia, (1997).

[5] Glover, G, & Sellars, C. M. Static recrystallization after hot deformation of α-iron. Metallurgical Transactions, (1972).

[6] Glover, G, & Sellars, C. M. Recovery and recrystallization during high temperature deformation of α-Iron. Scripta Materialia, (1973).

[7] Hongyan WuLinxiu Du, Xianghua Liu. Dynamic recrystallization and precipitation behavior of Mn-Cu-V weathering steel. Journal of Materials Science & Technology, (2011).

[8] Gourdet, S, & Montheillet, F. A model of continuous dynamic recrystallization. Acta Materialia, (2003).

[9] Smallman, R. E, & Ngan, A. H. W. Physical Metallurgy and Advanced Materials (seventh edition), (2007). Elsevier Ltd., Burlington.

[10] Bert VerlindenJulian Driver, Indradev Samajdar, Roger D. Doherty. Thermo-Mechanical Processing of Metallic Materials, Pergamon Materials Series), (2007). Elsevier Ltd., New York., 11

[11] Guo-Zheng QuanYuan-ping Mao, Gui-sheng Li, Wen-quan Lv, Yang Wang, Jie Zhou. A characterization for the dynamic recrystallization kinetics of as-extruded 7075 aluminum alloy based on true stress-strain curves. Computational Materials Science, (2012).

[12] Guo-Zheng QuanYu Shi, Yi-Xin Wang, Beom-Soo Kang, Tae-Wan Ku, Woo-Jin Song. Constitutive modeling for the dynamic recrystallization evolution of AZ80 magnesium alloy based on stress-strain data. Materials Science and Engineering: A, (2011).

[13] Guo-Zheng QuanTae-Wan Ku, Woo-Jin Song, Beom-Soo Kang. The workability evaluation of wrought AZ80 magnesium alloy in hot compression. Materials & Design, (2011).

[14] Quan Guo-zhengTong Ying, Zhou Jie. Dynamic softening behaviour of AZ80 magnesium alloy during upsetting at different temperatures and strain rates. Proceedings of the Institution of Mechanical Engineers. Part B-Journal of Engineering Manufacture, (2010).

[15] Quan GuozhengLi Guisheng, Chen Tao, Wang Yixin, Zhang Yanwei, Zhou Jie. Dynamic recrystallization kinetics of 42CrMo steel during compression at different temperatures and strain rates. Materials Science and Engineering: A, (2010).

[16] Guo-zheng QuanLei Zhao, Tao Chen, Yang Wang, Yuan-ping Mao, Wen-quan Lv, Jie Zhou. Identification for the optimal working parameters of as-extruded 42CrMo high-strength steel from a large range of strain, strain rate and temperature. Materials Science and Engineering: A, (2012).

[17] Quan Guo-zhengTong Ying, Luo Gang, Zhou Jie. A characterization for the flow behavior of 42CrMo steel. Computational Materials Science, (2010).

[18] Lin, Y. C, & Chen, X. M. A critical review of experimental results and constitutive descriptions for metals and alloys in hot working. Materials & Design, (2011).

[19] Lin, Y. C, Chen, X. M, & Zhong, J. Study of static recrystallization kinetics in a low alloy steel. Computational Materials Science, (2008).

[20] Kim, S. I, & Yoo, Y. C. Dynamic recrystallization behavior of AISI 304 stainless steel. Materials Science and Engineering: A, (2001).

Section 2

Recrystallization Involving Metals

Chapter 3

Simulation of Dynamic Recrystallization in Solder Interconnections during Thermal Cycling

Jue Li, Tomi Laurila, Toni T. Mattila, Hongbo Xu and
Mervi Paulasto-Kröckel

Additional information is available at the end of the chapter

http://dx.doi.org/10.5772/53820

1. Introduction

Solder alloys are widely used bonding materials in electronics industry. The reliability concerns for solder interconnections, which provide both mechanical and electronic connections, are rising with the increasing use of highly integrated components in portable electronic products [1-6]. As shown in Fig. 1, a typical ball grid array (BGA) component board usually consists of a silicone die, molding compound, solder interconnections, and printed wiring board (PWB). In service, all products are subjected to thermal cycles as a result of temperature changes due to component internal heat dissipation or ambient temperature changes. The existence of coefficient of thermal expansion (CTE) mismatches between dissimilar materials (about 16 ppm/°C for PWB and 2.5 ppm/°C for Si die [7]) is the source of deformation and thermomechanical stress in the solder interconnection, which leads to the cracking of the interconnections and failures of the electronic devices.

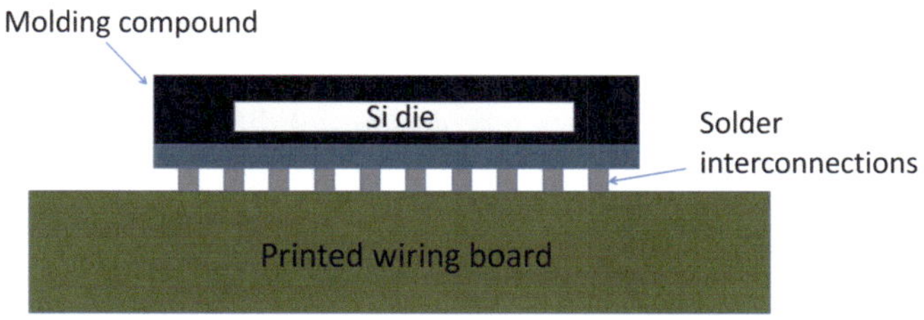

Figure 1. Schematic show of a typical BGA component board.

Since failure of solder interconnections is a typical failure mode in many electronic devices, the reliability and life time prediction of solder interconnections become crucial. Various reliability test and computer-aided simulations have been carried out to study the solder interconnection reliability [8-17]. With the experimental and simulation results, a number of lifetime prediction models have been established and they can be classified into two main categories: strain-based and energy-based. For instance, the Engelmaier model is based on the total shear strain range, the Coffin-Manson model on the plastic strain, and the Darveaux model on the energy density [18-20]. However, microstructural changes in the bulk solder have not yet been included in any of the popular prediction models. Especially the microstructural changes associated with recrystallization and grain growth are of importance because they can significantly affect the mechanical properties and can cause recrystallization induced failure of solder interconnections [21-26]. A new approach for lifetime prediction needs to be developed that takes into account the microstructural changes. In order to achieve this, the first step is to quantitatively study the recrystallization and grain growth occurring in solder interconnections.

In this chapter, the current understanding of the microstructural changes in solder interconnections is introduced, followed by a brief review of the Monte Carlo simulations of grain growth and recrystallization. Then, a new algorithm for predicting dynamic recrystallization in solder interconnections during thermal cycling tests is presented.

2. Microstructural changes of Sn-rich solder interconnections

The microstructure of a solder alloy has a very significant effect on its material properties. A brief introduction of the microstructural changes of tin (Sn) rich solder interconnections is addressed in this section. With the implementation of lead (Pb) free technology in microelectronics [27], tin based lead-free solder alloys have replaced the traditional SnPb alloys. Currently, the three-component tin-silver-copper (SnAgCu) alloy with near-eutectic composition is the most widely used solder alloy. For simplicity, the following microstructural study focuses on one solder alloy with the composition Sn3.0Ag0.5Cu.

2.1. As-solidified microstructures of Sn-rich solder interconnections

After reflow the solder interconnections normally consist of relatively few solidification colonies (less than five) [28]. Micrographs of a typical as-solidified SnAgCu solder interconnection are presented in Fig. 2. The boundaries between the contrasting areas, as seen in polarized light image in Fig. 2b are high-angle boundaries (larger than 15°) between matrixes of solidification colonies composed of Sn cells and Cu_6Sn_5 and Ag_3Sn particles. Within the colony boundaries, a uniformly oriented cellular solidification structure of tin is enclosed [29-33]. The cellular structure of tin is clearly distinguishable as cells are surrounded by eutectic regions. Besides the cellular structure, there are also some large Cu_6Sn_5 and Ag_3Sn bulk intermetallic compound (IMC) precipitates.

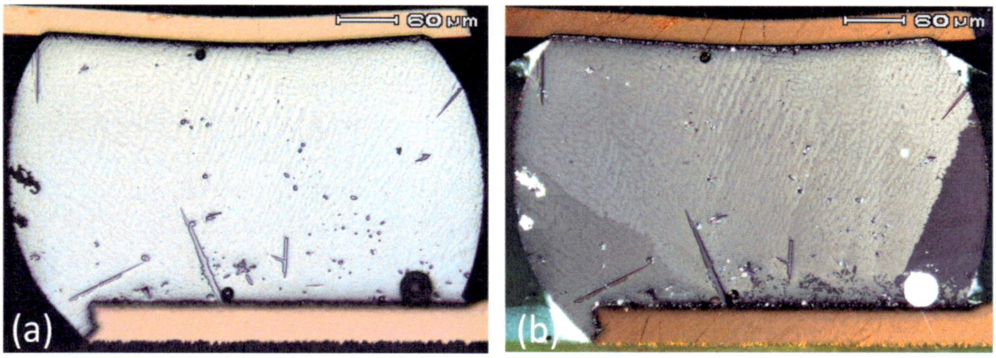

Figure 2. The as-solidified microstructure of a SnAgCu solder interconnection; (a) optical bright field image, (b) cross-polarized light image.

2.2. Recovery and recrystallization of Sn-rich solder interconnections

For decades, the industry has used recrystallization to control microstructures, and static recrystallization of structural metals after deformation is probably the best understood recrystallization process [22]. On the other hand, dynamic recrystallization during cyclic deformation, which occurs in solder interconnections, has received much less attention and is still poorly understood. This is because the related microstructural events are highly complex from the microstructural point of view. The major understanding of this subject is briefly summarized as follows.

Thermal cycling tests with extreme temperatures in the range of about -40 °C to +125 °C are usually carried out to assess the reliability of electronic devices [35, 36]. During thermal cycling, the solder interconnections are under cyclic loading conditions. The induced thermomechanical stresses are often higher than the yield strength of the material, which leads to plastic deformation. A fraction of the energy associated with the plastic deformation of solder interconnections is stored in the metal, mainly in the form of dislocations. The stored energy is subsequently released during restoration, which can be divided into three main processes: recovery, primary recrystallization and grain growth. Recovery and recrystallization are two competing processes, which are driven by the increased internal energy of the deformed solder. Recovery decreases the driving force for recrystallization and thus hinders the initiation of recrystallization. In high stacking fault energy metals such as Sn, the release of stored energy takes place so effectively by recovery that recrystallization will not practically take place [22, 23]. Studies have shown that after a single deformation static recrystallization rarely occurs in Sn-rich solders [37]. However, under dynamic loading conditions such as in thermal cycling tests, recrystallization often takes place in the high stress concentration regions of solder interconnections [28, 38, 39].

Experimental observations indicate that the microstructure of solder interconnections may change significantly during thermal cycling tests. The as-solidified microstructure can transform locally into a more or less equiaxed grain structure by recrystallization. An example is presented in Fig. 3 where the cross-section images of a solder interconnection after 6000 thermal cycles are shown. Part of the solder interconnection was recrystallized near the component side after 6000 thermal cycles (see Fig. 3b). It is noteworthy that optical microscopy with polarized light shows the areas of different orientations with different colors and it is an excellent tool for observing the recrystallized region. In the recrystallized region a continuous network of high angle grain boundaries provides favorable sites for cracks to nucleate and to propagate intergranularly, which can lead to an early failure of the component. This kind of failure mode is regarded as the recrystallization-assisted cracking. More details can be found in the references [14, 26, 28].

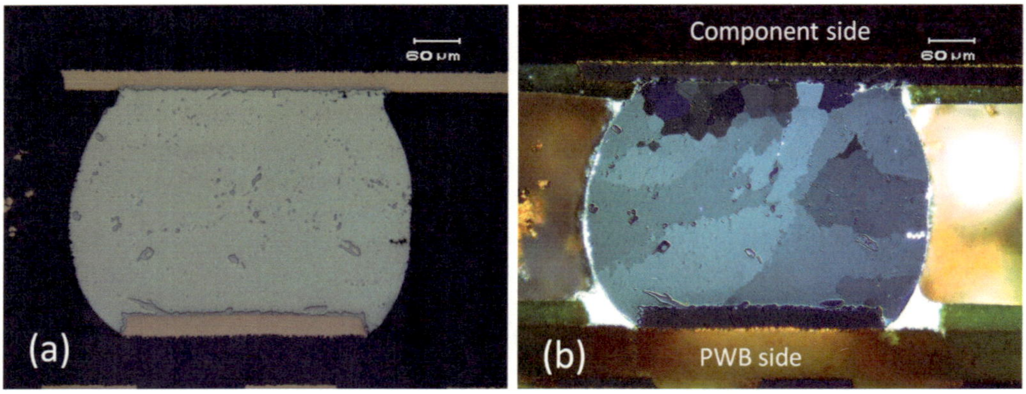

Figure 3. Micrographs of the solder interconnection after 6000 thermal cycles; (a) optical bright field image, (b) cross-polarized light image.

2.3. Effect of intermetallic compound precipitates

In the near-eutectic SnAgCu alloys, mainly two kinds of IMC precipitates, Cu_6Sn_5 and Ag_3Sn, can form upon solidification. The size of intermetallic particles (IMPs) varies a lot: the small and finely distributed IMPs are located at the boundaries of tin cells (eutectic structure) while the relatively large IMPs are randomly distributed in the bulk solder. Fine particles usually prevent the motion of grain boundaries by exerting a pinning force, and therefore, suppress the progress of recrystallization. The influence of fine particles on recrystallization has been studied in earlier work, e.g. [40]. It is believed that fine particles do not remarkably affect the distribution of stored energy within the grains. However, coarse particles exert localized stress and strain concentrations due to the mismatch of mechanical properties and thermal expansion coefficients during thermal cycling. Dislocation density is increased in the particle-affected deformation regions, which provide favorable sites for nucleation of recrystallization.

3. Monte Carlo simulations

Many models have been developed to simulate microstructural evolution, such as vertex model, Monte Carlo (MC) Potts model, phase field model, and cellular automata (CA) model [22, 23, 41, 42]. For modeling recrystallization, the MC Potts model and CA model are perhaps the two most popular candidates. In general, the MC Potts model and CA model are similar to each other since both models include a lattice, use discrete orientations and describe stored energy in terms of a scalar stored energy term. The MC Potts model provides a convenient way to simulate the changes in microstructures and it has been successfully applied to simulate the recrystallization process in solder interconnections as well as in many other applications, e.g. [38, 39, 43-45].

3.1. Monte Carlo simulation of grain growth

The MC grain growth simulation originates from Ising and Potts models for magnetic domain evolution [46]. The Ising model consists of two spin states, namely up and down, and the Potts model allows multiple states (Q states) for each particle in the system. The Potts model has been widely used for modeling material behaviors, such as grain growth and texture evolution, e.g. [47-49].

During the MC grain growth simulation, a continuum microstructure is mapped onto a 2D MC lattice, which can be either triangular or rectangular lattice [50]. In order to initialize the lattice, an integer number S_i (between 1 and Q) is assigned to each lattice site, where Q represents the total number of orientations in the system. Two adjacent sites with different grain orientation numbers are regarded as being separated by a grain boundary and each pair of unlike neighboring sites contributes a unit of grain boundary energy, J, to the system. A group of sites having the same orientation number and surrounded by grain boundaries are considered as a grain. The total energy of the system, E, is calculated by the grain boundary energy contributions throughout all the sites.

$$E = J \sum_{<ij>} \left(1 - \delta_{S_i S_j}\right) \quad (1)$$

where the sum of i is over all N_{MC} sites in the system, the sum of j is over all the nearest-neighbor sites of the site i, and δ_{ij} is the Kronecker delta.

The Monte Carlo method iteratively simulates the grain growth process by the following key steps [51-53].

a. Choose a lattice site i in random

b. If the selected site is an interior site, no reorientation will be tried. Go back to 'step a'

c. If the selected site is at the grain boundary, its neighboring sites are checked. Assign a new orientation number to the site. The new orientation is limited to those orientations of the neighboring grains, and is weighted by the number of neighbors with the same orientation

d. Calculate the energy change ΔE associated the site orientation change

e. If ΔE is non-positive, the attempted reorientation is accepted, otherwise, the old orientation of the site is recovered

f. Increment time regardless of whether the reorientation attempt is accepted or not

g. Go to 'step a' until the end of the simulation

3.2. Monte Carlo simulation of recrystallization

The major differences between the simulation of recrystallization and grain growth are the bulk stored energy and the nucleation process. A fraction of the energy associated with the deformation of material is stored in the metal, mainly in the form of dislocations. The distribution of stored energy is heterogeneous, and therefore each site contributes an amount of stored energy, $H(S_i)$, to the system. The total energy of the system, E, is calculated by summing the volume stored energy and the grain boundary energy contributions throughout all the sites.

$$E = J\sum_{<ij>}(1-\delta_{S_iS_j}) + \sum_i H(S_i) \tag{2}$$

In static recrystallization simulations, the stored energy of each site is positive for unrecrystallized sites and zero for recrystallized sites. However, in the case of dynamic recrystallization the stored energy is a function of both time and position as new energy is added to the lattice continuously.

The nucleation process is modeled by introducing nuclei (small embryos with zero stored energy) into the lattice at random positions. Embryos have orientations that differ from all the other grains of the original microstructure. Embryos can be added to the lattice at the beginning of the simulation or at a regular interval during the simulation. In the reorientation process, if the randomly selected site is unrecrystallized, it will be recrystallized under the condition that the total energy of the system is reduced. If the selected site is recrystallized, the reorientation process is a simulation of the nucleus growth process or the grain growth process.

3.3. Monte Carlo simulation of recrystallization with the presence of particles

The particles are normally introduced to the MC sites at the beginning of the simulation, and those sites have an orientation different from any other grains. The particles do not react, dissolve or grow during the simulation, and thereby are named as inert particles [50]. These assumptions have been proved to be effective, especially for small particles. However, the influence on nucleation stimulation should be considered for large particles. Large particles exert localized stress and strain concentrations and cause the increase of dislocation density in the particle-affected deformation regions, which provide favorable sites for nucleation of recrystallization.

http://dx.doi.org/10.5772/53820

4. Multiscale simulation algorithms

The treatment of heterogeneous nucleation and inhomogeneously deformed material remains a challenge for the MC method. Hybrid methods are needed to perform the task. As steps in this direction, Rollett et al. developed a hybrid model for mesoscopic simulation of recrystallization by combining the MC and Cellular Automaton methods [43]. Song et al. presented a hybrid MC model for studying recovery and recrystallization of titanium at various annealing temperatures after inhomogeneous deformation [44]. Yu et al. combined the MC method with the finite element (FE) method in order to simulate the microstructure of structural materials under forging and rolling [45]. However, most of the early studies are about the simulations of static recrystallization. Recently, Li et al. presented a hybrid algorithm for simulating dynamic recrystallization of solder interconnections during thermal cycling [38, 39].

The details of the multiscale simulation of dynamic recrystallization of solder interconnections are presented as follows. The finite element method is utilized to model macroscale inhomogeneous deformation, and the Monte Carlo Potts model is utilized to model the mesoscale microstructural evolution. Compared to the in situ experimental observations, a correlation between real time and MC simulation time is established. In addition, the effects of intermetallic particles (Cu_6Sn_5 and Ag_3Sn) on recrystallization in solder matrix are included in the simulation.

4.1. Thermal cycling test and model assumptions

Thermal cycling (TC) tests are accelerated fatigue tests, which subject the components and solder interconnects to alternating high and low temperature extremes [35, 36]. The tests are conducted to determine the ability of the parts to resist a specified number of temperature cycles from a specified high temperature to a specified low temperature with a certain ramp rate and dwell time. A typical temperature profile for a TC test is shown in Fig. 4.

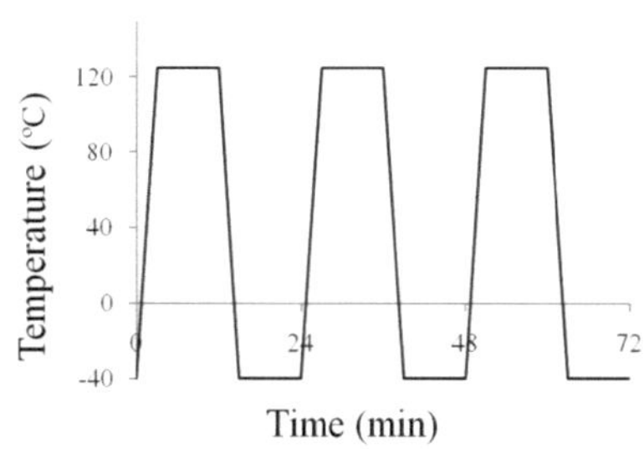

Figure 4. A typical temperature profile with temperature range from -40 °C to 125 °C, with a 6 minute ramp time and an 18 minute dwell time.

Each thermal cycle can be regarded as 'deformation + annealing' and TC tests normally last several thousand cycles before failures of the solder interconnections are detected. The algorithm of dynamic recrystallization is based on the principle that the stored energy of solder is gradually increased during each thermal cycle. Even though recovery consumes a certain amount of the energy, the net change of the energy is always assumed to be positive due to the fact that newly recrystallized grains appear after a certain number of thermal cycles. When a critical value of the energy is reached, recrystallization is initiated. The stored energy is released through the nucleation and growth of new grains, which gradually consume the strain-hardened matrix of high dislocation density.

4.2. Multiscale simulation process

In order to schematically describe the simulation process, a flow chart is shown in Fig. 5. There are three major steps and all the key inputs for the simulation are listed in the boxes, which are on the left side of each step. In Step I, the finite element method is employed to calculate the inelastic strain energy density of the solder interconnections under thermal cycling loads. As discussed above, it is assumed that the net increase of the stored energy takes place after every thermal cycle. In Step II (scaling processes), the stored energy, as the driving force for recrystallization, is mapped onto the lattice of the MC model, and moreover, a correlation is established to convert real time to MC simulation time with the help of the in situ test results. In Step III, the grain boundary energy and the volume stored energy are taken into consideration in the energy minimization calculations to simulate the recrystallization and grain growth processes. Furthermore, intermetallic particles are treated as inert particles and their influence on the distribution of stored energy is included.

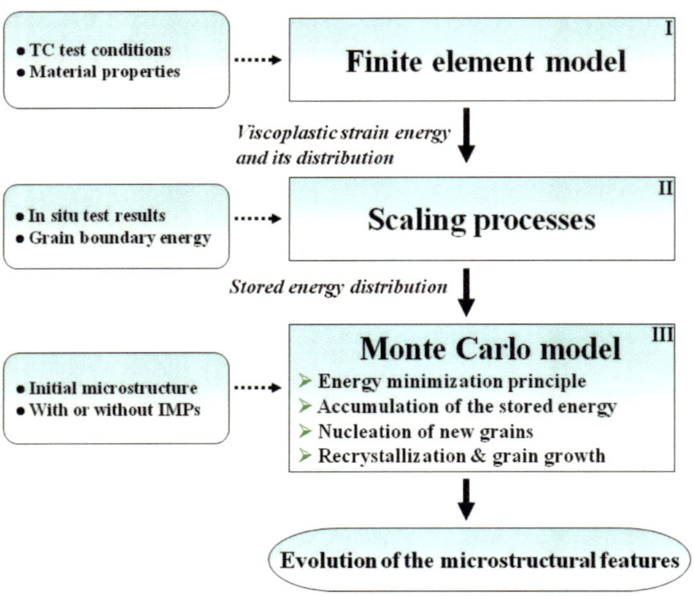

Figure 5. Flow chart for the simulation of microstructural changes in solder interconnections [39].

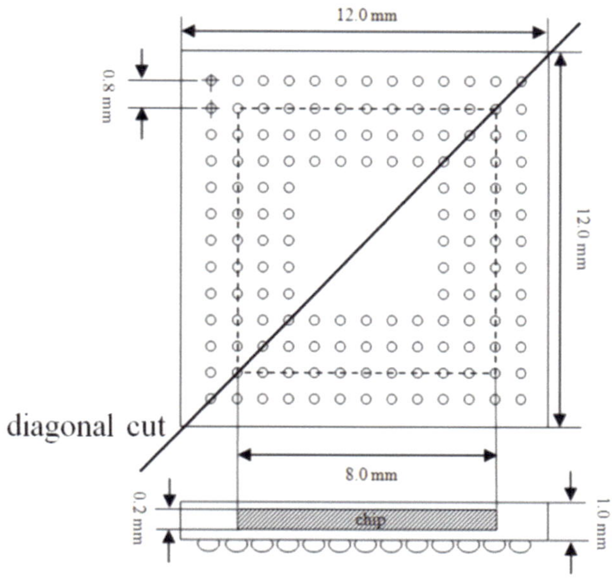

Figure 6. Schematic show of the BGA component under study [39].

4.3. Finite element model

The 3-D finite element analysis (FEA) was performed with the help of the commercial finite element software ANSYS v.12.0. The model was built up according to the experimental set-up where the ball grid array (BGA) components were cut along the diagonal line before the in situ test. A schematic drawing of the BGA component is shown in Fig. 6.

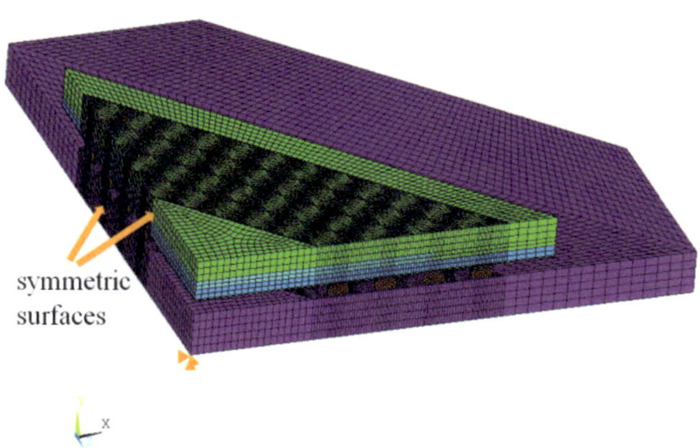

Figure 7. Finite element model for the thermomechanical calculation [39].

Symmetrical design of the component board enabled the employment of the one-fourth model of the package during FE calculation (see Fig. 7). The symmetry boundary conditions were applied to the symmetric surfaces as mechanical constraints, and the central node of the bottom of the PWB was fixed to prevent rigid body motion. Each solder interconnection was meshed with 540 SOLID185 elements as SOLID185 has plasticity, viscoplasticity, and large strain capabilities. The rest of the model was meshed with SOLID45 elements. The total number of nodes and elements of the model was 158424 and 134379, respectively. The SnAgCu solder was modeled by Anand's constitutive model with the parameters provided by Reinikainen et al. [54]. The Anand model is often used for modeling metals' behaviors under elevated temperature when the behaviors become very sensitive to strain rate, temperature, history of strain rate and temperature. The model is composed of a flow equation and three evolution equations that describes strain hardening or softening during the primary stage of creep and the secondary creep stage [55]. The inelastic strain energy is calculated by the integral of the stress with respect to the plastic strain increment.

4.4. In situ experiments

The in situ thermal cycling tests were carried out for the verifications. During the tests, the components were taken out of the test vehicle after every 500 cycles. The solder interconnections were repolished, examined, and then put back to the test vehicle to continue the test. The microstructures of the solder cross sections were examined by optical microscopy with polarized light, which shows the areas of different orientations with different colors (see Fig. 8). The boundaries between the areas of different contrast are the high angle grain boundaries. Fig. 8 (a) and Fig. 8 (b) present the typical microstructures of an outermost solder interconnection after solidification and after 1000 in situ thermal cycling test respectively, and Fig. 8 (c) is the close-up view of the recrystallized region. The dashed rectangle in Fig. 8 (b) shows the domain for the microstructural simulations.

4.5. Scaling processes

Generally speaking, there is no physically meaningful time and length scale in MC simulation, and therefore, it is difficult to compare the simulation results to experimental observations. Some calibration procedures are necessary in order to establish a relationship between real time and MC simulation time. Furthermore, the calculated inelastic strain energy needs to be converted to stored energy via a scaling process before being mapped onto the MC lattice. In the following, length scaling, time scaling and energy scaling are addressed respectively.

4.5.1. Length scaling

MC simulation does not model the behavior of single atoms and accordingly is performed at the mesoscale level. An MC lattice site represents a large cluster of atoms with the typical size being in the order of micrometers. The domain of the 2-D MC simulation covers the chosen region, which belongs to the cross section of the solder interconnection. For instance, the size of the MC domain in this case is a 200×200 square lattice, which covers the 185×185 μm^2 region. Thus, the unit boundary length of the MC model, *s*, equals 0.925μm.

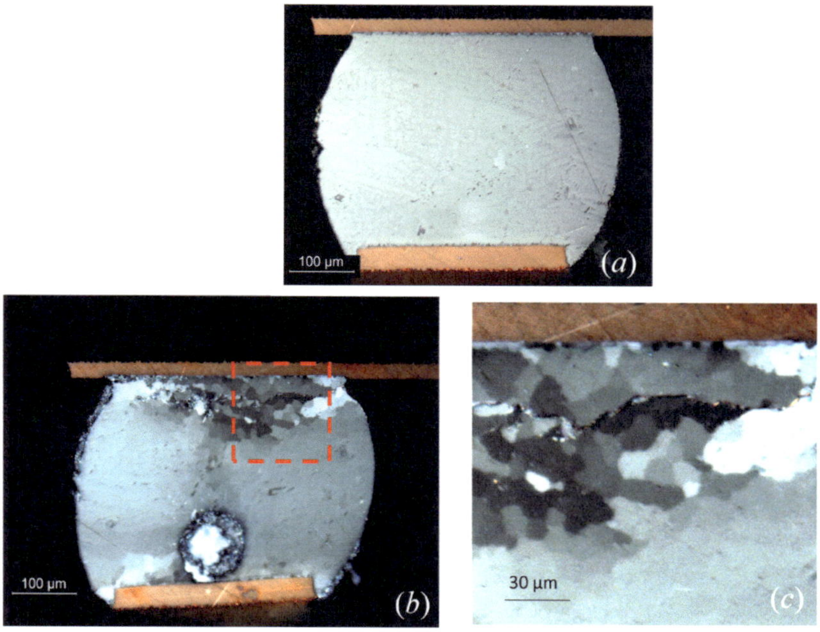

Figure 8. Microstructure of the outermost solder interconnection observed with polarized light (*a*) after solidification, (*b*) after 1000 thermal cycles observed with polarized light, (*c*) close-up view of the top right corner of the cross section [39].

4.5.2. *Time scaling*

One Monte Carlo time step (MCS) is defined as N_{MC} reorientation attempts, where N_{MC} is the total number of sites in the MC lattice. This means that each site is given an opportunity to change its orientation. A correlation between the simulation time t_{MC} [MCS] and real time t [s] is usually expressed in the following form including an apparent activation energy factor (Q_m) as well as an atomic vibration frequency (around 10^{13} Hz).

$$t_{MC} = v \exp(-\frac{Q_m}{RT})t \quad (3)$$

where R is the universal gas constant, T is the temperature, and t is the time.

However, this time scaling process is not employed in the current study due to three main concerns. Firstly, the value of activation energy factor (Q_m) for tin is seldom reported in the literatures. Secondly, the time scaling is extremely sensitive to the value of Q_m. As shown in

Table 1, a possible error of Q_m (within the range from 20 to 107 kJ/mole) leads to a significant difference in time scale, which will finally result in unreliable simulation results. Thirdly, during each thermal cycle, the temperature alternates between a low temperature and a high temperature (e.g. from -40 ºC to 125 ºC, see Fig. 1), which makes it difficult to use Eq. (3).

Q_m (kJ/mole)	20	40	64	70	90	107
t_{MC} /t (MCS/s)	1.9e+10	4.7e+7	3.3e+4	5.4e+3	1.3e+1	7.5e-2

Table 1. Time scales with different activation energy factors

Besides the scaling approach in Eq. (3), other real time scaling approaches have also been developed and successfully applied to various applications. For instance, Safran et al. [56] set the time scale by multiplying the transition probability with a basic attempt frequency, and Raabe [57] scaled the real time step by a rate theory of grain boundary motion.

In order to improve both the accuracy and the efficiency, a new correlation between the simulation time t_{MC} [MCS] and real time t [TC] is established as follows.

$$t_{MC} = \frac{c_1}{c_2} t \tag{4}$$

where c_1 and c_2 are model parameters with the units [MCS] and [TC], respectively.

One thermal cycle is defined as the unit of time instead of using seconds. In this way, the relatively complicated temperature change within a thermal cycle is simplified and included only in the FE simulation and not in the MC simulation. The physical meaning of c_1 is the number of Monte Carlo time steps required for the growth of the newly recrystallized grains during each simulation time interval (STI). A certain amount of external energy is added to the MC lattice at the beginning of each STI and the amount of energy is calculated according to a certain number of thermal cycles, i.e. c_2. Therefore, the parameter c_2 can be considered as a time compression factor and the number of simulation time intervals is equal to t/c_2.

A series of numerical experiments were carried out and the simulated microstructures were compared to Fig. 8 (c) in order to decide the suitable model parameters, c_1 and c_2, in Eq. 4 for the time scaling process. In theory, the value of the time compression factor, c_2, can range from 1 to N_{TC}, where N_{TC} is the total number of thermal cycles. A larger value of c_2 leads to a more efficient calculation with less accuracy. One extreme case is when c_2 equals N_{TC}, and then, there will be only one simulation time interval and the process will be similar to a static recrystallization simulation. Considering efficiency and accuracy, c_2 was assumed to be 100 thermal cycles, making 10 simulation time intervals for the case N_{TC} =1000. Besides c_2, a number of different c_1 were studied. The microstructures of several typical values of c_1 are shown in Fig. 9. A small value of c_1, e.g. $c_1 = 10$ (see Fig. 9 (a)) results in a small average grain size and immature microstructures, where newly introduced embryos do not have enough time to grow up. A large value of c_1, for instance $c_1 = 100$ (see Fig. 9 (d)), leads to a relatively large average grain size as well as long and narrow grain shapes. Fig 9 (c) is regarded as a good representative of the studied microstructure in terms of the similar average grain size and more or less equiaxed grain shapes. Hence, 'c_1 = 50' was used for the rest of the simulations.

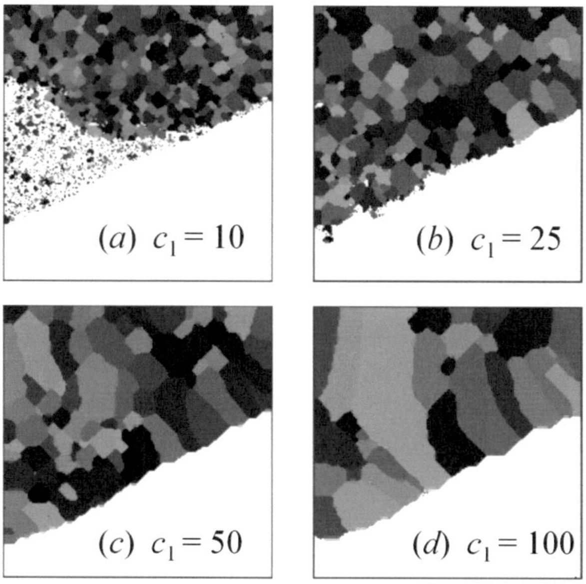

Figure 9. Simulated microstructures with four typical c_1 values [39].

4.5.3. Energy scaling

In contrast to the situation in a static recrystallization process, the stored energy of the MC sites accumulates during dynamic loading. The correlation between the external work and the stored energy is discussed as follows. It is well known that a fraction of the external work, typically varying between 1% and 15%, is stored in the metallic material during deformation [58]. According to the parameter study carried out in [38] with considering the effect of recovery, the most suitable retained fraction, 5%, is used for the unrecrystallized region in the current MC simulation. Zero is assumed to be the retained fraction for the recrystallized region. This assumption is valid due to the fact that only primary recrystallization has been found in the experimental observation of solder interconnections. It should be mentioned that if non-zero retained fraction is used for the recrystallized region, the present model is capable of predicting secondary recrystallization, which may be applicable in other cases.

The energy scaling is based on the principle that the ratios of the volume stored energy to the grain boundary energy should be equated in the MC model and the physical system [51]. The driving force due to the grain boundary energy, P^{grgr}, equals $\gamma/\langle r\rangle$, where γ is the grain boundary energy per unit area and $\langle r\rangle$ is the mean grain radius. Since high angle grain boundaries are of interest to the present investigation, the value γ =0.164 J/m^2 is used for the high angle grain boundary energy of Sn [22].

In the MC model, the driving force due to the stored energy density is given by $P_{MC}^{vol} = H / A$, where H is the volume stored energy of a site and A is the area of that site ($A = s^2$ in the 2-D square lattice). The driving force due to the grain boundary energy is given by

$$P_{MC}^{grgr} = \frac{\gamma_{MC}}{<r>_{MC}} = \frac{J}{s<r>_{MC}} = \frac{J}{s<r>} \tag{5}$$

where γ_{MC} and $<r>_{MC}$ are the grain boundary energy per unit and the mean grain radius in the MC model, respectively. Each unlike pair of nearest neighboring sites contributes a unit of grain boundary energy J to the system. Because of the existence of the length scale factor, s, an ideal prediction should satisfy the requirement that the mean grain radiuses in the model and in the physical system are equal, $<r>_{MC} = <r>$.

In the physical system, the ratio of the volume stored energy to the grain boundary energy per unit volume is

$$\frac{P^{vol}}{P^{grgr}} = \frac{P^{vol} <r>}{\gamma} \tag{6}$$

In the MC model, the ratio is given by

$$\frac{P_{MC}^{vol}}{P_{MC}^{grgr}} = \frac{H<r>}{J\,s} \tag{7}$$

Equating the ratios of the model and the physical system, and rearranging, gives

$$H = \left(\frac{P^{vol}\,s}{\gamma} \right) J \tag{8}$$

The increment of H can be easily calculated by Eq. (8), where s and γ are known parameters. P^{vol} distribution is obtained by scaling and mapping the energy density distribution calculated by FEM onto the MC lattice. J is a unit of grain boundary energy in the MC model. It is noteworthy that the absolute value of J is not essential and knowing the ratio, H/J, is sufficient for the MC simulation.

4.6. Nucleation

Nucleation stage is very crucial for the simulation of recrystallization. The recrystallization process is modeled by introducing nuclei (small embryos with zero stored energy) into the lattice at a constant rate, i.e. continuous nucleation. According to the locations of the valid nuclei, the nucleation is non-uniform. Although the locations of nuclei are randomly chosen, the volume stored energy of the chosen sites has to be larger than the critical stored energy, H_{cr}, before the nuclei are placed in their sites. In this way, the sites with high stored energy (e.g. near interfaces, grain boundaries, and IMPs) will have a high probability of nucleation. The critical stored energy, H_{cr}, is set to be $2J$ so that a single-site isolated embryo can grow as a new grain. Detailed discussion about the critical stored energy and critical embryo size were presented in the reference [51].

4.7. Treatment of large intermetallic particles

The intermetallic particles in SnAgCu solder are mainly Cu_6Sn_5 and Ag_3Sn. The size of IMPs varies a lot and only coarse IMPs (particle size of 1 μm or above) are studied in the MC simulation. Zener-type particle pinning is not considered in the current model for the sake of simplicity. It is believed that fine particles do not remarkably affect the distribution of stored energy within the grains; however, coarse particles exert localized stress concentrations due to the mismatch of mechanical properties and thermal expansion coefficients. The large particles have a significant influence on stimulating nucleation.

The IMPs are introduced into the MC simulation as inert particles. They are assigned an orientation different from any of the surrounding grains and are not allowed to be reoriented during the simulation. Thus, the inert particles do not grow or move. To include the IMPs in the FE model is not realistic and too computationally expensive in view of the fact that the size and shape of IMPs vary a lot and the locations of IMPs are randomly distributed in the bulk solder. Instead, the size of the particle-affected deformation region is estimated and an energy amplification factor (EAF) distribution is introduced in order to consider the effects of IMPs.

A 2-D FE simulation was carried out to study the particle-affected deformation region and the energy amplification factor distribution. The model was composed of a solder matrix (25×25 μm^2) and a round IMP (radius = 5 μm). The material properties of the relatively soft solder matrix and the hard IMP were from the reference [59]. The loading was defined by applying the displacement on the top and right edges, i.e. bi-axial tension. The von Mises stress at the edges was not influenced by the IMP, and thereby, it was defined as one unit for the sake of normalization. The calculated stress contour and the 'EAF vs. distance' curve are shown in Fig. 10. According to the FE simulation results, the following assumptions are made with the purpose of treating IMPs in the MC simulation. Within a distance of approximately one particle radius, the calculated stored energy is amplified by a certain energy amplification factor before mapped onto the Monte Carlo lattice. The EAF for the site close to the IMP is about '1.12' and the EAF decreases linearly to '1'for the site more than one radius distance away from the IMP. In order to realize this, a new matrix storing the EAF distribution is introduced to the Monte Carlo algorithm. Every time the stored energy of a site is updated during the recrystallization simulation, the energy increment is multiplied by the associated EAF before being added to the site. By introducing the EAF, stored energy density close to IMPs is higher than usual, leading to a higher driving force for nucleation and growth of recrystallized grains. Thus, the particle stimulated nucleation is well taken into consideration in the MC simulation.

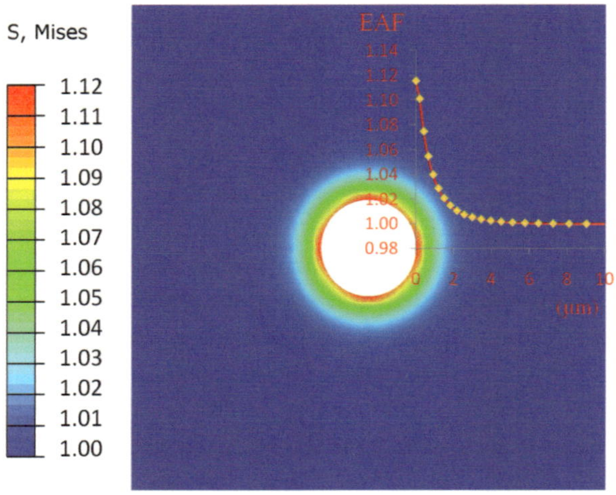

Figure 10. von Mises stress contour and 'EAF vs. distance' curve [39].

5. Simulation results and experimental verifications

5.1. Simulation with no presence of IMPs

A solder interconnection was selected to verify the performance of the presented algorithm. The interconnection was the second diagonal solder interconnection from the right end as shown in Fig. 7. The heterogeneous deformation of the interconnection after 1000 thermal cycles is shown in Fig. 11 (a). The image was taken with bright light before repolishing. The persistent slip bands are visible in the image, which show the severe plastic deformation near the interface on the component side. The distribution of the heavily deformed regions agrees well with the calculated inelastic strain energy density distribution (see Fig. 11 (b)). This agreement verifies that the energy input for the microstructural simulation is valid. The dashed rectangle in Fig. 11 (b) shows the domain of the following microstructural simulation.

Different from the experimental observations, the Monte Carlo simulation results offer a continuing process of the microstructural evolution. Three snapshots from the MC simulation with a time interval, 500 thermal cycles (TCs), are shown in Fig. 12. On the left side of the simulated microstructures, the related micrographs are presented. According to the simulation results, the incubation time for the recrystallization is about 400 TCs. During the incubation period, the stored energy is accumulated, but the magnitude remains below the critical value. As a result, no new grains are formed before 400 TCs. The upper right corner of the solder interconnection is the location where the highest inelastic strain energy is concentrated (see Fig. 11 (b)). It is this very same location where the magnitude of the stored energy first exceeds the critical value and recrystallization is initiated (see Fig. 12 (a) and Fig. 12 (d)). Then, as shown in Fig. 12 (e), the recrystallized region expands towards the lower left of the interconnection, which is in good agreement with the experimental finding (see Fig. 12 (b)). By comparing Fig. 12 (e) and 12 (*f*), it is found that the migration rate of the recrystallization front slows down during the period from 1000 TCs to 1500 TCs due to the decreasing driving force in the lattice. In the micrograph, Fig. 12 (c), cracks and voids are obvious, meaning that the continuity assumption of the finite element model is no longer valid. Therefore, the difference between the experimental finding and the simulated microstructure, Fig. 12 (f), is understandable. A possible solution is to simulate the behaviors of cracks and voids in the Monte Carlo model, output the microstructures to the finite element model, and then, use the calculated results as the inputs for the next round Monte Carlo simulation.

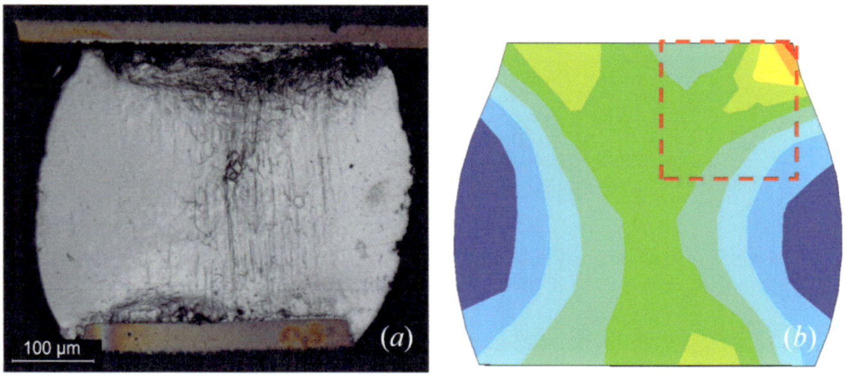

Figure 11. *a*) Plastic deformation of the solder interconnection after 1000 thermal cycles, (*b*) FEM-calculated inelastic strain energy density distribution, dashed rectangle shows the domain of the microstructural simulation [39].

5.2. Simulation with presence of IMPs

There was no obvious IMP-affected recrystallization in any of the in situ samples. Most of the observed recrystallized microstructures were located close to the interface region where the stored energy density was the highest. In order to focus on the influence of the IMPs and exclude the effects of the heterogeneous energy distribution, a uniform stored energy density distribution was assumed during the simulation. The assumption is valid when the calculation domain is located in the center part of the solder interconnection, where the energy magnitude is relatively low and energy distribution is quite uniform. Furthermore, the energy amplification factors introduced in Section 4.5.3 were used to increase the energy around the IMPs.

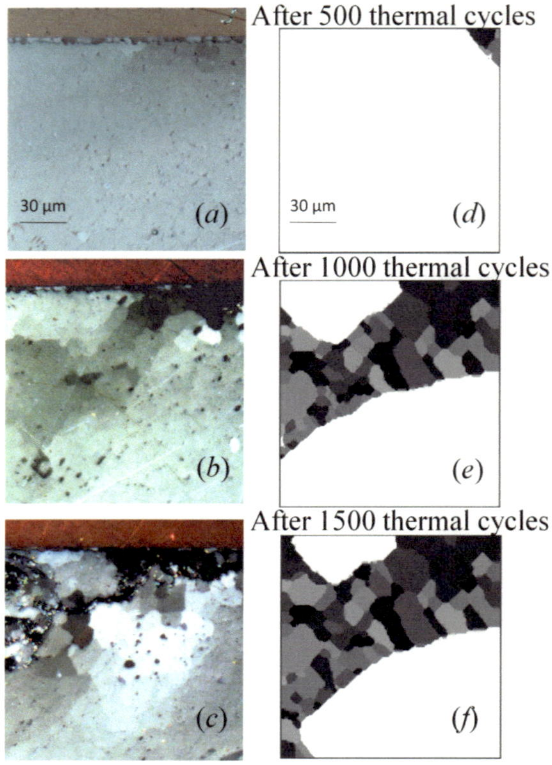

Figure 12. *a*), (*b*), and (*c*) are experimentally observed microstructures of the same location; (*d*), (*e*), and (*f*) are simulated microstructures after 500, 1000, and 1500 thermal cycles respectively [39].

A micrograph from a normal thermal cycling test was used to verify the simulation results (see Fig. 13 (a)). The sample was examined after 5000 TCs and the micrograph was taken from the center of the cross section. The major IMPs were highlighted in Fig. 13 (a) for easy recognition and Fig. 13 (b) was used as the initial microstructure for the microstructural simulation. As compared with in situ samples, solder interconnections in normal thermal cycling tests experience moderate plastic deformation, and thereby, require long incubation time for recrystallization.

Four snapshots (after 1500, 3000, 4000, and 5000 TCs respectively) of the simulated microstructural evolution are presented in Fig. 14. Since there are no interfaces and pre-existing grain boundaries in the calculation domain, the intermetallic particles are the most favorable sites for nucleation in this case. The particle stimulated nucleation is shown in the simulation results and the initiation of recrystallization near the IMPs is clearly visible in Fig. 14 (a). The growth of the new grains at the expense of the strain-hardened matrix is presented in Fig. 14 (b)-(*d*). After 4000 TCs, since the whole matrix is consumed by the recrystallized grains and most of the stored energy is released, there is practically no difference between Fig. 14 (c) and Fig. 14 (d). Furthermore, it is found that the IMPs tend to located at the grain boundaries or triple junctions of the new grains as a result of the energy minimization calculation, which is consistent with the experimental results (see Fig. 13 (a)).

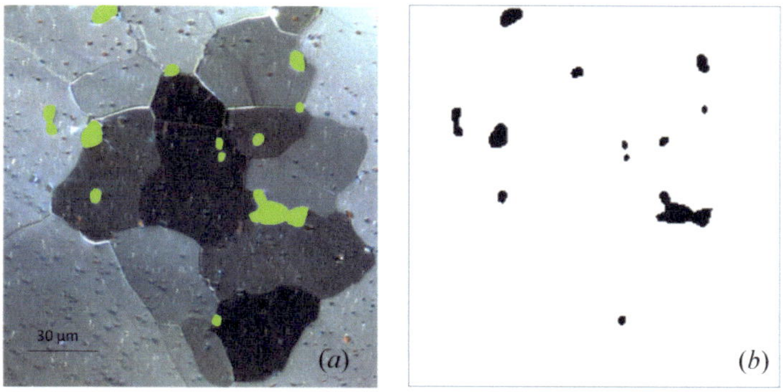

Figure 13. *a*) Micrograph shows IMP-affected recrystallization, (*b*) initial microstructure for the Monte Carlo simulation [39].

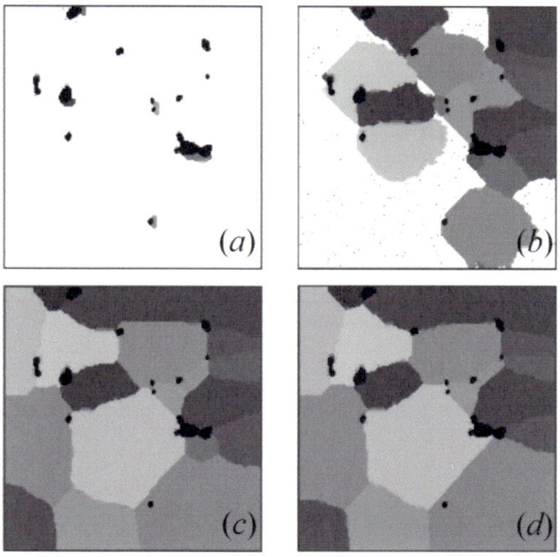

Figure 14. Simulated microstructural evolution with the present of IMPs, (*a*) after 1500 TCs, (*b*) after 3000 TCs, (*c*) after 4000 TCs, (*d*) after 5000 TCs [39].

6. Conclusions

In this chapter, the current understanding of the microstructural changes in solder interconnections was introduced, followed by a brief review of the Monte Carlo simulations of grain growth and recrystallization. A new algorithm for predicting dynamic recrystallization in solder interconnections during thermal cycling tests was presented. The algorithm was realized by combining a Potts model based Monte Carlo method and a finite element method. The correlation between real time and MC simulation time was established with the help of the in situ test results. Recrystallization with the presence of intermetallic particles in solder matrix was simulated by introducing the energy amplification factors in the particle-affected deformation regions. The algorithm predicts the incubation period of the recrystallization as well as the growth tendency of the recrystallized region, which are in good agreement with the experimental findings. Although the research for the microstructural simulation of solder interconnections is still at its primary stage, the presented algorithm shows potential for better reliability assessment of solder interconnections used in the electronics industry.

Acknowledgement

The authors would like to acknowledge Academy of Finland for financial support.

Author details

Jue Li, Tomi Laurila, Toni T. Mattila, Hongbo Xu and Mervi Paulasto-Kröckel

Department of Electronics, Aalto University School of Electrical Engineering, Espool, Finland

References

[1] R. R. Tummala, *Fundamentals of Microsystems Packaging*. New York: McGraw-Hill, 2001.

[2] J. H. Lau (Ed.), *Solder Joint Reliability: Theory and Applications*. New York: Van Nostrand Reinhold, 1991.

[3] J. H. Lau and Y. -H. Pao, *Solder joint reliability of BGA, CSP, flip chip, and fine pitch SMT assemblies*. New York: McGraw-Hill, 1997.

[4] G. Q. Zhang, W. D. Van Driel, and X. J. Fan, *Mechanics of Microelectronics*. Netherlands: Springer, 2006.

[5] J. K. Shang, Q. L. Zeng, L. Zhang, and Q. S. Zhu, "Mechanical fatigue of Sn-rich Pb-free solder alloys," *Journal of Materials Science: Materials in Electronics*, vol. 18, pp. 211-227, 2007.

[6] E. H. Wong, S. K. W. Seah, and V. P. W. Shim, "A review of board level solder joints for mobile applications," *Microelectronics Reliability*, vol. 48, pp. 1747-1758, 2008.

[7] Y. S. Touloukian and C. Y. Ho, *Thermal Expansion: Metallic Elements and Alloys*, New York: IFI/Plenum, 316 p, 1975.

[8] T. T. Mattila, T. Laurila, and J. K. Kivilahti, "Metallurgical factors behind the reliability of high density lead-free interconnections," in E. Suhir, C. P. Wong, and Y. C. Lee, *Micro-and Opto-Electronic Materials and Structures: Physics, Mechanics, Design, Reliability, Packaging*, New York, Springer, pp. 313-350, 2007.

[9] J. Karppinen, T. T. Mattila, and J. K. Kivilahti, "Formation of thermomechanical interconnection stresses in a high-end portable product," *The Proceedings of the 2nd Electronics System Integration Technology Conference*, London, UK, September 1-4, 2008, IEEE/EIA CPMT, (2008), pp. 1327-1332.

[10] J. –P. Clech, "Acceleration factors and thermal cycling test efficiency for lead-free Sn-Ag-Cu assemblies", in *Proc SMTA International 2005*, Chicago, pp. 902-917.

[11] R. Darveaux, "Effect of simulation methodology on solder joint crack growth correlation and fatigue life prediction", *Journal of Electronic Packaging*, vol. 124, pp. 147-154, 2002.

[12] M. Dusek, M. Wickham, and C. Hunt, "The impact of thermal cycling regime on the shear strength of lead-free solder joints", *Soldering & Surface Mount Technology*, vol. 17, no. 2, pp. 22-31, 2005.

[13] J. Gong, C. Liu, P. P. Conway, and V. V. Silberschmidt, "Micromechanical modelling of SnAgCu solder joint under cyclic loading: Effect of grain orientation", *Computational Materials Science*, vol. 39, pp. 187-197, 2007.

[14] J. Li, J. Karppinen, T. Laurila, and J. K. Kivilahti, "Reliability of Lead-Free solder interconnections in Thermal and Power cycling tests", *IEEE Transactions on Components and Packaging Technologies.*, vol. 32, no. 2, pp. 302-308, 2009.

[15] T. T. Mattila, V. Vuorinen, and J. K. Kivilahti, "Impact of printed wiring board coatings on the reliability of lead-free chip-scale package interconnections," *Journal of Materials Research*, vol. 19, no. 11, pp. 3214-3223, 2004.

[16] G. Zeng, S. B. Xue, L. Zhang, Z. Sheng, and L. L. Gao, "Reliability evaluation of SnAgCu/SnAgCuCe solder joints based on finite element simulation and experiments", *Soldering & Surface Mount Technology*, vol. 22, no. 4, pp. 57-64, 2010.

[17] C. J. Zhai, Sidharth, and R. Blish, "Board level solder reliability versus ramp rate and dwell time during temperature cycling", *IEEE Transactions on Device and Materials Reliability*, vol. 3, no.4, pp. 207-212, 2003.

[18] W. Engelmaier, "Fatigue life of leadless chip carrier solder joints during power cycling," *IEEE Transactions on Components, Hybrids, and Manufacturing Technology*, vol. 6, pp. 232-237, 1983.

[19] L. F. Coffin, "Low cycle fatigue – A review," *Applied Materials Research*, vol. 1, pp. 129-141, 1962.

[20] R. Darveaux, "Effect of simulation methodology on solder joint crack growth correlation," *The Proceedings of the 50th Electronic Component and Technology Conference*, Chandler, AZ, May 21-24, 2000, pp. 1048-1058.

[21] L. Zhang, S. B. Xue, L. L. Gao, Y. Chen, S. L. Yu, Z. Sheng, and G. Zeng, "Microstructure and creep properties of Sn-Ag-Cu lead-free solders bearing minor amounts of the rare earth cerium", *Soldering & Surface Mount Technology*, vol. 22, no. 2, pp. 30-36, 2010.

[22] F. J. Humphreys and M. Hatherly, *Recrystallization and Related Annealing Phenomena*, 2nd ed., Oxford: Elsevier Ltd, 2004

[23] R. D. Doherty, D. A. Hughes, F. J. Humphreys, J. J. Jonas, D. Juul Jensen, M. E. Kassner, W. E. King, T. R. McNelley, H. J. McQueen, and A. D. Rollett, "Current issues in recrystallization," *Materials Science and Engineering A*, 238, pp. 219 – 274, 1997.

[24] S. Terashima, K. Takahama, M. Nozaki, and M. Tanaka, "Recrystallization of Sn grains due to thermal strain in Sn-1.2Ag-0.5Cu-0.05N solder," *Materials Transactions, Japan Institute of Metals*, vol. 45, no. 4, pp. 1383-1390, 2004.

[25] S. Dunford, S. Canumalla, and P. Viswanadham, "Intermetallic morphology and damage evolution under thermomechanical fatigue of lead (Pb)-free solder interconnections," *The Proceedings of the 54th Electronic Components and Technology Conference*, June 1-4, 2004, Las Vegas, NV, USA, IEEE/EIA/CPMT, (2004), pp. 726-736.

[26] J. J. Sundelin, S. T. Nurmi, and T. K. Lepistö, "Recrystallization behavior of SnAgCu solder joints," *Materials Science and Engineering A*, vol. 474, pp. 201-207, 2008.

[27] Directive 2002/95/EC of the European Parliament and of the Council on the Restriction of the Use of Hazardous Substances in Electrical and Electronic Equipment (RoHS), Jan. 27th, 2003.

[28] T. T. Mattila and J. K. Kivilahti, "The role of recrystallization in the failure mechanism of SnAgCu solder interconnections under thermomechanical loading," *IEEE Transactions on Components and Packaging Technologies*, vol. 33, no. 3, pp. 629-635, 2010.

[29] A. LaLonde, D. Emelander, J. Jeannette, C. Larson, W. Rietz, D. Swenson, and D. W. Henderson, "Quantitative metallography of β-Sn dendrites in Sn3.8Ag0.7Cu ball grid array solder balls via electron backscatter diffraction and polarized light microscopy," *Journal of Electronic Materials*, vol. 33, no. 12, pp. 1545-1549, 2004.

[30] D. Henderson, J. J. Woods, T. A. Gosseling, J. Bartelo, D. E. King, T. M. Korhonen, M. A. Korhonen, L. P. Lehman, E. J. Cotts, S. K. Kang, P. Lauro, D.-Y. Shih, C. Goldsmith, and K. J. Puttliz, "The microstructure of Sn in near eutectic Sn-Ag-Cu alloy solder joints and its role in thermomechanical fatigue," *Journal of Materials Research*, vol. 19, no. 6, pp. 1608-1612, 2004.

[31] S. Terashima and M. Tanaka, "Thermal fatigue properties of Sn-1.2Ag-0.5Cu-xNi Flip Chip interconnects," *Materials Transactions*, vol. 45, no. 3, pp. 681-688, 2004.

[32] S. K. Kang, P. A. Lauro, D.-Y. Shih, D. W. Henderson, and K. J. Puttlitz, "Microstructure and mechanical properties of lead-free solders and solder joints used in microelectronic applications," *IBM Journal of Research and Development*, vol. 49, no. 4/5, pp. 607-620, 2005.

[33] A. U. Telang, T. R. Bieler, J. P. Lucas, K. N. Subramanian, L. P. Lehman, Y. Xing, and E. J. Cotts, "Grain-boundary character and grain growth in bulk tin and bulk lead-free solder alloys," *Journal of Electronic Materials*, vol. 33, no. 12, pp. 1412-1423, 2004.

[34] L. P. Lehman, S. N. Athavale, T. Z. Fullem, A. C. Giamis, R. K. Kinyanjui, M. Lowenstein, K. Mather, R. Patel, D. Rae, J. Wang, Y. Xing, L. Zavalij, P. Borgesen, and E. J. Cotts, "Growth of Sn and intermetallic compounds in Sn-Ag-Cu solder," *Journal of Electronic Materials*, 33, 12, pp. 1429-1439, 2004.

[35] JESD22-A104C, "Temperature Cycling," Jedec Solid State Technology Association, (2005), 16 p.

[36] IPC-TM-650 rev. A, "Thermal Shock and Continuity, Printed Board," The Institute for Interconnecting and Packaging Electronic Circuits, (1997), 2 p.

[37] S. Miettinen, *Recrystallization of Lead-free Solder Joints under Mechanical Load*, Master's Thesis (in Finnish), Espoo, (2005), 84 p.

[38] J. Li, T. T. Mattila, and J. K. Kivilahti, "Multiscale simulation of recrystallization and grain growth of Sn in lead-free solder interconnections," *Journal of Electronic Materials*, vol. 39, no. 1, pp. 77-84, 2010.

[39] J. Li, H. Xu, T. T. Mattila, J. K. Kivilahti, T. Laurila, and M. Paulasto-Kröckel, "Simulation of dynamic recrystallization in solder interconnections during thermal cycling," *Computational Materials Science*, vol. 50, pp. 690-697, 2010.

[40] A. D. Rollett, D. J. Srolovitz, M. P. Anderson, and R. D. Doherty, "Computer simulation of recrystallization — III. Influence of a dispersion of fine particles," *Acta Metallurgica*, vol. 40, no. 12, pp. 3475-3495, 1992.

[41] M. A. Miodownik, "A review of microstructural computer models used to simulate grain growth and recrystallisation in aluminum alloys," *Journal of Light Metals*, vol. 2, no. 3, pp. 125-135, 2002.

[42] N. Yazdipour, C.H.J. Davies, and P.D. Hodgson, "Microstructural modeling of dynamic recrystallization using irregular cellular automata," Computational Materials Science, vol.44, 566-576, 2008.

[43] A. D. Rollett and D. Raabe, "A hybrid model for mesoscopic simulation of recrystallization," *Computational Materials Science*, vol. 21, no. 1, pp. 69-78, 2001.

[44] X. Song and M. Rettenmayr, "Modelling study on recrystallization, recovery and their temperature dependence in inhomogeneously deformed materials," *Materials Science and Engineering A*, vol. 332, no. 1-2, pp. 153-160, 2002.

[45] Q. Yu and S. K. Esche, "A Multi-scale approach for microstructure prediction in thermo-mechanical processing of metals," *Journal of Materials Processing Technology*, vol. 169, pp. 493-502, 2005.

[46] R. B. Potts, "Some generalized order-disorder transformations," *Proceedings of the Cambridge Philosophical Society*, vol. 48, pp. 106-109, 1952.

[47] M. P. Anderson, D. J. Srolovitz, G. S. Grest, and P. S. Sahni, "Computer simulation of grain growth I — Kinetics," *Acta Metallurgica*, vol. 32, no. 5, pp. 783-791, 1984.

[48] D. J. Srolovitz, M. P. Anderson, G. S. Grest, and P. S. Sahni, "Computer simulation of grain growth — III. Influence of a particle dispersion," *Acta Metallurgica*, vol. 32, no. 9, pp. 1429-1438, 1984.

[49] A. D. Rollett, D. J. Srolovitz, and M. P. Anderson, "Simulation and theory of abnormal grain growth — Variable grain boundary energies and mobilities," *Acta Metallurgica*, vol. 37, no. 4, pp. 2127-1240, 1989.

[50] A. D. Rollett, "Overview of modeling and simulation of recrystallization" *Progress in Materials Science*, vol. 42, no. 1-4, pp 79-99, 1997.

[51] A. D. Rollett, P. Manohar, *Chapter 4 in Continuum Scale Simulation of Engineering Materials: Fundamentals-Microstructures-Process Applications*, ed. D. Raabe, et al., Weinheim: WILEY-VCH, 2004.

[52] Q. Yu and S. K. Esche, "A Monte Carlo algorithm for single phase normal grain growth with improved accuracy and efficiency," *Computational Materials Science*, vol. 27, no. 3, pp. 259-270, 2003.

[53] E. A. Holm and C. C. Battaile, "The computer simulation of microstructural evolution," *JOM*, vol. 53, no. 9, pp. 20-23, 2001.

[54] T. O. Reinikainen, P. Marjamäki, and J. K. Kivilahti, "Deformation characteristics and microstructural evolution of SnAgCu solder joint," *in Proc. 6th EuroSim Conference*, Berlin, Germany, April, 2005, pp. 91-98.

[55] L. Anand, "Constitutive Equations for Hot Working of Metals," *Journal of Plasticity*, vol. 1, pp. 213–231, 1985.

[56] S. A. Safran, P. S. Sahni, and G. S. Grest, "Kinetics of ordering in two dimensions. I. Model systems," *Physical Review B*, vol. 28, no. 5, pp. 2693-2704, 1983.

[57] D. Raabe, "Scaling Monte Carlo kinetics of the Potts model using rate theory," *Acta Materialia*, vol. 48, no. 7, pp. 1617-1628, 2000.

[58] M. B. Bever, D. L. Holt, and A. L. Titchener, "The stored energy of cold work," *Progress in Materials Science*, vol. 17, pp.1-190, 1973.

[59] R. R. Chromik, R. P. Vinci, S. L. Allen, and M. R. Notis, "Measuring the mechanical properties of Pb-free solder and Sn-based intermetallics by nanoindentation," *JOM*, vol. 55, no. 6, pp. 66-69, 2003.

Chapter 4

Texturing Tendency in β-Type Ti-Alloys

Mohamed Abdel-Hady Gepreel

Additional information is available at the end of the chapter

http://dx.doi.org/10.5772/53588

1. Introduction

1.1. Textures

Preferred orientation of crystal is an intrinsic feature of metals and has an influence on physical properties such as strength, electrical conductivity and wave propagation, particularly in the anisotropy of these properties [1]. For example, in the single crystals of many metals it is well known that the main mechanism of plastic deformation, on a microscopic scale, is a simple shear parallel to certain planes and directions. Slip will occur in a certain direction on a crystallographic plane when the shear stress in that direction attains a critical value [2]. So, the observed strength might depend on the loading direction of the crystal. Some other physical properties of the crystal vary depending on the measuring direction.

On the other hand, in a polycrystalline metal, each grain normally has a crystallographic orientation different from that of its neighbors resulting in isotropy of the properties of the metal. Considering a polycrystalline as a whole, the orientations of the grains may tend to cluster about some particular orientations. Such polycrystalline is said to have a preferred orientation, or texture, which may be defined simply as anon-random distribution of crystal orientations [3]. In this case, a polycrystalline performs in a way close to the single crystal depending on the strength of texture formed in it.

The appearance of preferred orientations (or texture) is very common. The texture produced by forming such as wire drawing and rolling, is called a deformation texture. The grains in a polycrystalline metal tend to rotate during plastic deformation, which results in the texture formation. Each grain undergoes slip and rotation in a complex way by the imposed force and by the restriction of slip and rotation of adjoining grains [3]. The preferred orientation also appear when cold-worked metal (show a deformation texture) is annealed. This is the so-called recrystallization texture (or annealing texture).

Texture may be influenced by a number of factors that can be divided into two major categories, material variables and process variables [4]. The material variables include crystal structure, solute content, second-phase particles, and initial grain size. The process variables include the amount of deformation, strain rate, stress and strain states [5], reduction in thickness and area, intermediate annealing conditions, heating and cooling rates, and annealing atmosphere [6].

1.2. Texture characterization techniques

The most common method of characterizing texture is the presentation of pole figures where X-ray diffraction is used to specify the orientation of the crystallographic planes in space using the stereographic projection. The two different representations of these textures are the inverse and direct pole figures [1]. The pole figure is a two-dimensional projection of the three-dimensional distribution and represents the probability of finding a pole to a lattice plane (*hkl*) in a certain sample direction. Pole figures are normalized to express this probability in multiples of a random distribution. Inverse pole figures are also projections of the orientation distribution factor (ODF), but in this vase the probability of finding a sample direction relative to crystal directions is plotted. In other words, the difference between these two representations is their reference frame; inverse pole figures are shown with respect to the specimen reference frame, while direct pole figures refer to the crystal reference frame [1].

Local orientation can also be measured with the Scanning Electron Microscope (SEM). Interaction of the electron beam with the uppermost surface layer of the sample produces electron back-scatter diffraction patterns (EBSPs or EBSD) that are analogous to kikuchi patterns in Transmission Electron Microscope (TEM). EBSD are captured on a phosphorus screen and recorded with a low intensity video camera or a CCD device [1].

The texture of sheet is in the most highly developed form, so, most of grains are oriented and the sheet surface becomes roughly parallel to a certain crystallographic plane (hkl). Also, a certain direction [uvw] in that plane becomes roughly parallel to the direction of deformation. Such a texture is described by the shorthand notation (hkl)[uvw], and is called an ideal orientation. Most sheet textures, however, have so much scatter in orientations, and hence they may be approximated symbolically only by the sum of number of ideal orientations or texture components [3].

1.3. Texture types

1.3.1. Deformation texture

The deformation of a polycrystal is a very complicated heterogeneous process. When an external stress is applied to the polycrystal, it is transmitted to individual grains. Dislocations move on slip systems, dislocations interact and cause 'hardening', grains change their shape and orientation, thereby interacting with neighbours and creating local stresses that need to be accommodated [1].

The weakest slip systems in fcc metals are{111}<110>, consisting of 12 symmetrically equivalent variants. Activity of these systems produce a characteristic texture pattern during rolling [1]. The large number of slip systems makes it easy to achieve compatibility and the rolling texture can be well explained with the Taylor theory, especially if individual slip systems are allowed to harden, according to their activity [8] and if allowance is made for some heterogeneity across small grain boundaries [9].

Also, the bcc metals have 12 variants of slipping system [1] and the most common deformation mode in bcc metals is {110}<111> slip, which is a transposition of slip plane and slip direction with respect to fcc metals. Also, bcc metals slip on other planes than {110} in the <111>direction [10] such as {112}<111>and {123}<111> [3]. Generally, the most preferred orientations in cold rolled bcc metals consist of two types. The first type is represented by {100}<110> or its rotations around<110> axis, resulting in {hkl}<110>. The other is represented by {111}<112> or its orientations around surface normal axis, resulting in {hkl}<uvw> [3].

1.3.2. Recrystallization texture

The relationship between slip and crystal rotations is straightforward. Other processes such as climb, grain boundary sliding, and diffusion in general may also affect orientation distributions [1]. Of particular importance is recrystallization. Recrystallization is the replacement of deformed grains by the nucleation and growth of recrystallized grains, both can lead to drastic changes in texture. The nuclei during recrystallization are regions exists in the deformed microstructure and at the same time the recrystallization does not lead to precise orientation relationship between deformed and recrystallized grains. Although there might be some approximate orientation relationships present in-between the recrystallized and parent grains, there has been no report in making quantitative predictions for recrystallization texture based on such approximate orientation relationships. The texture evolution is still highly sensitive to almost the entire spectrum of metallurgical variables and sometimes difficult to fully predict. However sometimes it is observed that the recrystallization texture component bear crystallographic relationships to the original deformation textures, which may be described by rotations about simple crystal direction which is often ~25° rotations around <110> directions in bcc alloys [11].

In recrystallization, the nuclei by the shear bands competes with the nuclei forming at other nucleation sites. Since stored energy represents the driving force for recrystallization, certain crystallographic orientations will be enhanced during annealing in comparison with others because of more favorable nucleation and/or growth kinetics [12]. Grains with higher stored energies may be consumed through boundary migration by grains with less stored energy (i.e., growth stage). Alternatively, dislocation-free nuclei may form grains with high dislocation density and then grow at expense of others. The texture which finally forms is believed to be controlled either by nucleation or by grain growth. It is possible that grain growth is 'oriented', i.e. for some reasons grains with certain crystallographic orientation grow faster than others. In that case the grain growth mechanism is likely to control the final texture. Otherwise, 'oriented' nucleation may control the final texture. The texture fresh nuclei may not be random and reflect the final texture. The last word has not been yet said on this problem. Not only orientation distribution, but also grain size distribution are important considerations [1].

The stored energy is the most important parameter characterizing the influence of micro-structure on the recryasallization process which is a driving force of it. The stored energy is proportional to the average critical shear stress for slip in the crystal [13].

1.4. Elastic anisotropy change with β-phase stability

The elastic anisotropy of bcc metals can be represented by the elastic anisotropy factor, A, which is calculated from elastic constants from the formula, $A= C_{44}/C'$. $C'= (C_{11}-C_{12})/2$ represents the resistance for the {110}<110> shear and C_{44} represents the resistance for the {001}<001> shear [13] and both of them increases with β-phase stability [3,10]. Further more, the ideal strength for tension σ_{max}, which is needed for tensile separation of bcc crystal on {100} plane, is proportional to C′ and ideal strength for shear σ_{max} which is needed for plastic shear along <111> on {011}, {112} or {123} [14]. As a result, the deformation mechanism in bcc alloys is related to the β-phase stability as mentioned before [15]. Figure 1 show the change in A and C′ with the Cr-content in Ti-Cr binary alloys. The Cr-content is related to β-phase stability [16]. So, it is expected that the anisotropy and texturing tendency are related to the β-phase stability in β Ti-alloys. So, a correlation between the β-phase stability and the texturing tendency will be presented below.

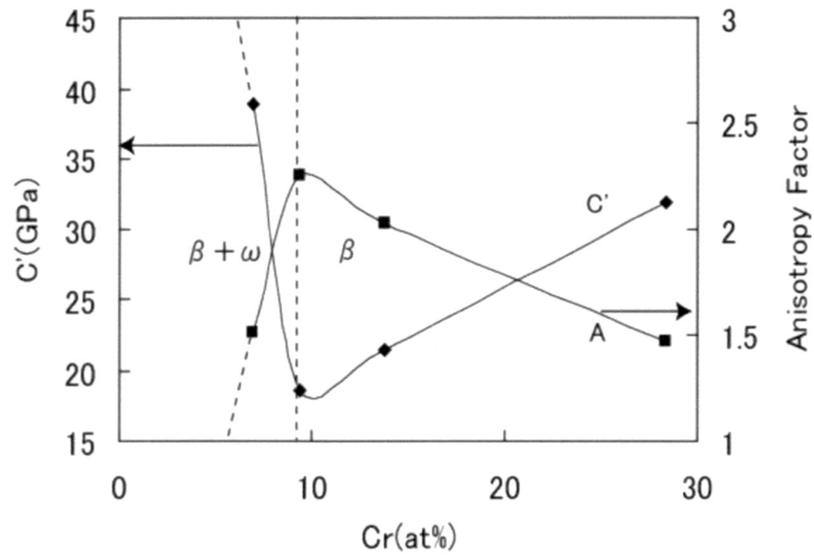

Figure 1. Changes in elastic anisotropy factor, *A*, and tetragonal shear constant, C', with the β –phase stability in Ti-Cr binary alloys. *A* shows maximum and C' shows minimum at the β/β+ω phase boundary

1.5. BCC β-type Ti alloys

Recently, considerable efforts have been devoted to exploring novel β-titanium alloys for different applications because of their superior properties such as the superelasticity, low Young's modulus, high strength-to-weight ratio, and better formability compared to the α and $\alpha + \beta$ titanium alloys [17-20]. The mechanical properties of the β-titanium alloys depend strongly on the presence of several phases (e.g., ω phase and martensitic α"-phase) in them. The appearance of these phases could be controlled by either the optimized alloy design [19,20] or the materials processing [18]. The phases present in the alloy are related to the alloying elements and the thermal history of the alloy. The change of the Ti-alloy type depending on the β-stabilizer content is shown schematically in figure 2.

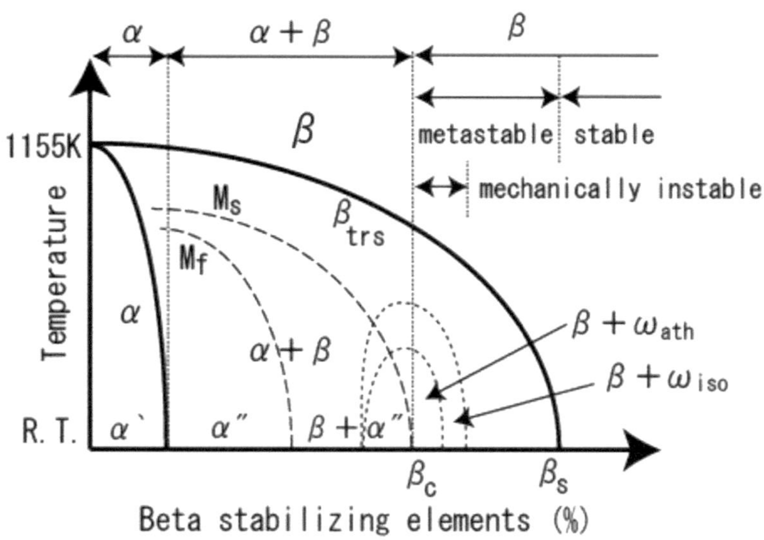

Figure 2. Schematic phase diagram of titanium alloys with the decomposition products of the β-phase. β_c is the critical minimum β stabilizer amount for near β or metastable β alloys to retain β completely on quenching from β phase field and β_s is the minimum amount of β stabilizer for stable β alloys; β_{trs}, M_s and M_f refer to β-transus, martensite start and finish temperatures, respectively.

Most of the β-Ti alloys possess good workability. It is possible to fabricate a cold-rolled sheet of the alloys by a reduction ratio higher than 90%. In this case strong deformation textures are developed and even recrystallization texture may be developed when subsequently heat treated. Therefore, the anisotropy in elastic and plastic properties is induced inevitably to the sheet, resulting in the modification of alloy properties such as the elastic modulus, elastic strain, Poisson's ratio, strength, ductility, toughness, magnetic permeability and the energy of magnetization [21]. In other words, the elastic and plastic properties of the alloy may be improved by using an orientation effect arising from the textures. It is, therefore, important to examine which kind of textures can be developed in the β-Ti alloys under the given conditions of thermo-mechanical treatment, and to investigate the texture effect on the elastic and plastic properties.

In this chapter, the effect of β-phase stability on the texturing tendency of β-type Ti-alloys are discussed. It is important to high light here that Zr has been known for many decades as neutral element on the stability of β-phase; however, the recent studies have proved that Zr shows a β-stabilizing effect in the β-type alloys [20,22]. Therefore, in this study, it was chosen to study two groups of alloys, one group is Zr-free β-type alloys (referred hereafter as A-alloys) and the other group is high Zr-containing alloys (referred herefater as Z-alloys). The design of these alloys is explained below.

2. Alloys design

The two series of alloys were designed across the single β-phase boundary, β/β+ω+(α"), with the aid of the $\overline{Bo}$- $\overline{Md}$ diagram. At this boundary, the elastic anisotropy factor, $A=C_{44}/C'$, is rather high since the value of the elastic shear modulus, $C' = (C_{11}-C_{12})/2$, is diminishing as the alloy approaches this boundary [23,24], as shown in figure 1. Also, it has been reported that the $(C_{11}-C_{12})/2$ is related to the electronic parameter *e/a* (electron-per-atom ratio) and its value approaches zero when the e/a value is about 4.24 [23]. This is the reason why, in this work, the *e/a* value was kept at 4.24 in almost all the designed alloys. Here, $\overline{Bo}$is the average bond order between atoms, and $\overline{Md}$is the average d-orbital energy level (eV) of the elements in the alloy.

In A-alloys, A00 alloy is designed to be located in the β+α"(+ω) phase zone in the $\overline{Bo}$-$\overline{Md}$dia-gram [20], as shown in figure 3. The composition of A-alloys is listed in table 1. The Fe and O were added to stabilize the β-phase of the alloys. Fe was chosen to stabilize β-phase due to its very strong β-stabilizing effect as obvious from its alloying vector in the $\overline{Bo}$-$\overline{Md}$dia-gram [22]. The oxygen was added to the alloys to suppress the ω and martensite phases, as discussed in ref.[20].

As for Z-alloys, four alloys were designed with the aid of the $\overline{Bo}$-$\overline{Md}$ diagram across the β/β +ω+(α") phase boundary in two steps; in the 1st step, four alloys were designed, consist mainly of β-stabilizers, Ta, Nb, Mo, Cr and V. These four alloys are shown across the β/β+ω+ (α") phase boundary with the dashed arrow in figure 3.

As discussed above, Zr works as β-stabilizer in the β-type Ti alloys and also raises the $\overline{Bo}$ value of the alloy. The 2nd step was to add Zr to the alloys with different amounts varies with the required β-phase stability of each alloy. So, the stability difference between the alloys will be much bigger and the properties difference would be clearer. Considering that the e/a value of the alloys did not change after Zr addition.

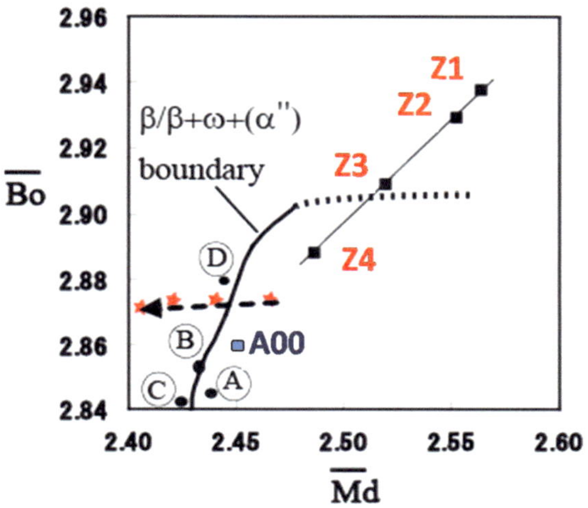

Figure 3. Extended $\overline{Bo}$-$\overline{Md}$ diagram showing the β/β+ω+(α") phase boundary and the location of the designed alloys, A00 and Z1-4. Also, the alloys A, B, C, and D are 35mass%Nb-4mass%Sn, Ti-24Nb-3Al, Ti-35mass%Nb-7.9mass% Sn, and Ti-22Nb-6Ta, respectively.

At last, the four designed Z-alloys, namely Z1 - 4, are located across the β/β+ω+(α") phase boundary in the high $\overline{Bo}$ region (i.e., high Zr-Containing alloys zone) in the extended $\overline{Bo}$-$\overline{Md}$ diagram shown in figure 3, and the chemical compositions are listed in Table 1.

3. Experimental procedure

As explained above, two series of alloys, namely; high Zr-containing [Z-alloys] and Zr-free [A-alloys], were designed across the β/β+ω+(α") phase boundary and the chemical compositions are listed in table 1. In this chapter, all the compositions are given in atomic percent units unless otherwise noted. These alloys were prepared by the arc-melting of an appropriate mixture of pure metals (purity: 99.99%) under a high purity argon gas atmosphere. The button-shaped specimens with average 7.5 mm in thickness were cut and homogenized at 1273K for 7.2ks, and then cold rolled to the plate with 4.5 mm thick, followed by the solution treatment at 1223K for 1.8 ks. Subsequently, the specimen was cold rolled by 60%, 90 % or 98% reduction in thickness. The cold rolled specimen is called CR specimen hereafter. The 90%CR specimen was then solution-treated at 1223K for 1.8ks. This finally solution treated specimen is called ST specimen hereafter.

The phases existing in the specimen and its pole figures were identified by the conventional X-ray diffraction (XRD) using a Ni-filtered Cu-Kα radiation. Electron back scattered diffraction (EBSD) analysis was also made using a HITACHI S-3000H scanning electron microscope (SEM) equipped with a OXFORD INCA Crystal EBSD detector, operated at an acceleration voltage of 20 kV and a tilt angle of 71°. The microstructural characterization was performed using the optical microscope (OM), the scanning electron microscope (SEM) and the transmission electron microscopy (TEM).

Alloy	O	Fe	V	Cr	Mo	Nb	Ta	Zr
Z1			4		2	9	7	30
Z2					3	15	3	25
Z3				1		8	14	15
Z4						4	20	5
A00						17	6	
A01	1					17	6	
A11	1	1				17	6	

Table 1. Chemical compositions of the designed A and Z-alloys, at.%.

4. Results and discussion

4.1. Change in β-phase stability with alloys' composition

The β-phase stability increases with increasing content of the β-stabilizing elements. Shown in figure 4.a are the X-ray diffraction patterns taken from the A-alloys in the ST condition. The α'' martensite phase is the predominant phase in A00 alloy beside small amount of β and ω phases. The α'' martensite phase is suppressed by O addition to the A00 β-phase alloys. As shown in figure 4.a, the addition of 1 mol.% O to A00 alloy was very effective in suppressing the α'' martensite phase as observed in A01 alloy. Also, it is clear from this figure that the addition of Fe to the A01 alloy resulted in stabilizing the β-phase as observed in the A11 alloy and the α'' martensite phase couldn't be observed by XRD at room temperature.

It can be deducted from the XRD results that the A11 is a single β-phase alloy. So, the least stable single β-phase alloys in the A series is A11 alloy. It is concluded that the co-addition of Fe and O to these β-type alloys is very effective in suppressing the α'' martensite phase and instantaneously stabilize β-phase of the alloy. An important observation from figure 4.a is that, the co-addition of only 1% O and 1% Fe could modify greatly the phase stability of alloy A00, with the α'' predominant phase, to alloy A11, with the single β-phase. Therefore, the β-phase stability in these alloys increases in the order, A11> A01> A00.

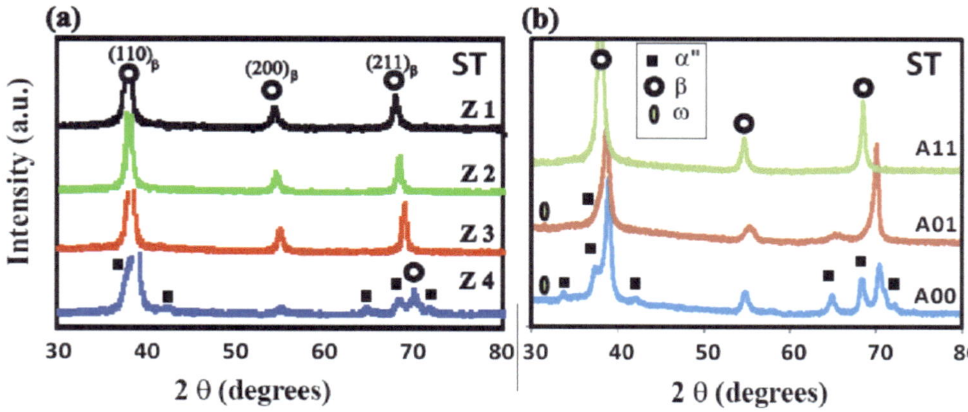

Figure 4. XRD profiles of the alloys Z1-4 (a), and A-alloys (b) in the solution treated at 1223K, ST, after 90% cold rolling.

	Z-alloys (60%CR)				A-alloys (90%CR)		
	Z1	Z2	Z3	Z4	A00	A01	A11
CR, $I_{Rcr} = I_{(200)\beta} / I_{(110)\beta}$	1.7	1.2	0.8	0.1	0.086	0.194	1.232
ST, $I_{Rst(1)} = I_{(110)\beta} / I_{(200)\beta}$	6.0	8.6	8.5	33.0			
ST, $I_{Rst(2)} = I_{(211)\beta} / I_{(200)\beta}$	1.6	2.0	2.2	3.6			

Table 2. X-ray peak intensity ratios of Z-alloys and A-alloy in the CR and ST conditions.

By the same way for Z-alloys, shown in figure 4.b are the X-ray diffraction patterns taken from the Z-alloys in the ST condition. A single β-phase was predominant in the alloys Z1, Z2, and Z3. Only in the alloy Z4, the martensite α''-phase coexisted with the β-phase, as shown in figure 4.b. Therefore, the$\beta/\beta+\alpha''$ boundary shown in figure 3 is located between the alloys Z3 and Z4 as indicated by a dotted curve. So, the alloy Z3 is the least stable single β-phase alloy which is defined as the alloy containing a least amount of the β-stabilizing elements to get a β single phase [20,25]. According to the $\overline{Bo}$-$\overline{Md}$ diagram shown in figure 3, the alloy closer to the $\beta/\beta+\alpha''$ boundary has the lower β-phase stability, so the β-phase stability in these alloys decreased in the order, Z1> Z2> Z3> Z4. The stability was highest in the alloy Z1 and lowest in alloy Z4.

4.2. Textures developed by cold rolling

Most of the annealed β-type Ti-alloys have random equiaxed grains and their X-ray peak intensity of the $\{110\}_\beta$ plane is the highest among the different atomic planes reflections. However, the $\{100\}_\beta$<110> rolling texture is formed normally after high reduction ratio of the β-type Ti-alloys due to dislocation slipping and grains rotation. In this cold rolling texture, the $\{200\}_\beta$ planes aligned parallel to the rolling plane preferentially. Figure 5.a shows the X-ray diffraction patterns taken from the Z-alloys in the 60% cold rolled (60CR) condition. It is known that the cold rolling texture is formed in conventional β- type Ti alloys. As a result, the measured X-ray peak intensity ratio of the cold rolled specimen, $I_{Rcr,}$ defined as I_{Rcr}= $I_{\{200\}\beta}$ / $I_{\{110\}\beta}$, changed with cold rolling. Here, $I_{\{200\}\beta}$ and $I_{\{110\}\beta}$ are the X-ray peak intensities of the $\{200\}_\beta$ and $\{110\}_\beta$ reflections, respectively. The I_{Rcr} usually increases with the reduction ratio of cold rolling. In addition, as is evident from figure 5.a and table 2, when the alloys were cold rolled by 60%, I_{Rcr} tended to increase with increasing β-phase stability. In other words, a cold rolling texture was developed in the way that the {200} planes were aligned parallel to the rolling plane preferentially. This texture was formed more readily in the order, Z1>Z2> Z3> Z4, in agreement with the order of the β- phase stability.

Figure 6 (a-c) shows {200}, {110} and {112} pole figures obtained from a 90% cold rolled (CR) specimen of the alloy Z2. The center of the pole figures corresponds to the direction normal to the specimen surface (ND). The right and the top of the pole figures correspond to the rolling direction (RD) and the transverse direction (TD), respectively. It was realized from these pole figures that typically {100}<110> rolling texture with a strength of 97.5 times larger compared to the random orientation is well developed in the 90% cold rolled (CR) specimen of the alloy Z2, as shown in the {100}<110> texture stereoprojection [26] in figure 6 (d). It is important to mention here that the alloy Z2 with the relatively high β-phase stability shows the{100}<110> texture after 90%CR more remarkably, as compared to the lower β-phase stability alloy undergone by the same or even severer cold rolling [18,27,28].

It has been reported that the {100}<110> texture is a main rolling texture formed in the β-type Ti-alloys [17,18,26-28]. Beside this texture, the {211}<110> texture forms in a Ti-35mass %Nb-4mass%Sn alloy [18,26] (location A in figure 3) or the {111}<112> texture forms in a Ti-24Nb-3Al alloy [26] (location B in Figure 3). With a little increase in the β-phase stability, the {100}<110> texture becomes dominant as observed in a Ti-35mass%Nb-7.9mass% Sn alloy [28] (location C in figure 3). As explained in Ref. [20], both Al and Sn work as the β-stabilizing elements in these β-phase alloys. In the much higher β-phase stability alloys such as Ti-22Nb-6Ta alloy (location D in figure 3), only the {100}<110> texture is developed after 99% cold rolling [27]. This was consistent with the present results that the $I_{Rcr,}$ which represents {100}<110> texture, increases monotonously with the β-phase stability. Therefore, it can be concluded that the {100}<110> rolling texture is developing in β-type Ti-alloys and its strength is increasing monotonously with increasing β-phase stability. However, in Ti alloys with low β-phase stability, other rolling textures such as {211}<110> and {111}<112> textures are developing more readily than {100}<110> texture in such low β stability alloys.

In the same way for A-alloys, as is evident from figure 5.b and table 2, when the A-alloys were cold rolled by 90%, I_{Rcr} tended also to increase with increasing β-phase stability. In other words, a rolling texture was developed, in the higher β-phase stability alloys, in the way that the {200} planes aligned parallel to the rolling plane preferentially, in accordance with that reported for the above Z-alloys. For example, the I_{Rcr} of A11 alloy is around 6 times of that of A01 alloy with relatively lower β-phase stability. However, I_{Rcr} (A01) / I_{Rcr} (A00) are around 2 times only. This can be interpreted as the addition of 1% O is less efficient to highly stabilize β-phase and therefore to develop the {100}<110> rolling texture if compared with the addition of 1% Fe in this A-alloy. However, the O addition seems to be very effective in suppressing the ω- and the α"-phases (in other words, stabilizing β-phase at room temperature), as shown in figure 5.b. It is well know that Fe is very strong β-stabilizer. Therefore, the co-addition of O and Fe in A11 alloy was enough to increase the β-phase stability to a level high enough to develop strong {100}<110> rolling texture, same as in Z-alloys with higher β-phase stability such as Z2 and Z3 alloys, as shown in figure 5.b and table 2.

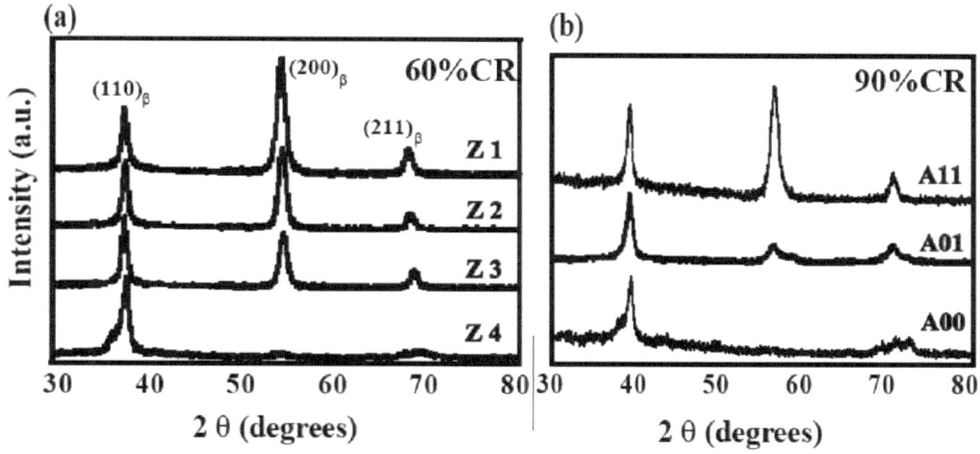

Figure 5. XRD profiles of Z-alloys after 60%CR, (a) and A-alloys after 90%CR, (b).

Figure 7 (a-c) shows {200}, {110} and {112} pole figures obtained from a 98% cold rolled (CR) specimen of the A01 alloy. The center of the pole figures corresponds to the direction normal to the specimen surface (ND). It seems from these pole figures that the {111}<112> rolling texture was developed predominantly compared to the other textures and random grains orientations. The {111}<112> rolling texture developed in A01 after 98%CR is moderate in strength if compared to that {100}<110> rolling texture developed in Z2 alloy after only 90%CR as evidenced from the inverse pole figures shown in Figure 8. This means that the rolling textures in β-type Ti-alloys become stronger with increasing the β-phase stability irrelative to the type of texture developed in them. Also, Z2 with the relatively high β-phase stability shows the{100}<110> texture after 90%CR more remarkably when compared to the lower β-phase stability alloys undergone by the same or even severer cold rolling as reported in Ref. [18,27,28]. Therefore, it is concluded that the β-phase stability is more effective than reduction ratio in developing rolling textures in β-type Ti-alloys.

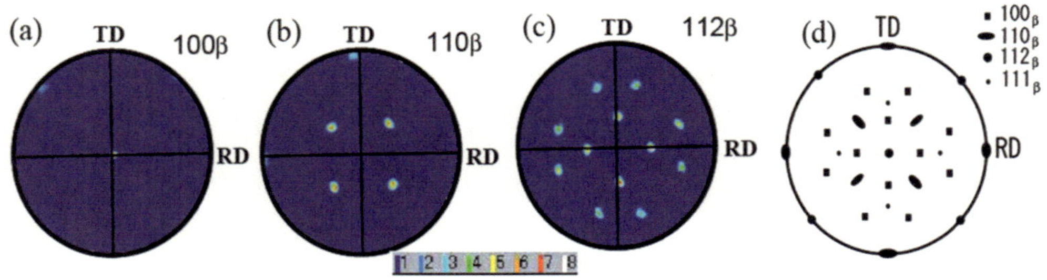

Figure 6. EBSD pole figures of alloy Z2 specimen after 90% cold rolling using 200_β, (a), 110_β, (b) and 112_β (c) and the stereoprojection of the $\{100\}_\beta$<110$>_\beta$-type texture, (d).

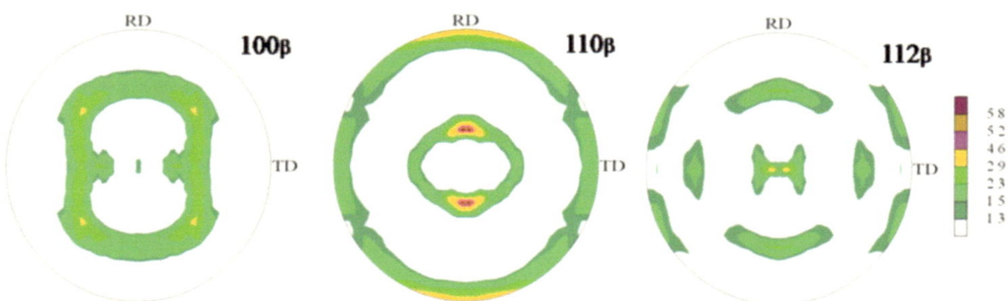

Figure 7. XRD pole figures of alloy A01 specimen after 98% cold rolling using 200β, 110β, and 112β.

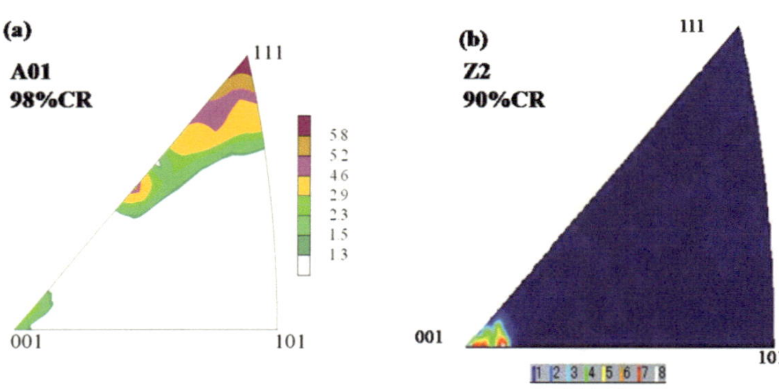

Figure 8. Inverse pole figures in the normal plane of alloy A01 specimen after 98% cold rolling calculated from the XRD ODF, (a), and alloy Z2 specimen after 90% cold rolling calculated from the EBSD ODF.

The β-type Ti-based alloys deform by either slip or twin mechanism [29,30]. The stress-induced martensitic transformation also takes place in some alloys upon applying external stress to them [30,31]. These phenomena emerge depending on the β-phase stability and hence will be controlled by alloying. Also, it is known that the slip/twin boundary is close to the $\beta/\beta+\omega+(\alpha")$ boundary [20].

As crystal deforms by slip, it undergoes crystal rotations. Such rotations lead to the development of preferred orientations in polycrystalline alloys [32]. The main reason why the {100}<110> texture develops by cold rolling in conventional bcc alloys may be attributable to the glide of dislocations along <111> on {011}, {112}or {123}[33] and the crystal rotations in them. It is well known that C' represents the resistance for the {011}<011> shear and C_{44} represents the resistance for the {001}<001> shear [34] and both of them decreases with decreasing β-phase stability [24,35]. The e/a value is kept at 4.24 in the present alloys and C' decreases with decreasing β-phase stability as discussed earlier, so the elastic softening will be enhanced in the order; Z1 < Z2 < Z 3 < Z4. This is a reason why the alloys locating at the $\beta/\beta+\alpha"$ boundary showed a very low shear modulus along both <011> on {011} and along <111> on {011}, {112} or {123} as reported in Ref. [36]. So it is expected that as the β-phase stability decreases, secondary slipping systems such as {011}<011> may be activated by deformation beside the main slipping system, i.e., <111> on {011}, {112}or {123}. As a result, a portion of the applied stress is consumed in slipping in such secondary slipping systems, which will make some disturbances in forming the main {100}<110> rolling texture. This can be a reason why the {100}<110> rolling texture was dominant in the high β-phase stability alloys. This was also seen in the steel in which the addition of a ferrite (bcc) stabilizer, Si, is enhanced to form the rolling texture [32].

4.3. Textures developed by recrystallization

The textures developed by severe cold rolling will diminish if the specimen is reheated for long time. For example, the $I_{H(cr)}$ of a 98%CR A01 specimen was decreased when reheated to different temperatures for 3 hrs, as shown in figure 9. $I_{H(cr)}$ is defined as $I_{Hcr}= 1/I_{Rcr}= I_{\{110\}\beta} / I_{\{200\}\beta}$. Hence, it is expected that the {100}<110> rolling texture is diminishing as a result of this reheating. The decrease in IH_{cr} was higher with increasing the reheating temperature till it reach a saturation temperature. This saturation temperature is most probably related to the completion of recrystallization process. Therefore, further increase in temperature will not affect much the value of IH_{cr}, as shown in figure 9. The rolling textures will not only diminish the heavily cold rolled specimens by reheating, but also recrystallization textures may develop as a result of this reheating.

In the present designed alloys, the recrystallization textures were developed in different strength in the ST specimen that was solution-treated after 90% cold rolling as explained bellow. In figure 4 (a) and table 2, the X-ray peak intensity and the intensity ratios are shown of the ST specimen solution-treated after 90% cold rolling. Here, $I_{Rst(1)}$ and $I_{Rst(2)}$ were defined as $I_{Rst(1)}= I_{\{110\}\beta} / I_{\{200\}\beta}$ and $I_{Rst(2)}= I_{\{211\}\beta} / I_{\{200\}\beta}$, respectively. Both of them decreased monotonously with increasing β-phase stability. Namely, in the recrystallization textures the high atomic density planes, i.e., {110} and {211}, aligned parallel to the rolling plane preferentially. This trend further increased with decreasing β- phase stability, as shown in table 2 for Z-alloys.

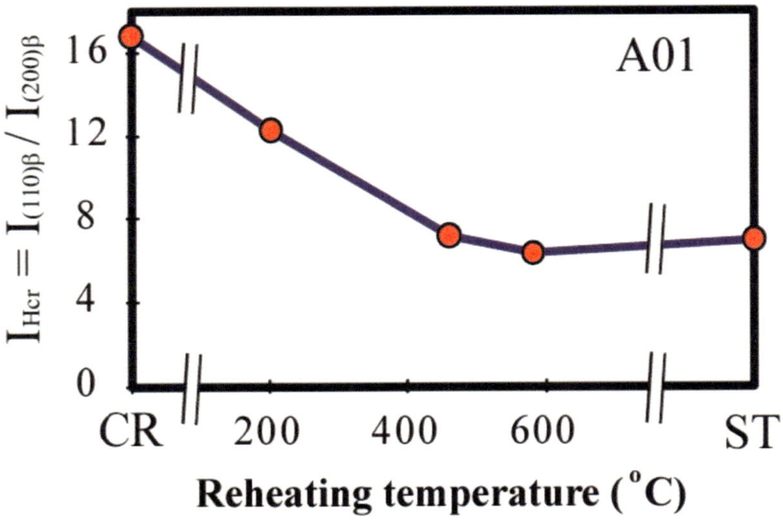

Figure 9. Effect of reheating temperature, for 3hrs, on the strength of formerly developed <100>{110} rolling texture after 98%CR, represented by *IHcr= 1/IRcr= I{110}* • */ I{200}* • of the A01 alloy.

In figure 10, the EBSD Euler space plots are shown of the ST specimen of the alloys Z2-4 which were subjected to solution treatment at 1223K for 1.8 ks after 90% cold rolling. The Euler space density maps showed that the recrystallization textures were well developed in the alloy Z4, followed by the alloy Z3 and then the alloy Z2. Thus, the tendency of forming the recrystallization texture in the alloys changed in the order, Z4> Z3 > Z2, which was the opposite order of the rolling texture formed by cold rolling.

In the solution treatment after cold rolling, it has been reported that {112}<110> recrystallization texture is mainly developed in the low β-phase stability alloys such as Ti-24Nb-3Al alloy, position B in figure 3, [29,37] and Ti-35mass%Nb-4mass% Sn alloy, position A in figure 3, [31,38]. Beside this {112}<110> recrystallization texture, the {110}<211> texture is developed in a little higher β-phase stability alloy, Ti-35mass%Nb-7.9mass%Sn, position C in figure 3, [38]. However, only the {112}<110> recrystallization texture appears in the Ti-22Nb-6Ta alloy, position D in figure 3, [24]. It is important to note here that the {211}<110> recrystallization texture is well developed in the alloy only after severe cold rolling (i.e., 95% and 99%) [24,39] and it tends to diminish with decreasing reduction ratio of cold rolling [24]. Also, recrystallization textures could be controlled by the temperature and time for the heat treatment [39].

Recrystallization is the replacement of deformed grains by the recrystallized grains [40]. The grains with certain crystallographic orientations will be nucleated and grown in the course of annealing [41]. The growth rate of the grains is also 'oriented', because some grains with certain crystallographic orientation will grow faster than others [32]. Otherwise, 'oriented' nucleation may control the final texture structure. As discussed earlier, both the elastic softening and the elastic anisotropy becomes more remarkable with the decrease in the β-phase stability. So, it was likely that the oriented nucleation and/or oriented growth was enhanced with decreasing β-phase stability, leading to the increase in the strength of the recrystallization texture.

From these discussions and the data given in table 2, it was concluded that, for the single β-phase alloys, the tendency of forming the {100}<110> texture by cold rolling increased with increasing β-phase stability, whereas, for low β-phase stability alloys (such as A00 and Z4 alloys in this study) other rolling textures may develop. On the other hand, the tendency of forming the recrystallization textures increased with decreasing β-phase stability. Thus, the β-phase stability was operating in the completely reverse way between the rolling and the recrystallization textures.

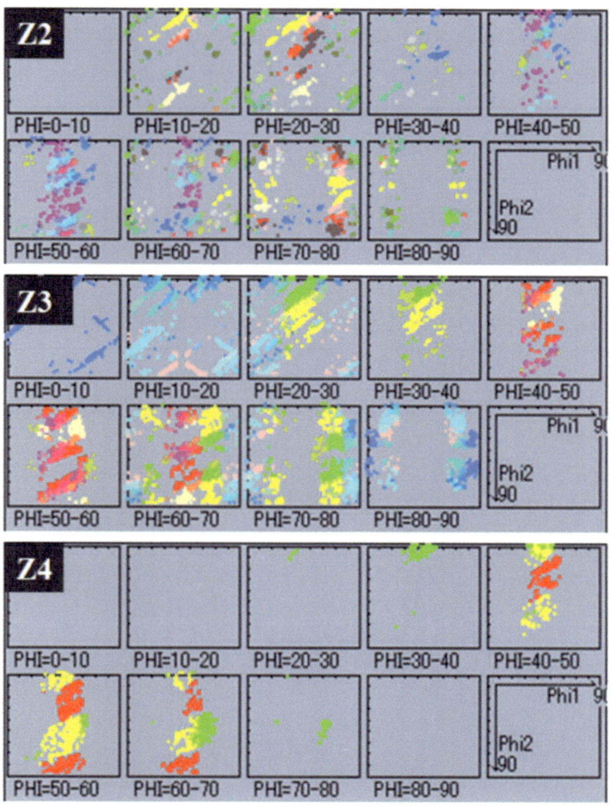

Figure 10. φ sections of the EBSD Euler space plot of alloys Z2-4 specimens solution treated at 1223K for 1.8 Ks after 90% cold rolling.

4.4. Textures developed in low β-phase stability alloys

The co-existence of α"- and/or ω-phase in the low β-phase stability alloys, such as Z4, A00 and A01 alloys in the present study, seems to affect much the deformation process and therefore the rolling texture developed in them when severly deformed. This is because in such low β-phase stability alloys, the deformation by twin and/or stress-induced martensitic transformation mechanisms are predominant. The deformation stress are consumed in forming twinning and inducing ω and/or α'' marensite phases. As a result, the grains rotation is expected to be much less in these low β-phase stability alloys resulting in low strength of the {100}<110> texture. For example, Z4 and A01 alloys show low I_{cr} and in the same time the stress induced α'' martensitic transformation occurred in them by cold rolling, as evedinced from the two-steps yielding during the tensile test. For further details, refere to Refs. [25,42].

It is interesting to mention here that the low β-phase stability alloys, with α'' martensite as the predominant phase, will show also a texture in the α''-phase, as shown in figure 11. The $\{200\}_{\alpha''}$, $\{012\}_{\alpha''}$ and $\{220\}_{\alpha''}$ XRD pole figures obtained from a 98%CR specimen of the low β-phase alloy,A00, are shown in Figure 11.a. The center of the pole figures corresponds to the direction normal to the specimen surface (ND). It seems from these pole figures that the α''-phase was textured in the specimen by cold rolling in the way that the $\{220\}_{\alpha''}$ planes aligned parallel to the rolling plane preferentially compared to the other textures and random grains orientations. In such low β-phase stability alloys, cold deformation induces aligned ω and/or α'' martensite phases with orientation relationships as explained else where. The three α'', ω, and β- phases are co-existed in A01 alloy after deformation as shown in figure 12. However, the ST specimen show less amount of α'' martensite and the recrystalliztion texture in the α'' –phase is also less as evidenced from figure 11.b for A00 alloy.

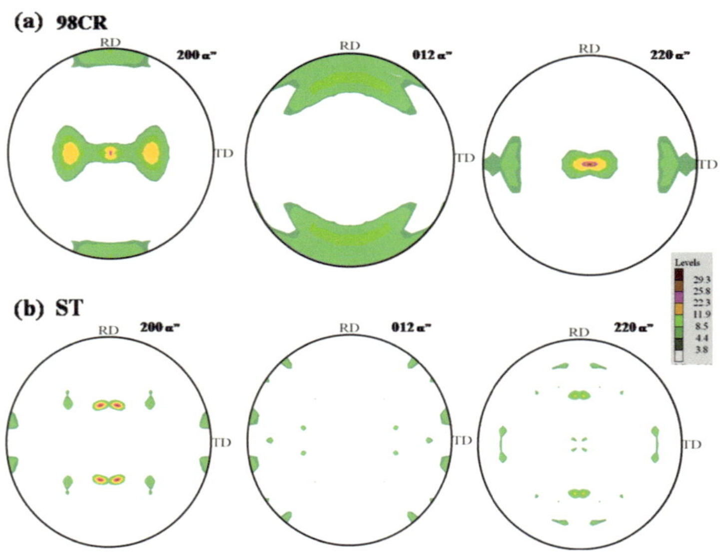

Figure 11. XRD pole figures of alloy A00 specimen after 98% cold rolling, (a) and after subsequent solution treatment at 1223K for 1.8 Ks, (b), using 200α", 012α", and 220 α".

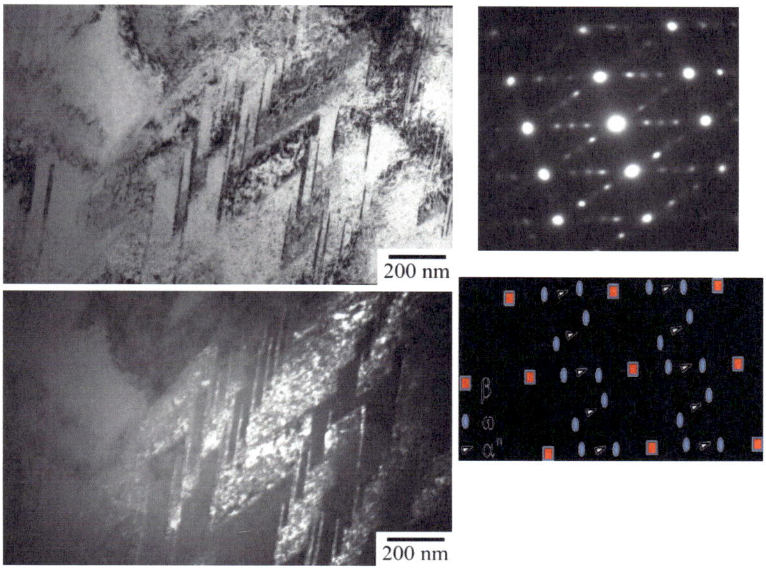

Figure 12. TEM analysis of a A01 specimen after 98CR. (a) and (b) present, respectively, bright-field and dark-field [imaged on $(-113)_{\beta}$] pair of micrographs showing stress induced of the ω and/or α"- phases; (c) shows the corresponding composite electron diffraction pattern and its key diagram. The ω-phase is with tow variants.

4.5. Microstructure change with β-phase stability

4.5.1. The microstructures after cold rolling

The microstructures shown in figure 13 are of the alloys Z2-4 after 90% cold rolled (90CR) in the RD and TD and the ND cross sections. The stream-like deformation bands were observed clearly along the RD and TD directions in both the RD and TD cross-sections. As the β-phase stability decreased, the deformation bands seemed to become finer. Also the density of the deformation bands seemed to be higher in the TD cross-section shown in figure 13.a than in the RD cross-section shown in figure 13.b.

The microstructure in an enlarged magnification is shown in figure 14 for the 90CR alloys. In the alloy 4, fine shear bands with an angle 42° inclined to the main deformation band were observed in the RD cross-section. However, no such shear bands were observed in the TD cross-section. Also in the alloy Z3, the secondary finer deformation bands with the angles of 23° and 37° inclined to the main deformation bands were observed in the RD and TD cross-sections, respectively. Neither the shear bands nor the secondary deformation bands were observed in the alloy Z2 in both the RD and TD cross-sections. These results can be interpreted, as above, due to the presence of the texturing systems other than {100}<110> developed by the cold rolling and the plastic deformation among these three alloys depending on the β-phase stability. Also, the similarity in the TD and RD microstructure in the alloy Z2 is attributable to the well developed {100}<110> rolling texture in which both the RD and the TD are parallel to <110> [35].

4.5.2. The micorstructure after solution treatment

The microstructures of the ST alloys Z2-4 solution treated at 1223K for 1.8ks after 90% cold rolling are shown in figure 15. Shear bands supposedly provide the majority of nucleation sites during recrystallization in the severely cold rolled alloys. In particular, the triple joints between shear bands sites will work as nucleation sites [43]. So, the number of the nucleation sites was supposed to increase with decreasing β-phase stability, judging from the former deformation microstructures shown in figures 13 and 14. As a result, the grain size in the ST condition seems to decrease as the β-phase stability decreases. As is evident from the slightly defocused micrographs shown in figure 15.a some deformation bands still existed in the TD cross-section, although the recrystallization process seemed to be completed in the rolling plane shown in figure 15.b. However, the equiaxed grains were observed in the TD cross-section of the alloy Z2 in the ST condition in the EBSD image quality map (IQM) as presented in Ref. [25]. Similar deformation bands were observed in a Ti-45Nb alloy in the recrystallized condition after severe cold working [44]. The reason why these deformation bands still remained after the recrystallization treatment is not clear at the moment.

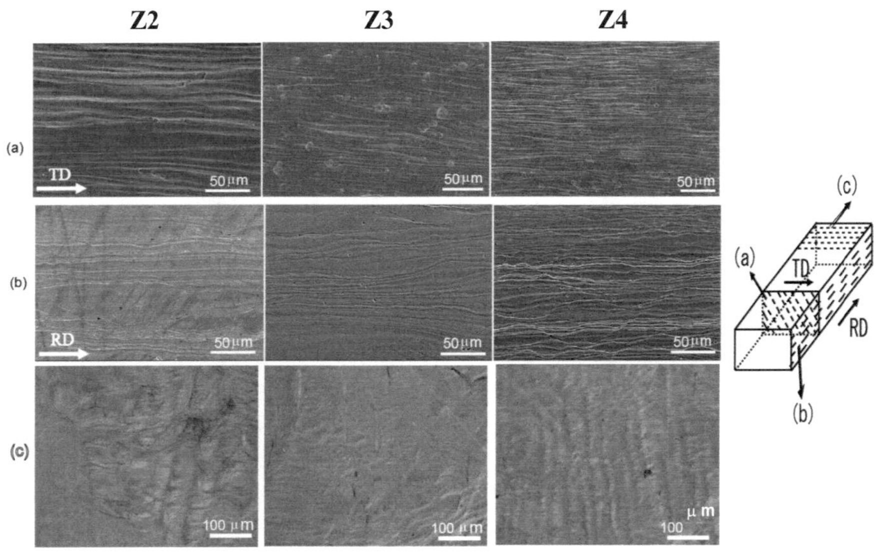

Figure 13. SEM micrographs of alloys Z2-4 after 90% cold rolling in the transverse, TD, cross-section (a), in the rolling, (RD), cross-section (b), and in the normal plane (c).

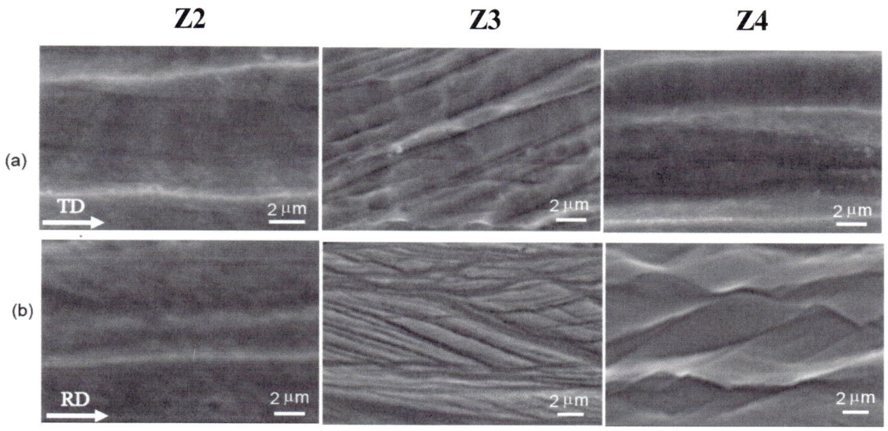

Figure 14. SEM micrographs of alloys Z2-4 shown in the enlarged scale of figure 13 (a) and (b).

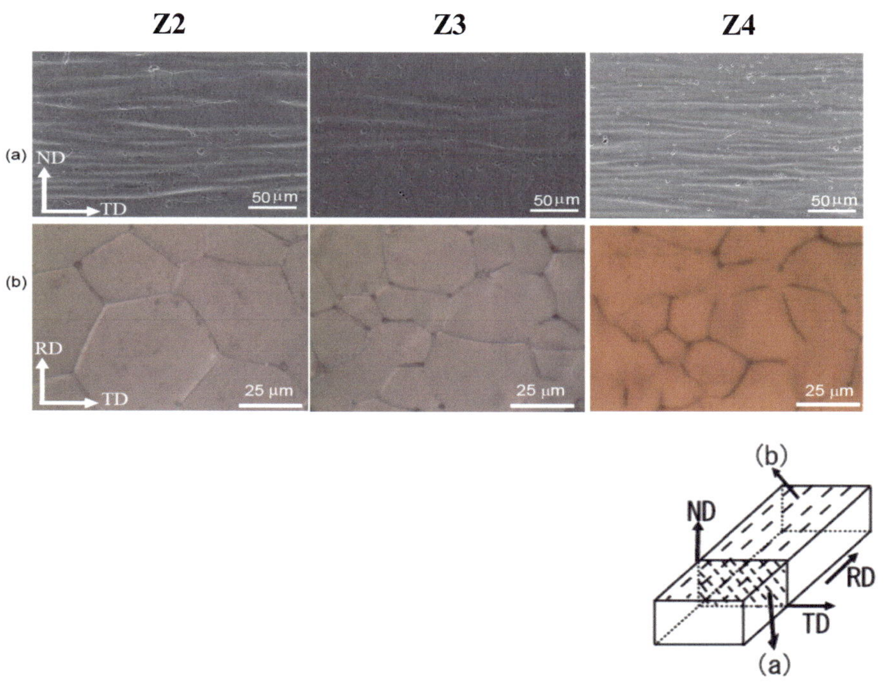

Figure 15. Micrographs of alloys Z2-4 after solution treatment at 1223K for 1.8 ks after 90% cold rolling in the transverse, TD, cross section, defocused SEM micrographs to show the deformation bands (a), and in the rolling plane, OM (b).

5. Conclusions

The $\overline{Bo}$-$\overline{Md}$ diagram was confirmed to be valid for the design of β-Type Ti alloys with irrelevant chemical compositions in the high $\overline{Bo}$ zone and with different β-phase stability. For the single β-phase alloys, the tendency of forming the {100}<110> texture by cold rolling increased with increasing β-phase stability, whereas, for low β-phase stability alloys (such as A00 and Z4 alloys in this study) other rolling textures may develop. On the contrary, the tendency of forming the recrystallization textures increased with decreasing β-phase stability and the recrystallization texture was more enhanced in the lower β-phase stability alloys resulting in the larger anisotropy of the properties.

Acknowledgments

The author thanks Prof. M. Morinaga of Nagoya University for his valuable comments and discussion, Prof. K. Sasaki of Nagoya University for his help with the TEM investigation, and Prof. H. Hosoda and Dr. T. Inamura of Tokyo Institute of Technology for their helpful discussions and XRD investigation. This study was supported by a Grant-in-Aid for Scientific Research from the Ministry of Education, Culture, Science, Sports and Culture of Japan, and by the Science and Technology Developing Fund (STDF) of Egypt.

Author details

Mohamed Abdel-Hady Gepreel

Department of Materials Science and Engineering, Egypt-Japan University of Science and Technology (E-JUST), Alexandria, Egypt

References

[1] H-R. Wenk, P.V. Houtte, Rep. Prog. Phys. 67 (2004) 1367.

[2] R. Sowerby, W. Johnson, Mater. Sci. Eng. 20 (1975) 101.

[3] M. Kawata, X-Ray analysis of residual stress and texture in ground carbon steels; Master theses; Toyohashi Univ. Tech., March (1982) pp. 45 and 47.

[4] E. Tenckhoff, A review of texture and texture formation in zircaloy tubing, Zirconium in the Nuclear Industry: Fifth Conference, ASTM STP 754 (1982) p. 5.

[5] J.C. Britt, K.L. Murty, Proceedings of Symposium (ZARC-91), BARC, Bombay (1991) p. 1.

[6] I.L. Dillamore, W.T. Roberts, Metals Reviews 10 (1965) 271.

[7] H.R Ogden, Ohio, in: Rare earth metals handbook, Physical properties of metals, C.A. Hampel (Edi), London, (1961), pp. 687-701.

[8] U.F. Kocks and H. Mecking, Prog. Mater. Sci. 48 (2003)171

[9] H. Honneff and H. Mecking, 6th Int. Conf. on Textures of Materials (Tokyo: The Iron and Steel Institute of Japan) (1981) p. 347.

[10] J.W. Christian, Proc. ICSMA 2: 2nd Int. Conf. on Strength of Metals and Alloys Vol. 1, ASM (1970) p. 29.

[11] Y.N. Wang, J.C.Huang, Mater. Chem. Phys. 81 (2003) 11.

[12] S.F. Castro, J.Gallego, F.J.G. Landgraf, H.J. Kestenbach, Mater. Sci. Eng. A 427 (2006) 301.

[13] K. Wierzbanowski, J. Tarasiuk, B. Bcroix, K. Sztwiertnia, P. Gerner, Recrystallization and grain growth, proceedings of the first joint international conference 12 (2000) 1075.

[14] S. Kuramoto, T. Furura, J.H. Hwand, K. Nishino, T. Saito, Metall. Mater. Trans. A 37 (2006) 657.

[15] S. Hanada, Sixth World Conf. on Titanium, France; Vol. 1 (1988) p. 105.

[16] S. Ankem, C.A. Greene, Mater. Sci. Eng. A 263 (1999) 127.

[17] S. Ishiyama, S. Hanada, O. Izumi, ISIJ Int. 31 (1991) 807–813.

[18] H. Matsumoto, S. Watanabe, S. Hanada, J. Alloys Compd. 439 (2007) 146–155.

[19] T. Saito, T. Furuta, J.H. Hwang, S. Kuramoto, K. Nishino, et al., Science 300 (2003) 464–467.

[20] M. Abdel-Hady, K. Henoshita, M. Morinaga, Scripta Mater. 55 (2006) 477–480.

[21] Y.N. Wang, J.C. Huang, Mater. Chem. Phys. 81 (2003) 11–26.

[22] M. Abdel-Hady, H. Fuwa, K. Henoshita, Y. Shinzato, M. Morinaga: Scripta. Mater. 57, 1000(2007).

[23] H. Ikehata, N. Nagasako, T. Furuta, A. Fukumoto, K. Miwa, T. Saito, Phys. Rev. B 70 (2004) 174113.

[24] E.S. Fisher, D. Dever, Acta Metall. 18 (1970) 265–269.

[25] M. Abdel-Hady, K. Henoshita, H. Fuwa, Y. Murata, M. Morinaga: Mater. Sci. Eng. A 480 (2008) 167.

[26] T. Inamura, Y. Fukui, H. Hosoda, S. Miyazaki, Mater. Sci. Forum. 475–479 (2005) 2323–2328.

[27] H.Y. Kim, T. Sasaki, H. Hosoda, S. Miyazaki, Acta Mater. 54 (2006) 423–433.

[28] H. Matsumoto, S. Watanabe, S. Hanada, Mater. Trans. 46 (2005)1070–1078.

[29] S. Ishiyama, S. Hanada, O. Izumi, ISIJ International 31 (1991) 807.

[30] W. Xu, K.B. Kim, J. Das, M. Calin, J. Eckert, Scripta Mater. 54 (2006) 1943.

[31] H. Matsumoto, S. Watanabe, S. Hanada, J. Alloy Compd. 439 (2006) 146.

[32] H.R. Wenk, P. Van Houtte, Rep. Prog. Phys. 67 (2004) 1368.

[33] T. Furuta, S. Kuramoto, J. Hwang, K. Nishino, T. Saito, Mater. Trans. 46 (2005) 3001.

[34] S. Kuramoto, T. Furura, J.H. Hwand, K. Nishino, T. Saito, Metall. Mater. Trans. A 37 (2006) 657.

[35] H.Y. Kim, T. Sasaki, H. Hosoda, S. Miyazaki, Acta Mater. 54 (2006) 423.

[36] T. Saito, T. Furuta, J. H. Hwang, S. Kuramoto, K. Nishino et al., Science 300 (2003) 464.

[37] T. Inamura, Y. Fukui, H. Hosoda, S. Miyazaki, Mater. Sci. Forum 475-479 (2005) 2323.

[38] H. Matsumoto, S. Watanabe, S. Hanada, Mater. Trans. 46 (2005) 1070.

[39] H. Hosoda, Y. Kinoshita, Y. Fukui, T.Inamura, M. Miyazaki, Mater. Sci. Eng. A 438-440 (2006) 870.

[40] Y.N. Wang, J.C. Huang, Mater. Chem. Phys. 81 (2003) 11.

[41] S.F. Castro, J. Gallego, F.J.G. Landgraf, H.J. Kestenbach, Mater. Sci. Eng. A 427 (2006) 301.

[42] M. A. Gepreel, Key Eng. Mater. 495 (2012) 62-66.

[43] O. Engler, Scripta Mater. 44 (2001) 229.

[44] Collings EW, Applied Superconductivity, Metallurgy and Physics of Titanium Alloys, Vol. 1. New York, Plenum Press (1986) pp. 464-469.

Chapter 5

Deformation and Recrystallization Behaviors in Magnesium Alloys

Jae-Hyung Cho and Suk-Bong Kang

Additional information is available at the end of the chapter

http://dx.doi.org/10.5772/55597

1. Introduction

Magnesium alloys have a great potential for application to lightweight structural components due to their low density, high specific strength and stiffness. In particular, wrought Mg alloys have attracted much attention due to their more advantageous mechanical properties compared to cast Mg alloys. The strong preferred orientations, mechanical planar anisotropy, and thus the poor formability of wrought magnesium alloys at room temperature, however, prevent their wider use in areas such as automotive and aerospace parts, electronic devices, and consumable products. These disadvantages are mainly attributed to their hcp (hexagonal close packed) structure and the associated insufficient independent slip systems.

Various studies of wrought magnesium alloys have focused on a microstructure modification of the fine grains and off-basal texturing to improve their formability by means of alloy designs and appropriate forms of thermo-mechanical processing. Alloying designs, which mainly incorporate rare-earth metals, have been used to obtain better off-basal texturing [1–4]. Deviation from the strong basal texture by extrusion [5, 6] and equal-channel angular extrusion (ECAE) processes [7] has been shown to enhance the formability. The weakening of the basal textures was also observed during an asymmetric rolling process [8–13], which resulted in better elongation.

Table 1 shows the frequently-found deformation modes in magnesium alloys, including both slip and twinning systems. The critical resolved shear stresses (CRSS) of non-basal slip systems such as prismatic and pyramidal slip systems show much higher values at room temperature than those of basal slip systems [14, 15]. These hinder the strong activation of the deformation modes enough to accommodate external loading or plastic deformation at room temperature, finally resulting in poor formability. The basal $\langle a \rangle$ system, which is mainly activated at room temperature, does not fully accommodate external elongation, although tensile twinning offers lattice reorientation and further adjusts the degree of deformation. The CRSS of magnesium alloys commonly varies with the temperature. Particularly, non-basal slip systems are more sensitive to temperature, and the CRSS in those cases quickly decreases with the temperature. The activation of non-basal slip systems at elevated temperatures improves the formability of magnesium alloys. The various misorientation relationships between the matrix and the twins commonly observed during the microstructural evolution of Mg alloys are summarized in Table 2.

Deformation mode	$\{hkil\}\langle uvtw\rangle$
Basal $\langle a\rangle$	$\{0002\}\langle 11\bar{2}0\rangle$
Prismatic $\langle a\rangle$	$\{1\bar{1}00\}\langle 11\bar{2}0\rangle$
Pyramidal $\langle c+a\rangle$	$\{11\bar{2}\bar{2}\}\langle 11\bar{2}3\rangle$
Tensile twin	$\{10\bar{1}2\}\langle 10\bar{1}\bar{1}\rangle$
Compressive twin	$\{10\bar{1}1\}\langle 10\bar{1}\bar{2}\rangle$

Table 1. Various deformation modes in magnesium alloys.

Twin type	Misorientation angle/axis	
$\{10\bar{1}1\}$	$56^\circ\langle 1\bar{2}10\rangle$	compressive twin
$\{10\bar{1}2\}$	$86^\circ\langle 1\bar{2}10\rangle$	tensile twin
$\{10\bar{1}3\}$	$64^\circ\langle 1\bar{2}10\rangle$	
$\{10\bar{1}1\}-\{10\bar{1}2\}$	$38^\circ\langle 1\bar{2}10\rangle$	double twin
$\{10\bar{1}3\}-\{10\bar{1}2\}$	$22^\circ\langle 1\bar{2}10\rangle$	

Table 2. Twinning misorientations commonly observed in magnesium alloys

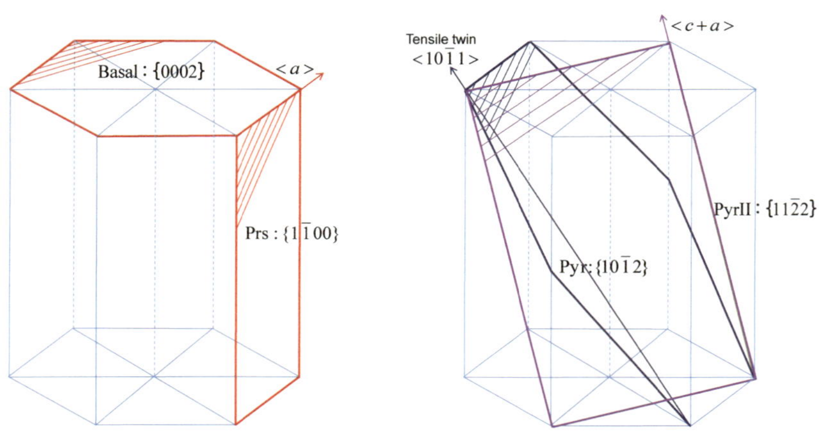

(a) (b)

Figure 1. Lattice planes and directions of the hexagonal crystal structure: (a) basal and prismatic planes, and (b) pyramidal planes

For magnesium alloys, the most dominant deformation mode is the basal ⟨a⟩ slip system regardless of the temperature. The basal ⟨a⟩ slip system is usually aligned with the deformation direction. It aligns the c-axis to the plane normal direction (ND) of the sheet during the rolling process [13]. Similarly, the c-axis vertically aligns to the extrusion direction of the billets during extrusion [6, 16]. The extrusion process contains axisymmetric deformation along the extrusion direction. Most thermo-mechanical processing methods of Mg alloys are carried out at elevated temperatures due to the limited slip system at room temperature. The warm- or hot-working processes frequently cause complex microstructural changes due to dynamic recrystallization in addition to deformation. The rolled sheets or extruded billets should also be annealed to relax the stored energy and to improve the degree of microstructural inhomogeneity for the next forming process.

Here, we present the textural and microstructural evolution during the deformation and recrystallization of various magnesium alloys. Mechanical responses corresponding to the microstructure are also discussed. Uniaxial compression and rolling processes followed by annealing were used as case studies.

2. Experiments

2.1. Materials

Four different magnesium alloys of extruded AZ31B (Mg-Al-Zn system), twin-roll casted AZ31B, ingot-casted ZK60 (Mg-Zn-Zr system) and ingot-casted AM31 (Mg-Al-Mn system) alloys were studied. The overall chemical compositions of the materials are presented in Table 3.

For uni-axial compression tests, cylindrical extruded AZ31B billets and ingot-casted ZK60 alloys were used. Through the compression tests, the deformation and mechanical responses were discussed. The extruded AZ31B billets were commercially fabricated and had an initial diameter of 9 mm. They were simply cut into compression samples with a length of 12 mm.

The other compression samples were prepared using ingot-casted ZK60 alloys, which were originally fabricated by conventional direct chill casting (DC) in a laboratory scale. Its initial thickness was approximately 20 mm and it was warm-rolled down to 15 mm, and then solution heat-treated at 673 K (400 °C) for 15 hours (T4). ZK60 billets with a thickness of 15 mm were machined into cylindrical samples for uniaxial compression. Figure 2 depicts the sample geometry obtained from the initial 15 mm-thick ingot. The character *V* referes to the vertical direction and *H*, denotes the horizontal direction. The V_0, V_{45} and V_{90} samples were machined from the rolling direction (RD) to the ND in each case. The numbers 0, 45, and 90 refer to the angles between the RD and the particular sample used. The ZK60 alloys are typical magnesium alloys with aging (precipitation) hardening, where their strength levels also change with aging [17–21]. Here, we focus on the solid solution state (T4) of the ZK60 alloys. A more detailed discussion of ZK60 alloys including both solid-solution (T4) and artificial-aging (T6) states will be given later.

The AZ31B strips fabricated by twin-roll casting and the ingot-casted AM31 alloys were used for warm-rolling and subsequent annealing. These were also fabricated at a laboratory scale. The initial thickness of the AZ31B strips was about 5 mm, and it was warm-rolled down to about 2 mm, as suggested in the literature [22]. The initial thickness of the ingot-casted

AM31 alloys was 20 mm. The discussion pertaining to the recrystallization behaviors was based on the warm-rolling and annealing processes of the AM31 and AZ31B samples.

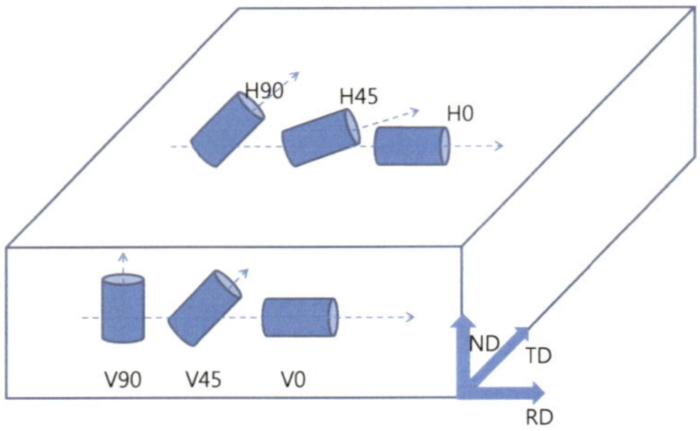

Figure 2. Uniaxial samples taken from the ZK60 billet along various directions with regard to the rolling direction (RD) and normal direction (ND). V_0, V_{45} and V_{90} represent the samples taken from the RD to the ND, while H_0, H_{45} and H_{90} represent the samples taken from the RD to the transverse direction (TD).

2.2. Thermo-mechanical processing

Uniaxial compression tests were carried out using the Thermecmastor-Z (Fuji Electronic Industrial Co.). Cylindrical AZ31B billets with a diameter of 9 mm and a length of 12 mm were used for the compression tests. The other alloys, ZK60 alloys, were machined into compression samples 12 mm in length and 8 mm in diameter, as shown in Fig. 2.

The deformation temperatures of the AZ31B billets were 473 K (200 °C), 523 K (250 °C), and 573 K (300 °C), while the strain rates were 0.00139/s and 0.139/s. Ex-situ experiments to determine the microstructural evolution were also carried out at a temperature of 523 K (250 °C) and a strain rate of 0.32/s. For the ex-situ compression, the sample was marked with micro-indentations and compression and EBSD measurements were then alternately repeated.

Uniaxial compression tests for the ZK60 alloys were also carried out at various temperatures and strain rates. Wider deformation temperatures were imposed on the ZK60 alloys than on the AZ31B samples, and the temperatures were 298 K (25 °C), 398 K (125 °C), 448 K (175 °C), 498 K (225 °C), 548 K (275 °C), 598 K (325 °C), and 698 K (425 °C). Two different strain rates were used, 0.0069/s and 0.139/s. At total strain values of 3% and 7% at a strain rate of 0.139/s, microstructural mapping was carried out using EBSD. Each cylindrical sample was prepared according to the specimen preparation sequence for EBSD mapping before compression, after which the uniaxial compression and EBSD mapping were repeated alternately to determine the microstructural evolution during compression.

Figure 3 shows a schematic diagram of the cylindrical sample and the EBSD measurement region. The extrusion direction is parallel to the compression direction. The orientation color code of the EBSD inverse pole figure maps is also shown.

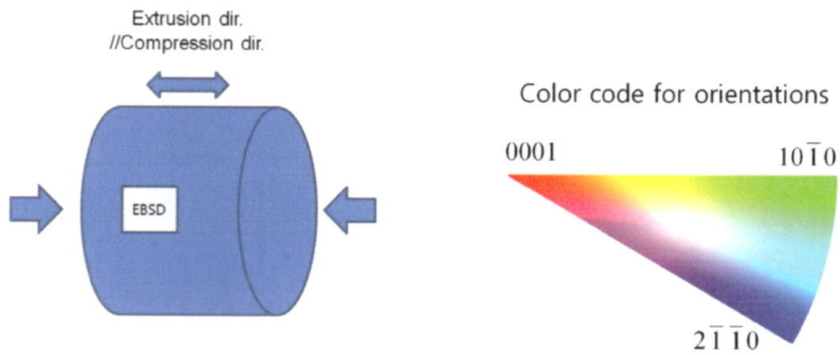

(a) (b)

Figure 3. Sample geometry for EBSD measurements (a), and orientation color code of the inverse pole figure maps (IPFs) (b)

Element	Al	Zn	Mn	Si	Cu	Ca	Fe	Ni	Zr	Mg
AZ31B	2.5-3.5	0.6-1.4	0.2-1.0	< 0.1	< 0.05	< 0.04	< 0.005	< 0.005	-	Bal.
ZK60	0.01	5.47	0.009	0.022	0.003	0.005	0.003	0.007	0.58	Bal.
AM31	3.3	-	0.8	-	-	-	-	-	-	Bal.

Table 3. Chemical composition of AZ31B, ZK60 and AM31 alloys used[wt%].

Ingot-casted AM31 alloys were hot-rolled at 673 K (400 °C) for a reduction in area of about 5% and the microstructure of the as-rolled state was measured using EBSD. AZ31B sheets with an initial thickness of 2 mm were warm-rolled down to 1 mm at a temperature of 498 K (225 °C). The average reduction in area per rolling pass was approximately 10%, and the total reduction in area was 50%. Static annealing was carried out on the warm-rolled AZ31B sheets for 10 min at 573 K (300 °C). The warm-rolling was carried out using a rolling mill with a diameter of 280 mm. The intermediate annealing time between each pass was about 5 min. No lubrication was applied.

2.3. Microstructure characterization

Characterization of the microstructure and texture was mainly carried out using EBSD (electron backscatter diffraction). An automated HR-EBSD (JEOL7001F) with a HKL Channel-5 and the generalized EBSD data analysis code REDS [23] were both used. EBSD samples were mechanically polished and then electropolished using a solution of butyl cellosolve (50 ml), ethanol (10 ml) and perchloric acid (5 ml) at a voltage of 10 V and at temperatures ranging from 253 K (−20 °C) to 258 K (−15 °C).

3. Deformation of Mg alloys

Textural and microstructural evolution and associated mechanical responses were investigated during the uniaxial compression of the Mg alloys at various temperatures and strain rates. Two different Mg alloys of extruded AZ31B billets and ingot-casted ZK60 alloys were prepared for the mechanical tests.

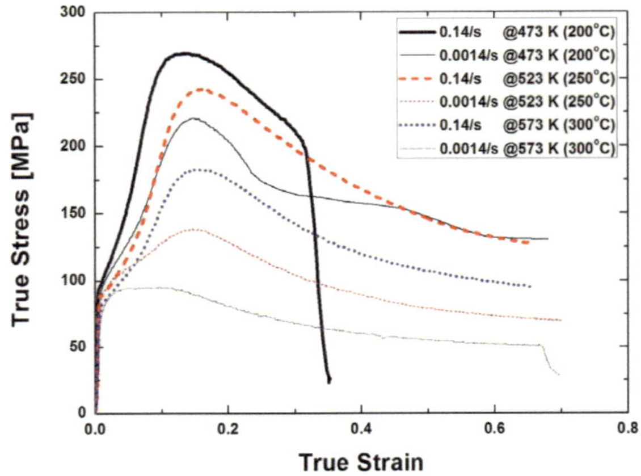

Figure 4. Uniaxial compression of extruded AZ31B billets at various temperatures and deformation rates

3.1. Extruded AZ31B billets

Figure 4 shows the flow curves of the extruded AZ31B billets during uniaxial compression. All three temperatures of 473 K (200 °C), 523 K (250 °C), and 573 K (300 °C) show the peak strength values, after which they decrease with the strain. With an increase in the compression temperature from 473 K (200 °C) to 573 K (300 °C), the peak strength also gradually decreases. The peak value in the strength implies some dynamic process during compression at that temperature and strain rate. The highest peak strength is found at a strain rate of 0.14/s and a temperature of 473 K (200 °C). Even at the same temperature, the lower deformation rate of 0.0014/s results in a low value of the peak strength. When considering the elongation during compression at 473 K (200 °C), the high deformation rate of 0.14/s results in a rapid failure at around 35%. In contrast, the low deformation rate of 0.0014/s reveals a more active dynamic process and the elongation continues, reaching more than 60% past the peak strength. As expected, the lowest strength is observed at a strain rate of 0.0014/s and a temperature of 573 K (300 °C). Overall, high temperature conditions of 523 K (250 °C) and 573 K (300 °C) enhance the elongation until it is greater than 60%.

Ex-situ EBSD measurements of the extruded AZ31B billets during uniaxial compression were carried out on the side, with the samples polished until they had a flat surface, as shown in Fig. 3. Figure 5 shows the flow curve with the ex-situ EBSD measuring points of the extruded AZ31B billets shown in Fig. 5. Open circles illustrate the various strains corresponding to the EBSD measurements. The EBSD measurements were taken of the sample before the peak strength.

Figure 6 presents the textural and microstructural evolution during the uniaxial compression of the extruded AZ31B billets. Micro-indentations were used to identify the mapping area during deformation. The initial microstructure reveals some mixture of large and small grains from the inverse pole figure (IPF) map shown in Fig. 6(a). The IPF maps were plotted based on the extrusion direction (ED), and thus both blue and green on the IPF imply that the $\langle 2\bar{1}\bar{1}0\rangle$ and the $\langle 10\bar{1}0\rangle$ directions align to the ED, respectively. Note the Fig. 3(b).

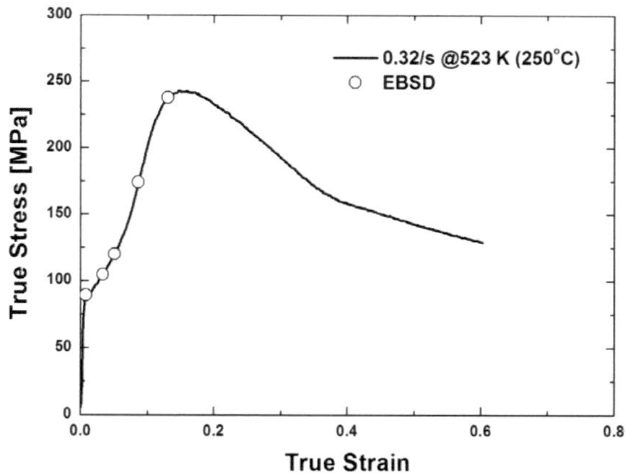

Figure 5. Flow curve with the ex-situ EBSD mapping points of extruded AZ31B billets

The extruded AZ31B billets contained typical extrusion textures. The texture and microstructure quickly changed upon deformation. Twinning activity was evident during compression. Approximately, a strain of 3% caused most of the blue and green grains to contain tensile twinning, which denotes a $\{10\bar{1}2\}\langle 10\bar{1}0\rangle$ or $86^{\circ}\langle 1\bar{2}10\rangle$ misorientation relationship. At a strain of 5%, twinning grew very fast. Nearly the entire region underwent twinning at a strain of 8%. As the deformation degree increased, the EBSD inverse pole figure contained more of a non-indexed region, as shown in black. Grain boundary maps are also presented in Fig. 7. A grain identification (GID) angle of 15° is denoted by the thick lines and the GID of 2° is denoted by the thin lines.

Reorientation of the grains during uniaxial compression can also be found easily using pole figures (PFs), as shown in Fig 8. The coordinates of X_0 and Y_0 correspond to the extrusion (ED) or compression (CD), and the transverse (TD) directions, respectively. As-extruded billets have a fiber texture along the TD, which is typical in the texturing of extruded samples. This stems from the grain reorientation during uniaxial compression, in which the basal planes are parallel to the billet surface. With an increase of the deformation, many tensile twins occurred and grew into parent regions. Strong intensity near the X_0 or ED direction reflects the twinning. The maximum intensity in the pole figures also increases with the deformation.

Misorientation angle distributions are shown in Fig. 9. The solid line represents the misorientation angle distribution of randomly orientated hexagonal polycrystals. The as-received billets show a distribution similar to that of a random distribution. At a strain of 1%, a strong peak arises around a misorientation angle of 86°, which is associated with tensile twinning. This was most frequently observed at a strain of 3%, after which the frequency of low misorientation angles of less than 15° gradually increases. At a strain of 13%, a great amount of low angle frequency develops.

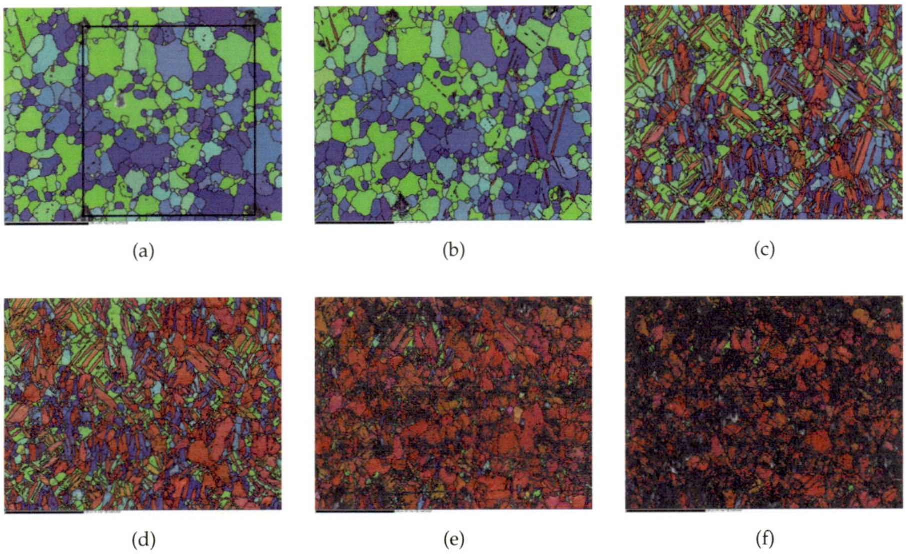

Figure 6. Inverse pole figure maps (IPFs) obtained along the extrusion direction (ED) or compression direction (CD) of the AZ31B billets. The grain identification angle (GID) is $15°$. The step size for the EBSD measurements is $2\ \mu m$ for as-extruded sample and $1\ \mu m$ for others, and the scale bar at the bottom of the IPF maps is $200\ \mu m$. (a) as-extruded, (b) 1%, (c) 3%, (d) 5%, (e) 8%, and (f) 13%

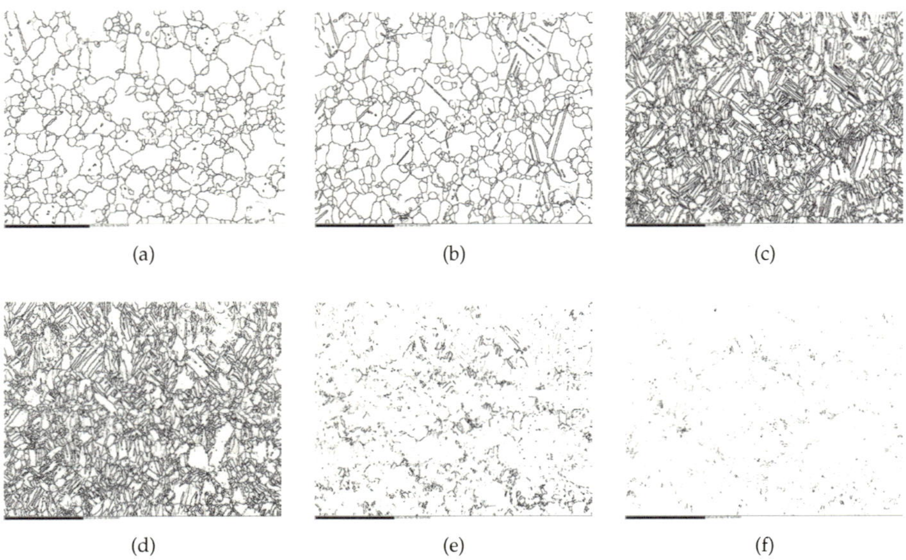

Figure 7. Grain boundary maps obtained from EBSD mapping (AZ31B billets). The grain identification angle (GID) is $15°$. The step size for the EBSD measurements is $2\ \mu m$ for as-extruded sample and for others $1\ \mu m$ and the scale bar at the bottom of the IPF maps is $200\ \mu m$. (a) as-extruded, (b) 1%, (c) 3%, (d) 5%, (e) 8%, and (f) 13%

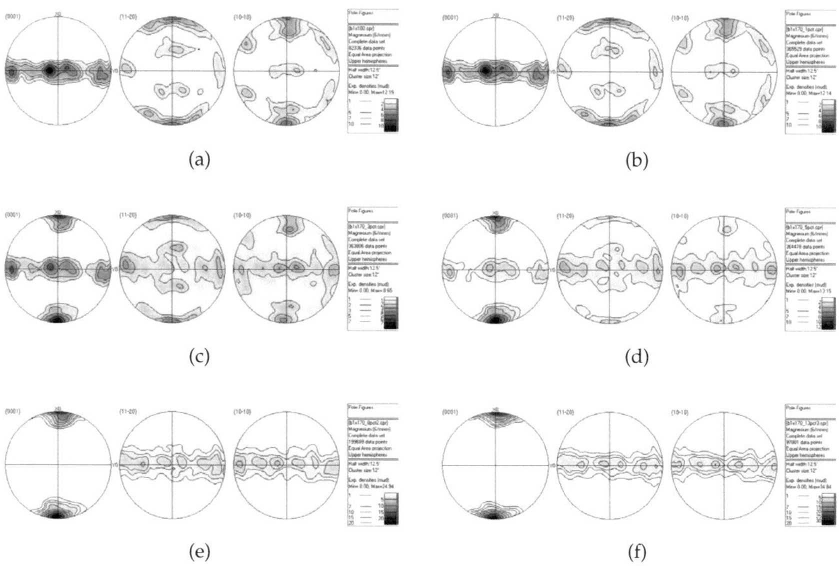

Figure 8. Pole figures obtained from EBSD mapping (AZ31B billets). The coordinates of X_0 and Y_0 correspond to the extrusion (ED) or compression (CD), and the transverse (TD) directions, respectively. (a) as-extruded, (b) 1%, (c) 3%, (d) 5%, (e) 8%, and (f) 13%

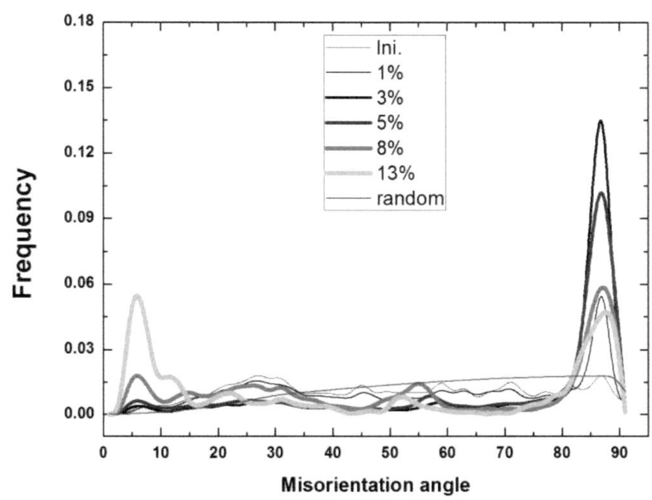

Figure 9. Misorientation distribution obtained from EBSD mapping (AZ31B billets).

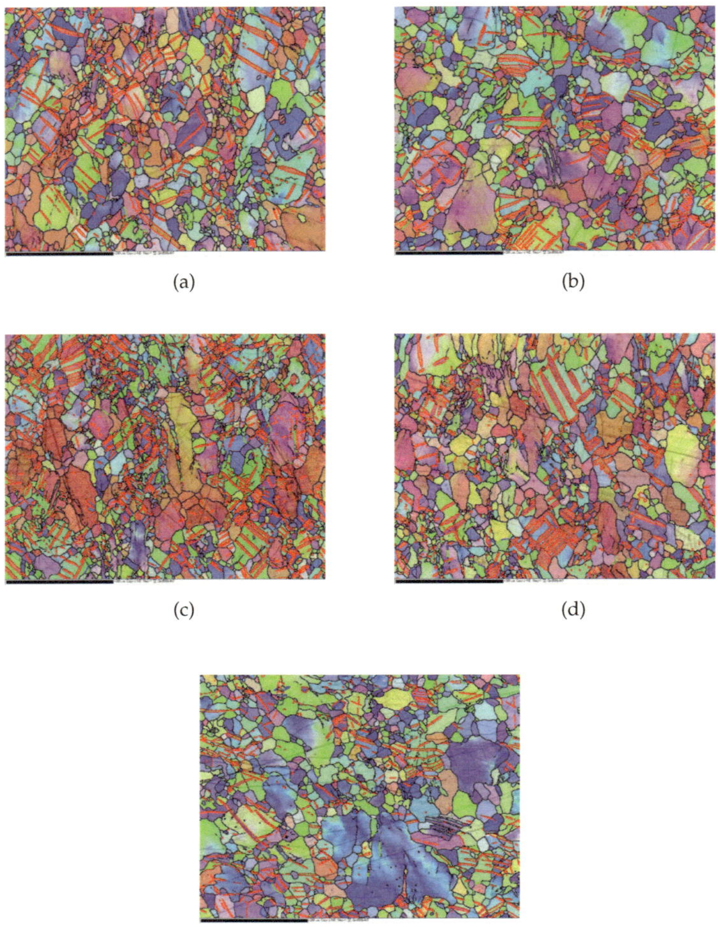

(a) (b) (c) (d) (e)

Figure 10. Inverse pole figure (IPF) maps obtained along the extrusion direction (ED) or compression direction (CD) of the ZK60 alloys ($\varepsilon = 3\%$, $\dot{\varepsilon} = 0.139$/s). The grain identification angle (GID) is $15\,^{\circ}$. The step size for the EBSD measurements is $1\ \mu m$, and the scale bar at the bottom of the IPF maps is $200\ \mu m$. (a) 448 K ($175\,^{\circ}C$), (b) 498 K ($225\,^{\circ}C$), (c) 548 K ($275\,^{\circ}C$), (d) 598 K ($325\,^{\circ}C$), and (e) 698 K ($425\,^{\circ}C$)

3.2. Ingot-casted ZK60 billets

The initial microstructure of the ingot-casted ZK60 alloys possessed an equi-axed grain structure with a weak basal texture.

During the uniaxial compression of the ZK60 alloys, the microstructural and textural evolution was investigated with the temperature. Figure 10 shows inverse pole figure maps of ZK60 at a strain and strain rate of 3% and 0.139/s, respectively. Tensile twinning boundaries are specified in red.

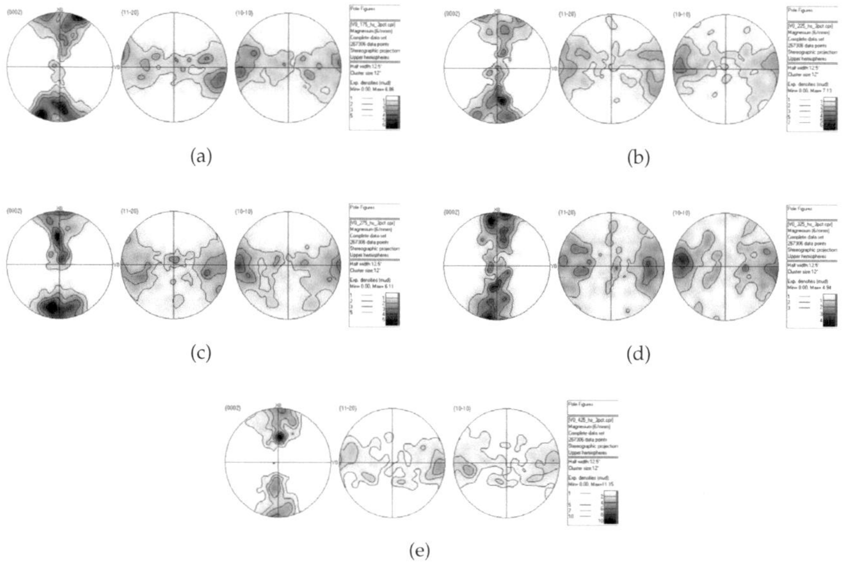

Figure 11. Pole figures of ZK60 alloys obtained from EBSD mapping (ε = 3%, $\dot{\varepsilon}$ = 0.139/s). The coordinates of X_0 and Z_0 correspond to the RD and extrusion direction (ED), respectively. (a) 448 K (175 °C), (b) 498 K (225 °C), (c) 548 K (275 °C), (d) 598 K (325 °C), and (e) 698 K (425 °C)

The variation of the basal intensity is more evident in Fig. 11, which shows pole figures computed from the EBSD mapping in Fig. 10. The initial sample showed weak basal intensity along the X_0 or the normal direction (ND) of the initial billet, as shown in Fig. 2. During compression, the grains were reoriented into the center of the pole figures, Z_0, or in the extrusion direction (ED). With an increase in the temperature, the overall basal intensity decreased. At a temperature of 698 K (425 °), the twinned region is small and the basal intensity in the center is comparatively low. It was noted that a high temperature affects twinning activation and propagation during compression, as less twinning was observed, as shown in Fig. 11(e). Note that the intensity in the center of the pole figure only appears slight in Fig. 11(e). Strong twinning activity results in a sharp increase in the basal intensity, which appears in the center of the (0002) pole figure. The maximum intensity found in Fig. 11(e) is associated with the initial large grains in Fig. 10(e), not with textural evolution due to compression. In fact, a texture analysis using EBSD mapping contains some statistical uncertainty coming from spatial limitations, and statistical uncertainty should therefore be assumed, unlike in the XRD texture analysis, which usually covers a large area and numerous grains.

The grain structure at a strain of 7% resulted in finer grains, as shown in Fig. 12, than that at a strain of 3 %, as shown in Fig. 10. Strong twinning activation and basal slip up to 598 K (325 °C) seem to increase the basal intensity further; thus, most grains appear in red, a result that is related to the basal fiber. In fact, more complicated activation of various deformation modes shown in Table 1 occurs during compression. Note that at a temperature of 698 K

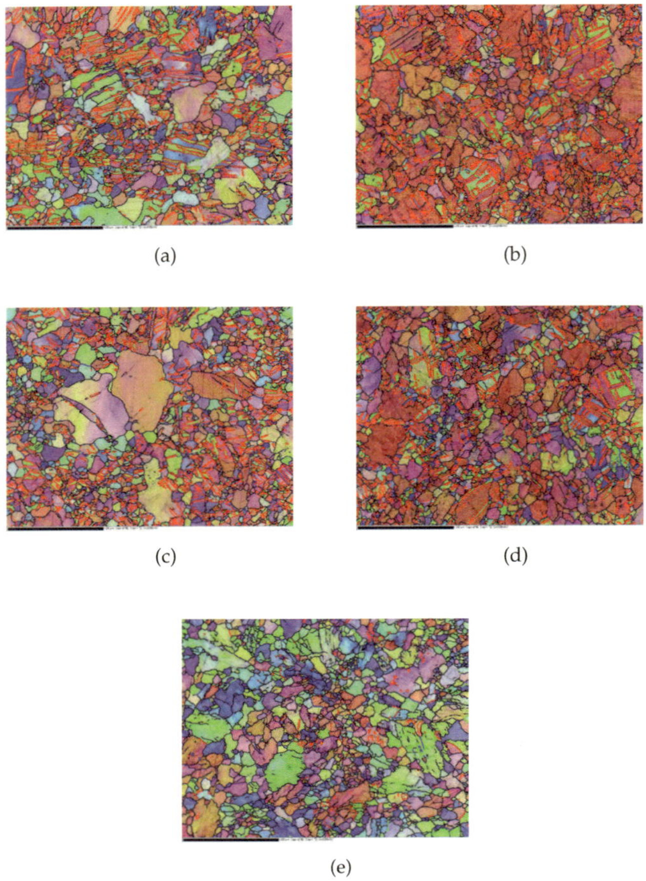

Figure 12. Inverse pole figure (IPF) maps obtained along the extrusion direction (ED) or compression direction (CD) of the ZK60 alloys ($\varepsilon = 7\%$, $\dot{\varepsilon} = 0.139/s$). The grain identification angle (GID) is 15°. The step size for the EBSD measurements is $1\ \mu m$, and the scale bar at the bottom of the IPF maps is $200\ \mu m$. (a) 448 K ($175^\circ C$), (b) 498 K ($225^\circ C$), (c) 548 K ($275^\circ C$), (d) 598 K ($325^\circ C$), and (e) 698 K ($425^\circ C$)

($425^\circ C$), twinning activation is also weakest among other temperature conditions, as shown in at a strain of 3% in Fig. 11(e). Pole figures for the EBSD mapping data in Fig. 12 are shown in Fig. 13. Compared to the PFs at a low strain of 3% in Fig. 11, basal intensity in the center of the pole figures is much more evident in Fig. 13. It was also found that some variation arises in the maximum intensity as the temperature changes. The greatest maximum intensity among the PFs is observed in Fig. 13(c), and this appears to be related to the initially large grains, as discussed in Fig. 11(e).

The misorientation angle distributions at strains of 3% and 7% are given in Fig. 14. A very typical angle distribution is found. The most dominant frequency is observed near tensile twinning. At a large strain of 7%, a higher frequency occurs with misorientation angles smaller than 15°. This is related to the subgrain boundaries due to increase in plastic strain.

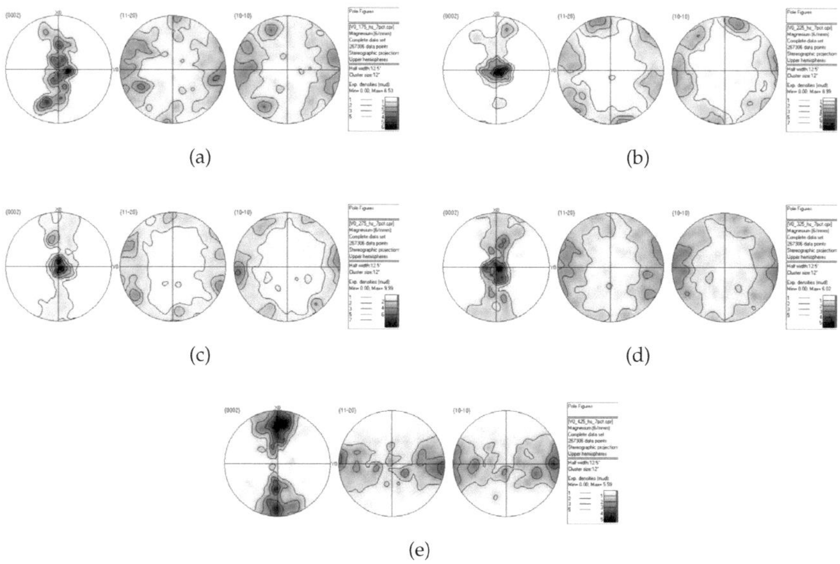

Figure 13. Pole figures of ZK60 alloys obtained from EBSD mapping ($\varepsilon = 7\%$, $\dot{\varepsilon} = 0.139/s$). The coordinates of X_0 and Z_0 correspond to the RD and extrusion direction (ED), respectively. (a) 448 K (175 °C), (b) 498 K (225 °C), (c) 548 K (275 °C), (d) 598 K (325 °C), and (e) 698 K (425 °C)

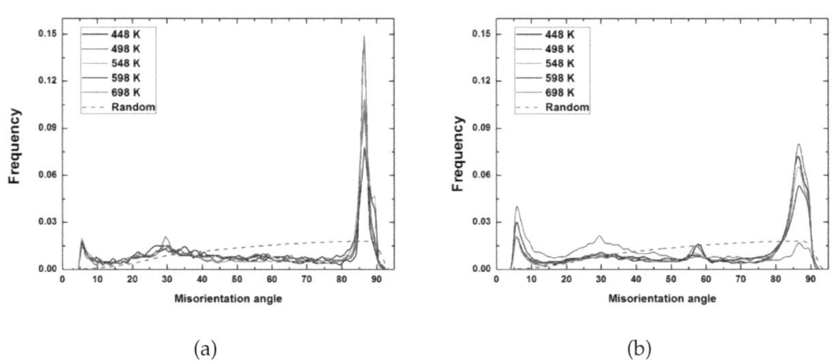

Figure 14. Misorientation distributions obtained from the EBSD mapping of ZK60. (a) $\varepsilon = 3\%$ (Fig. 10) and (b) $\varepsilon = 7\%$ (Fig. 12)

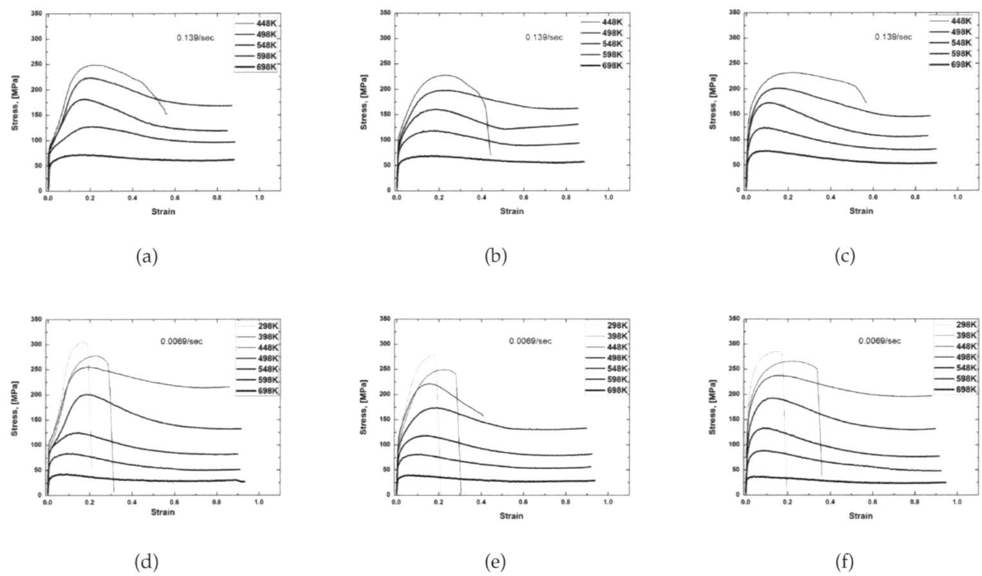

Figure 15. Flow curves obtained from ZK60 alloys. (a) V0, (b) V45, and (c) V90 at a deformation rate of 0.139/s. (d) V0, (e) V45, and (f) V90 at a deformation rate of 0.0069/s.

The flow curves obtained during the uniaxial compression of the ZK60 alloys along the vertical direction are shown in Fig. 15. There are some differences between a high strain rate (0.139/s) and a low strain rate (0.0069/s). For a low strain rate of 0.0069/s, low temperatures of 298 K (25 °C) and 398 K (125 °C) were additionally assessed. The low temperatures cannot provide enough deformation modes, and the elongation at the temperatures is less than it is at other temperatures. The high strain rate of 0.139 imparts higher strength than that at the low strain rate of 0.0069. At a temperature of 448 K (175 °C), it is clear that a strain of 0.0069 shows more extended elongation than a strain of 0.139. The flow curves in Fig. 15(a) correspond to the microstructures examined above in Figs. 10 and 12. The samples taken from the different direction respond differently to external loading or compression. The samples referred to as V_0 show the most dominant twinning; thus, some stress relaxation occurs during hardening (Figs. 15(a) and 15(d)). The samples termed V_{45} revealed some linear strain-hardening behavior (Figs. 15(b) and 15(e)). The samples referred to as V_{90} demonstrated typical non-linear strain-hardening behavior of the type usually found during the plastic deformation of polycrystalline materials (Figs. 15(c) and 15(f)). Note the curvature of the non-linear strain-hardening regions in the flow curves between V_0 and V_{90}. The former shows negative curvature, while the latter shows positive curvature.

Other flow curves obtained along the horizontal direction are shown in Fig. 16. At a high temperature of 498 K, the elongation is better than it is at a low temperature of 448 K. The strain-hardening behaviors of all samples measured from the horizontal directions, H_0, H_{45}, and H_{90}, are similar to those of V_0, as shown in Fig. 15(a). All show negative curvature during strain hardening, which implies that strong twinning occurred during warm compression.

http://dx.doi.org/10.5772/55597

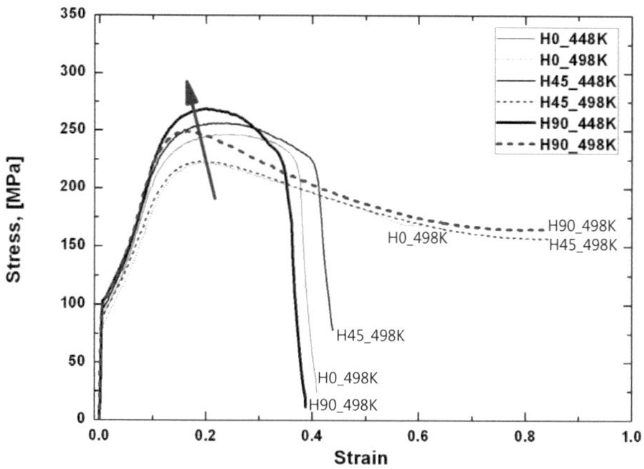

Figure 16. Flow curves obtained from ZK60 alloys along the horizontal direction (see Fig. 2).

4. Recrystallization of Mg alloys

The recrystallization behavior of magnesium alloys was investigated during warm rolling and the subsequent annealing processes. Usually, plastic works of magnesium alloys, including rolling, extrusion, and forming processes, are performed at elevated temperatures. In the previous section, uniaxial compression carried out at various elevated temperatures was discussed. During the warm processing of the samples, dynamic recrystallization can occur depending on the temperature and the total strain. The as-rolled sheets are usually annealed before the next forming process; thus, the texture and microstructure are expected to change. In this section, two different cases of recrystallization - dynamic recrystallization and static recrystallization - are presented. Static recrystallization at the shear band region is also discussed.

4.1. Dynamic recrystallization of hot-rolled AM31 alloys

Large and small grains are mixed as shown in Fig. 17. Particularly, small grains are located between the large grains, which show similar orientations based on the orientation colors and pole figures. It was noted that a shear band forms inside large grains during hot deformation. The small grains are dynamically-recrystallized. Inside the large grains, numerous low-angle grain boundaries with thin black lines and tensile twins, as specified by the red grain boundaries, are also observed.

4.2. Static recrystallization of warm-rolled AZ31B alloys

The warm-rolled sheets show a mixture of large and small grains, as shown in Fig. 18. It appears that the initial large grains are shattered into various sizes of grains during the warm rolling process. During static annealing for 10 min at 573 K (300 °C), the as-rolled structure becomes a fully-recrystallized grain structure, with all grain shapes equi-axed. There exists some variation in the grain sizes of the annealed sheets. The overall textures of both the as-rolled and annealed sheets revealed similar basal fibers. The basal intensity of the as-roll

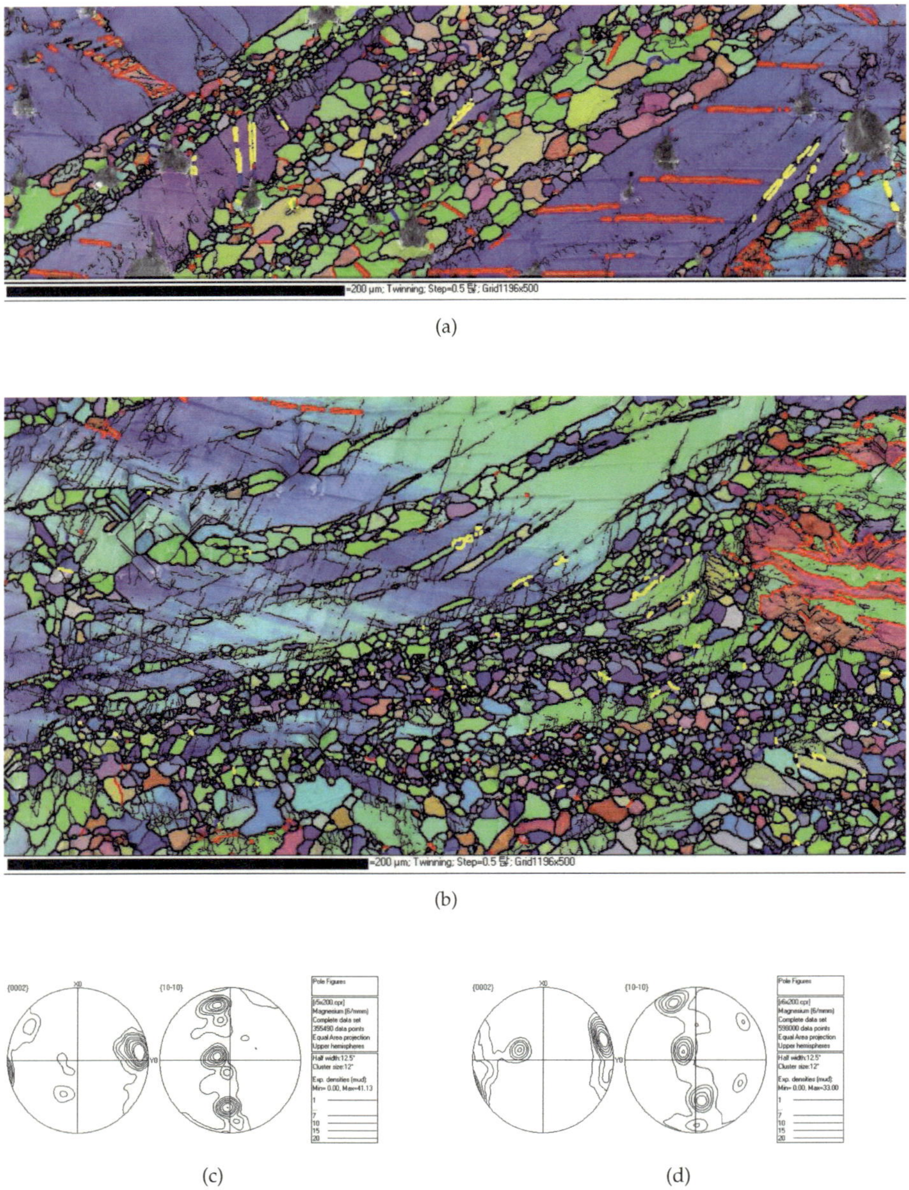

Figure 17. Inverse pole figure (IPF) maps of hot-rolled AM31 alloys. Two different regions were measured using EBSD. The grain identification angle (GID) is 15° with thick black lines, and the GID is 2° as denoted by the thin black lines. The step size for the EBSD measurements is 0.5 μm. The thick red, blue and yellow lines represent the tensile, compressive, and double twin boundaries, respectively. (a) inverse pole figure map and (c) pole figures for the first region. (b) inverse pole figure map and (d) pole figures for the second region.

http://dx.doi.org/10.5772/55597

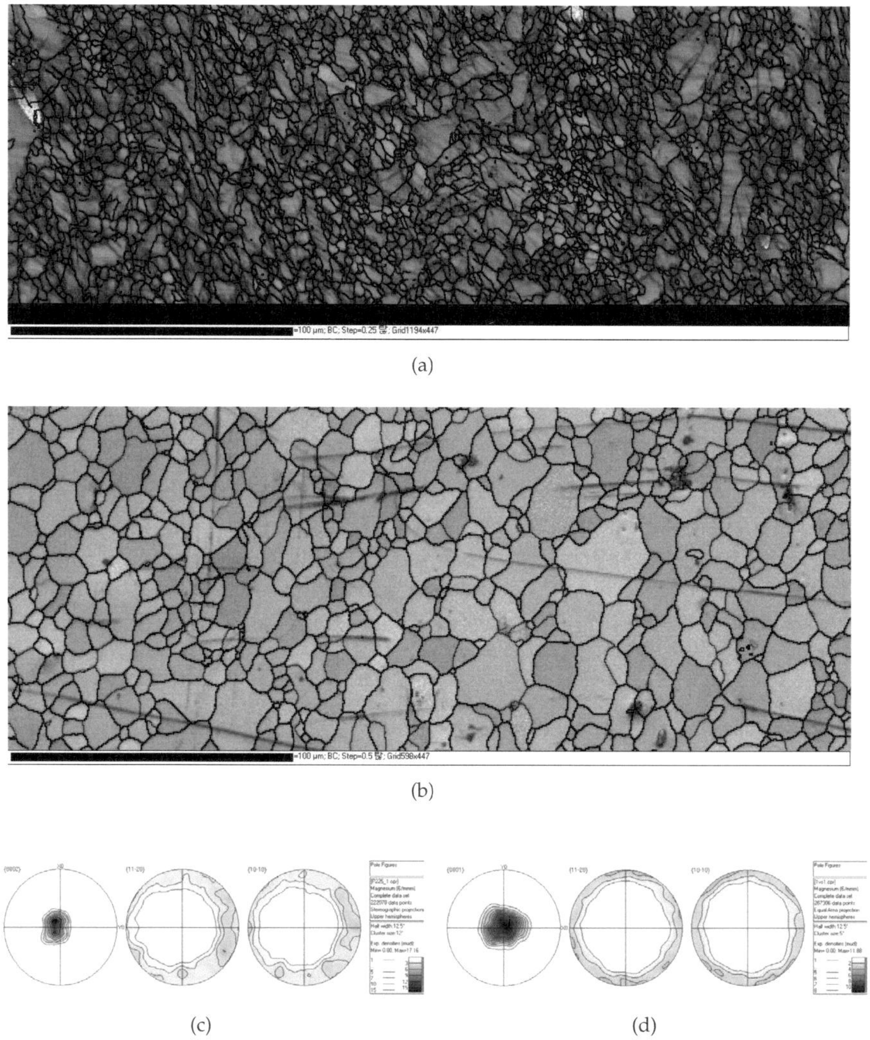

Figure 18. Band contrast maps of warm-rolled AZ31B sheets obtained by means of EBSD measurements. The grain identification angle (GID) is 15° with black lines. The step sizes of the as-rolled and annealed sheets are 0.25 μm and 0.5 μm, respectively. (a) band contrast, and (c) pole figures of the as-rolled sheets. (b) band contrast, and (d) pole figures of the annealed sheets.

sheet confirmed in the (0002) pole figures was stronger than that of the annealed sheet. The basal intensity distribution of the as-rolled sheet also illustrates a more compact distribution than that of the annealed sheet.

Figure 19(a) shows the shear band of the AZ31B alloy, which formed during warm rolling at a temperature of 498 K (225 °C). The shear band region contains numerous twins, and its pole figure in Fig. 19(b) reveals a second strong area of intensity near the X_0 direction. The entire region shows a strong basal texture in Fig. 19(c) of the type usually found during the warm rolling of Mg alloys. The as-rolled sample was annealed for 20 minutes at 573 K (300 °C) and

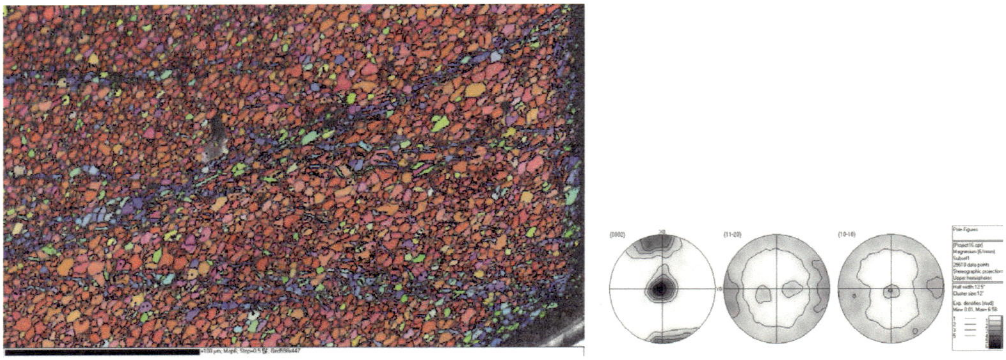

(a) (b)

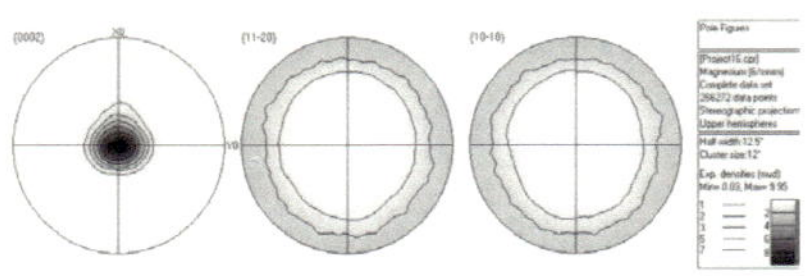

(c)

Figure 19. Inverse pole figure maps of warm-rolled AZ31B sheets obtained by means of EBSD measurements. The grain identification angle (GID) is $15°$ with black lines. The step size is $0.5\ \mu m$. (a) inverse pole figure map, (b) pole figures of the shear bands only, and (c) pole figures of the whole region.

was then remeasured using EBSD (Fig. 20). Marked scratch lines were used for the ex-situ mapping. Although the overall region still revealed a basal texture after annealing, a much wider distribution in the basal texture was observed after annealing, as shown in Fig. 20(b), due to the off-basal orientations in the recrystallized shear band region.

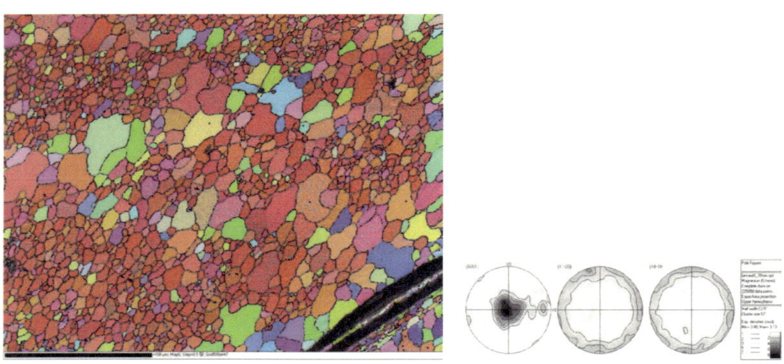

(a) (b)

Figure 20. Inverse pole figure masp of warm-rolled AZ31B sheets obtained by means of EBSD measurements. The grain identification angle (GID) is $15°$ with black lines. The step size is $0.5\ \mu m$. (a) inverse pole figure map, and (b) pole figures of the whole region.

5. Conclusion

The textural and microstructural evolution of wrought magnesium alloys of AZ31B, ZK60 and AM31 was investigated during deformation and recrystallization.

- As-extruded AZ31B billets with extrusion fiber orientations show strong twinning during compression along the extrusion direction. Most of the grains went through twinning at a strain of approximately 3%, after which all became a twinned region at a strain of about 8%. As-warm rolled ZK60 alloys with a weak basal texture also show strong twinning during compression along the rolling direction. Both the as-extruded AZ31B and the warm-rolled ZK60 compressed samples were designed to experience c-axis tension during compression, allowing easy activation of tensile twinning. The flow curves revealed that tensile twinning clearly relaxed the strain-hardening behavior, as evidenced by negative curvature.
- The as-casted AM31 alloys with an initial random orientation show that the shear band can provide a useful mechanism to accommodate external loading. Inside large grains, many shear bands form and are recrystallized during the hot rolling process. The as-rolled AZ31B strips fabricated by twin roll casting possessed initial basal fibers that changed into identical basal fibers after full annealing. During the static annealing process, the deformed microstructure became an equi-axed and recrystallized grain structure. The shear band region with off-basal texturing provides off-basal textures during static annealing and contributes to lowering the basal texturing.

Acknowledgements

The authors would like to thank Lili Chang, Yinong Wang, Shou-ren Wang, Sang Su Jeong and Hyoung-Wook Kim for their comments and help.

Author details

Jae-Hyung Cho and Suk-Bong Kang

Korea Institute of Materials Science (KIMS), Light Metals Division, South Korea

References

[1] S. R. Agnew, M. H. Yoo, and C. N. Tome. Application of texture simulation to understanding mechanical behavior of Mg and solid solution alloys containing Li and Y. *Acta Mater.*, 49:4277–4289, 2001.

[2] J. Bohlen, M.R. Nurnberg, J.W. Senn, D. Letzig, and S.R. Agnew. The texture and anisotropy of magnesium-zinc-rare earth alloy sheets. *Acta Mater.*, 55:2101–2112, 2007.

[3] L.W.F. Mackenzie and M. Pekguleryuz. The influences of alloying additions and processing parameters on the rolling microstructures and textures of magnesium alloys. *Mater. Sci. Eng., A*, 480:189–197, 2008.

[4] A.C. Hanzi, F.H. Dalla Torre, A.S. Sologubenko, P. Gunde, R. Schmid-Fetzer, M. Kuehlein, J.F. Loffler, and P.J. Uggowitzer. Design strategy for microalloyed ultra-ductile magnesium alloys. *Philos. Mag. Lett.*, 89(6):377–390, 2009.

[5] L. L. Chang, Y. N. Wang, X. Zhao, and J. C. Huang. Microstructure and mechanical properties in an az31 magnesium alloy sheet fabricated by asymmetric hot extrusion. *Mater. Sci. Eng., A*, 496:512–516, 6 2008.

[6] N. Stanford and M. R. Barnett. The origin of "rare earth" texture development in extruded mg-based alloys and its effect on tensile ductility. *Mater. Sci. Eng., A*, 496:399–408, 5 2008.

[7] J. Koike, T. Kobayashi, T. Mukai, H. Watanabe, M. Suzuki, K. Maruyama, and K. Higashi. The activity of non-basal slip systems and dynamic recovery at room temperature in fine-grained AZ31B magnesium alloys. *Acta Mater.*, 51:2055–2065, 13 2003.

[8] S. H. Kim, B. S. You, C. D. Yim, and Y. M. Seo. Texture and microstructure changes in asymmetrically hot rolled AZ31 magnesium alloy sheets. *Materials letters*, 59:3876–3880, 2005.

[9] X. Gong, S. B. Kang, S. Li, and J. H. Cho. Enhanced plasticity of twin-roll cast ZK60 magnesium alloy through differential speed rolling. *Mater. Des.*, 30:3345–3350, 4 2009.

[10] B. Beausir, S. Biswas, D. Kim, L. Toth, and S. Suwas. Analysis of microstructure and texture evolution in pure magnesium during symmetric and asymmetric rolling. *Acta Mater.*, 57:5061–5077, 13 2009.

[11] X. Gong, L. Hao, S. B. Kang, J. H. Cho, and S. Li. Microstructure and mechanical properties of twin-roll cast Mg-4.5Al-1.0Zn sheets processed by differential speed rolling. *Mater. Des.*, 31:1581–1587, 13 2010.

[12] J. H. Cho, H. M. Chen, Shi-Hoon Choi, Hyoung-Wook Kim, and S.-B. Kang. Aging effect on texture evolution during warm rolling of ZK60 alloys fabricated by twin-roll casting. *Metall. Mater. Trans. A*, 41:2575–2583, 2010.

[13] Jae-Hyung Cho, Hyoung-Wook Kim, Suk-Bong Kang, and Tong-Seok Han. Bending behavior, and evolution of texture and microstructure during differential speed warm rolling of az31b magnesium alloys. *Acta Mater.*, 59:5638–5651, 2011.

[14] A. Jain and S.R. Agnew. Modeling the temperature dependent effect of twinning on the behavior of magnesium alloy AZ31B sheet. *Mat. Sci. and Engr. A*, A462:29–36, 2007.

[15] A. Chapuis and Julian H. Driver. Temperature dependency of slip and twinning in plane strain compressed magnesium single crystals. *Acta Mater.*, 59:1986–1994, 2011.

[16] S.J. Liang, Z.Y. Liu, and E.D. Wang. Microstructure and mechanical properties of Mg-Al-Zn alloy sheet fabricated by cold extrusion. *Mater. Letters*, 62:4009–4011, 13 2008.

[17] L. Y. Wei, G. L. Dunlop, and H. Westengen. The intergranular microstructure of cast Mg-Zn and Mg-Zn-Rare Earth alloys. *Metall. Trans. A*, 26A:1947–1955, 1995.

[18] L. Y. Wei, G. L. Dunlop, and H. Westengen. Precipitation hardening of Mg-Zn and Mg-Zn-RE alloys. *Metall. Trans. A*, 26A:1705–1716, 1995.

[19] X. Gao and J. F. Nie. Characterization of strengthening precipitate phases in a Mg-Zn alloy. *Scripta Materialia*, 56(8):645–648, 2007.

[20] J. H. Cho, Y. M. Jin, H. W. Kim, and S. B. Kang. Microstructure and mechanical properties of ZK60 alloy sheets during aging. *Materials Science Forum*, 558-559:159–164, 2007.

[21] H. Chen, S. B. Kang, H. Yua, J. H. Cho, H. W. Kim, and G. Mina. Effect of heat treatment on microstructure and mechanical properties of twin roll cast and sequential warm rolled ZK60 alloy sheets. *Journal of Alloys and Compounds*, 476:324–328, 2009.

[22] Jae-Hyung Cho, Hyoung-Wook Kim, Suk-Bong Kang, and Sang Soo Jeong. Texture and microstructure evolution during the symmetric and asymmetric rolling of AZ31B magnesium alloys. *Materials science and engineering A*, in press, 2013.

[23] J. H. Cho, A. D. Rollett, and K. H. Oh. Determination of a mean orientation in electron backscatter diffraction measurements. *Metallurgical and materials transactions A*, 36A(12):3427–3438, 2005.

Section 3

Recrystallization in Natural Environments

Chapter 6

Recrystallization Processes Involving Iron Oxides in Natural Environments and *In Vitro*

Nurit Taitel-Goldman

Additional information is available at the end of the chapter

http://dx.doi.org/10.5772/53735

1. Introduction

Research of nano-size phases through electron microscopy and especially through high resolution electron microscopy enables to observe the morphology of each nano-sized crystal or short range ordered phase, that are usually not detected by other methods. Moreover, lattice fringes of the precursor and the product differ from each other but the original crystal size and morphology remain. The main object of this chapter is to demonstrate the advantages of using electron microscopy in detecting recrystallization processes and possible identification of the precursors.

Nano sized iron oxides are very common minerals in various environments. Various phases of iron oxides crystallize in natural environments of rocks, sediments and soils. Some of the iron oxides form directly either from melts or from solutions; others are formed by recrystallization processes of a precursor through dehydroxylation, dissolution/reprecipitation, oxidation or aggregation involving internal rearrangement within the structure of the precursor. Another formation pattern involves recrystallization of iron-bearing minerals crystallized under anaerobic conditions which are exposed to air. The Fe^{2+} of these iron bearing minerals is then oxidized and hydrolyzed into iron oxyhydroxides.

By using a high resolution electron microscopy, the morphology of newly formed iron oxides can be observed. Moreover, in some cases it is also feasible to detect the precursor's morphology. The newly formed end products are identified by electron diffraction and their chemical compositions are obtained by point analyses; hence, impurities that result from the initial phases can be detected. In most of the samples presented in this chapter, the fine fraction was checked with a High Resolution Transmission Electron Microscopy (HRTEM) using a JEOL FasTEM 2010 electron microscope equipped with a Noran energy dispersive spectrometer (EDS) for microprobe elemental analyses. Other pictures were obtained using a scanning electron microscope JEOL, JXA-8600 and a High Resolution Scanning Electron Microscope Sirion.

The iron oxides studied were found in various natural environments (Figure 1) including:

- recent precipitates in hyper-saline sediments in the Dead Sea Area;
- coating quartz grain in sand dunes and soils under the east Mediterranean climate in Israel;
- sediments of hydrothermal hyper-saline environment of the Red Sea;
- marl layers exposed in the Judean hills;

Other phases were synthesized in a NaCl solution under varying conditions.

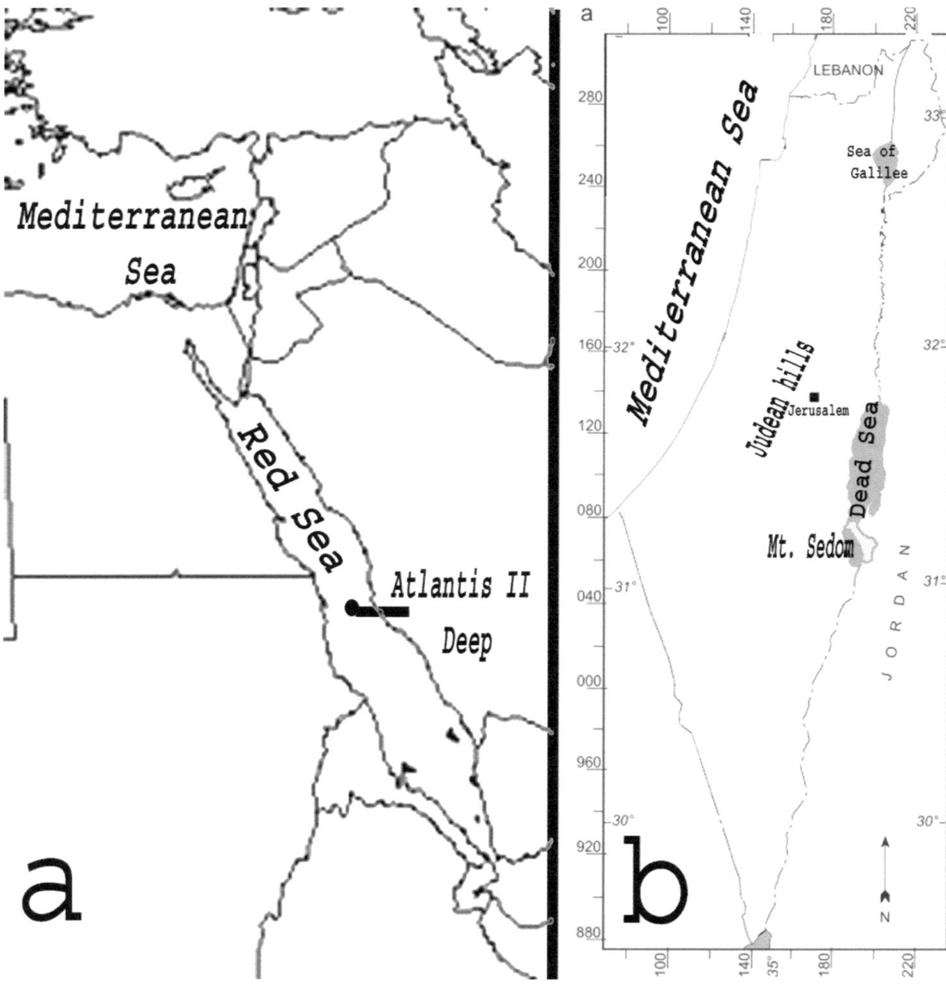

Figure 1. a) Middle East map with Atlantis II Deep located in the Red Sea. b) Israel map with location of Mt. Sedom near the Dead Sea and Judean hills. Samples were also collected along the Mediterranean Sea coast.

2. Results and discussion

Short range ordered 2-line ferrihydrite ($Fe_5HO_8*4H_2O$) is one of the precursors of other iron oxides. It initially precipitates due to Fe^{2+} oxidation and its crystal growth is hindered by the presence of silicate or soil organic matter. Its structural inner order can clearly be visible in HRTEM images. A selected area electron diffraction pattern shows 2 bright rings at 0.15 and 0.25nm. (Figure 2).

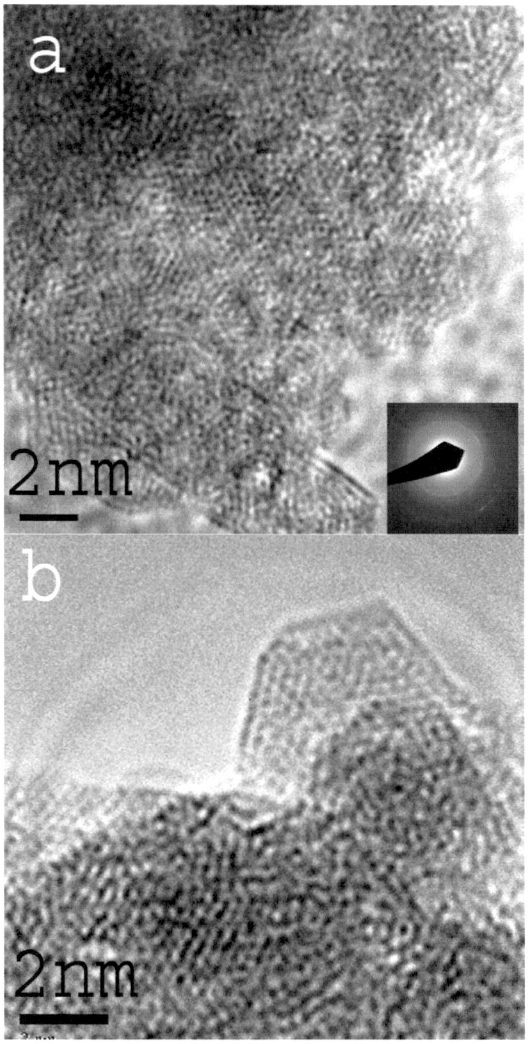

Figure 2. a) High resolution images of short range ordered ferrihydrite preserved within halite crystals in the hypersaline environment at the Dead Sea. The initial stage of recrystallization into a stable phase can be observed at the lower part of the image. Two bright rings at 0.15 and 0.25nm in SAED were obtained. b) A short range ordered pattern was observed in other crystals of ferrihydrite.

2.1. Recrystallization of ferrihydrite into akaganéite in a hyper-saline environment

Akaganeite (β-FeOOH) usually precipitates directly at acidic conditions with the presence of Cl (Cornell & Schwertmann, 2003). In the Dead Sea area close to Mount Sedom salt diaper, a hyper-saline brine discharges in a small spring. As the dissolved Fe^{2+} ions of the spring are exposed to air, ferrihydrite precipitates as an initial iron oxide (Figure 2) and the crystallites are preserved within halite crystals. Dissolution of the halite crystals and its reprecipitation at acidic conditions and elevated Cl concentration enables a recrystallization process into akaganéite crystallites, and they are preserved again within the newly formed halite crystals. Tiny crystallites of akaganéite preserve their initial ferrihydrite precursor's morphology and exhibit a well-crystallized pattern observed through a high resolution transmission electron microscope (Figure 3).

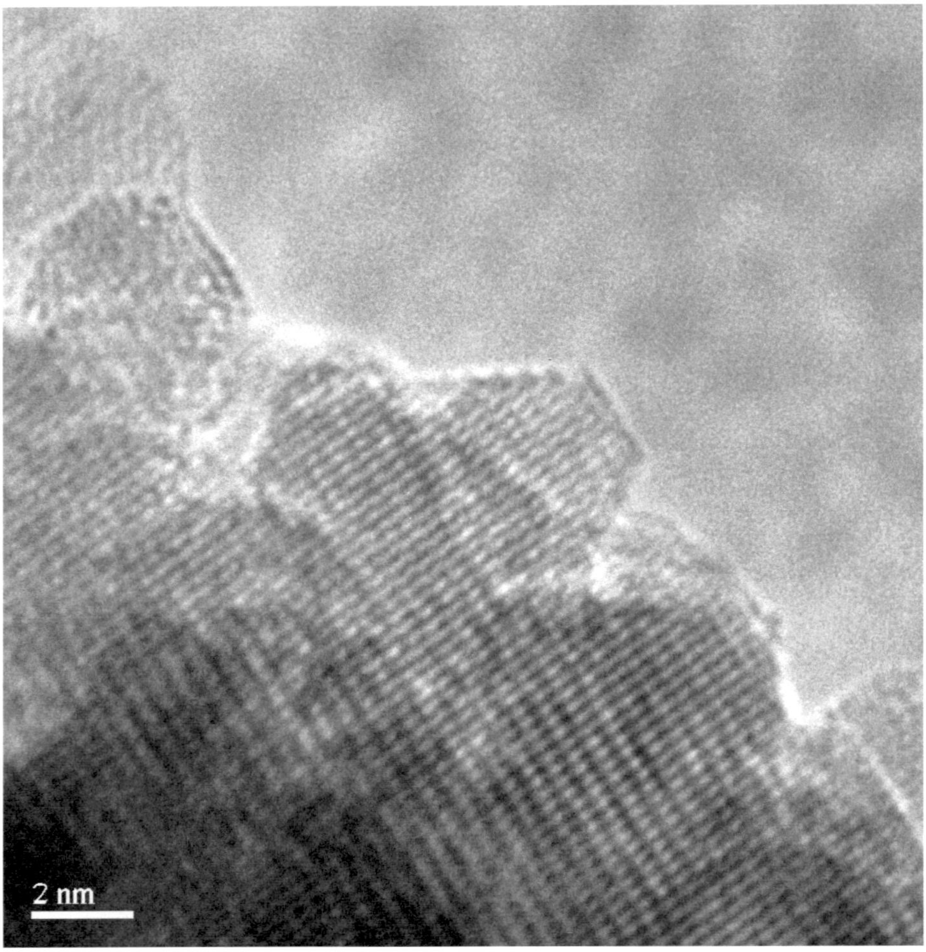

Figure 3. A high resolution image of akaganéite crystallites that present a recrystallization product of ferrihydrite in a hyper-saline environment in the Dead Sea area.

2.2. Recrystallization of ferrihydrite into hematite and goethite in sand dunes and soils

Quartz grains are blown inland along the eastern Mediterranean coastline and form sand dunes. A Mediterranean climate, namely a long hot dry season and a mild winter with a November–March annual precipitation of 400-600 mm (Saaroni et al., 2010) enables the piling up of additional clay minerals that are not washed away but adhere to the quartz grains and then serve as a surface for additional precipitation and recrystallization of iron oxides. Precipitation of these iron oxides causes reddening (rubification) of the quartz grains.

Ferrihydrite is the initial phase formed and it serves as the precursor for hematite (α-Fe_2O_3) or goethite (α-FeOOH) (Figure 4). Recrystallization of ferrihydrite into other iron oxides requires aerobic conditions in a warm climate with a wet and dry season that enables formation of more stable iron oxides. Hematite is formed during the dry season due to aggregation process that includes short range crystallization within a ferrihydrite aggregate (Cornell & Schwertmann, 2003). Dissolution and reprecipitation processes are usually involved in goethite formation. Yet the presence of clay minerals prevented complete drying of the iron oxides, and precipitation of goethite preserved the initial morphology of the ferrihydrite crystallites.

2.3. Recrystallization into hematite from other iron bearing minerals

Small grains of ilmenite ($FeTiO_3$) are transported along with quartz grains to the eastern coast of the Mediterranean Sea. As they are exposed to the Mediterranean climate, they recrystallize into tiny grains of hematite with Ti impurities resulting from their initial precursors. The recrystallization process is feasible since both minerals ilmenite and hematite share the same crystallographic system and even the same space group. Sometimes ilmenite and hematite even form a solid solution. By using electron microscopy, it is possible to observe the initial ilmenite grain along with the secondary hematite crystals that result from the recrystallization process. The precursor and the recrystallized tiny hematite crystals remain close to each other since they are all kept within clay minerals. SA-ED detects ilmenite crystal, and point analyses obtained from the large crystal and the surrounding tiny crystals show that the large ilmenite grain lost most of its iron to the hematite crystallites; however, some of the Ti precipitated within the hematite crystals as an impurity. Other elements shown in the point analyses resulted from the surrounding clay minerals (Figure 5).

Pyrite (Fe_2S) crystals precipitated within Cretaceous marl layers of the Judean hills probably under anaerobic conditions. Oxidation of these crystals yielded pseudomorphic recrystallization into tiny crystals of hematite preserving the initial pyrite large cubic crystal. Using a higher resolution image enabled the observation of small hexagonal plates of hematite crystallites (Figure 6); hence, pyrite served as a precursor for hematite crystals.

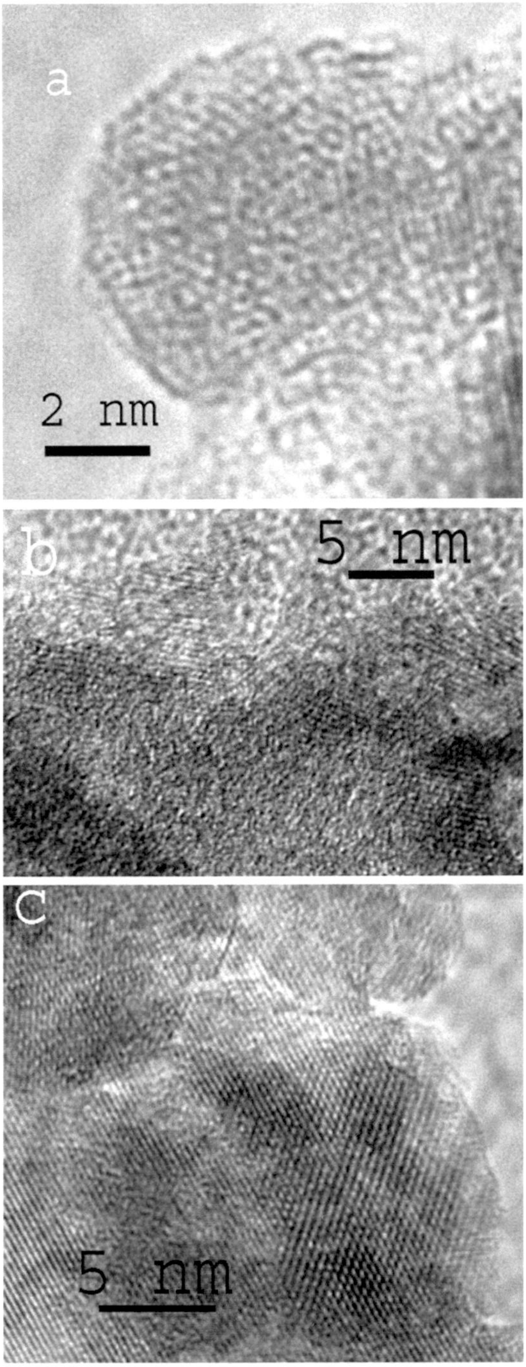

Figure 4. High resolution images of a) A short range ordered ferrihydrite; b) goethite crystals preserving ferrihydrite morphology; c) Hematite tiny crystals preserving the initial morphology of ferrihydrite.

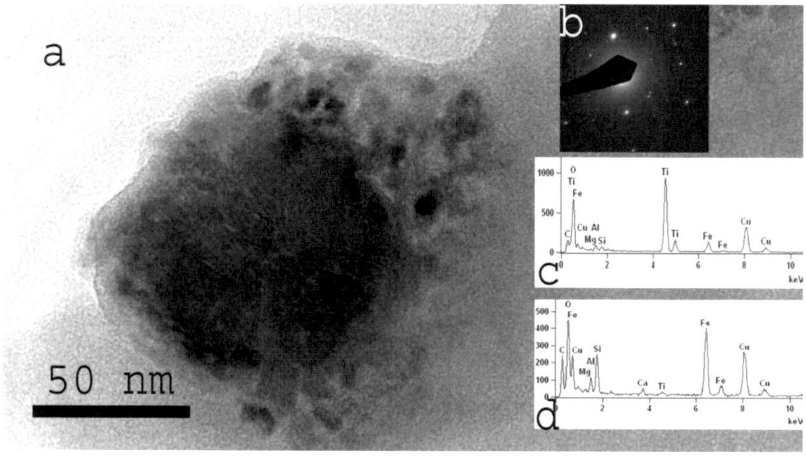

Figure 5. a) An ilmenite grain surrounded by tiny hematite crystals kept within clay minerals; b) SAED of ilmenite crystal; c) Point analysis of ilmenite crystal that went through some dissolution causing Fe removal from the grain into surrounding hematite crystallites; d) Point analysis of hematite crystallites showing a Ti impurity.

2.4. Recrystallization processes in a hydrothermal hyper-saline environment in the Atlantis II Deep, Red Sea

Iron oxides and short range ordered Si-Fe phases precipitate within the hydrothermal brine of the Atlantis II Deep, in the Red Sea (Taitel-Goldman, 2009). Multi-domain goethite is usually formed at elevated Na^+ concentration (Cornell & Giovanoli, 1986). Most of the goethite crystals found in the Atlantis II Deep exhibit multi-domainic character. Appearance of multi-domainic hematite crystals in the Atlantis II Deep was quite rare, yet, some were found in the sediments, probably resulting from the recrystallization process of a multi-domainic goethite precursor (Figure 7).

Rounded particles of short range ordered Si-Fe phase (suggested name: singerite) ($SiFe_4O_6(OH)_4H_2O$) were identified for the first time in the sediments of the Atlantis II Deep (Taitel-Goldman et al., 1999 ; Taitel-Goldman & Singer, 2002). This short range ordered phase is usually metastable and transforms into a more stable phase of iron rich clay mineral like nontronite (Figure 8).

2.5. Recrystallization in vitro of green rust into magnetite

Magnetite ($Fe^{2+}OFe^{3+}{}_2O_3$) mainly crystallizes in magmatic rocks or precipitates in a mixed Fe^{2+}/Fe^{3+} solution in aqueous alkaline systems. Formation of magnetite involves an initial stage of either green rust ($Fe^{3+}{}_xFe^{2+}{}_y(OH)_{3x+2y-z}A_z$; A=Cl, 1/2$SO_4$) or hexagonal flakes of $Fe(OH)_2$ that gradually oxidizes into green rust and then recrystallizes into magnetite crystals (Cornell & Schwertmann, 2003). At elevated temperatures and salinities, less oxygen is available for the oxidation of the dissolved Fe^{2+}, leading to magnetite (mixed Fe^{2+} and Fe^{3+} phases) precipitation. Synthesis was performed in NaCl matrix solutions (4M and 5M) that were kept in a water bath at 70ºC and 80ºC. N_2 was bubbled through the solutions for 20 minutes to remove dissolved oxygen. $FeCl_2$ $4H_2O$ salt was chosen for the Fe^{2+} solutions to

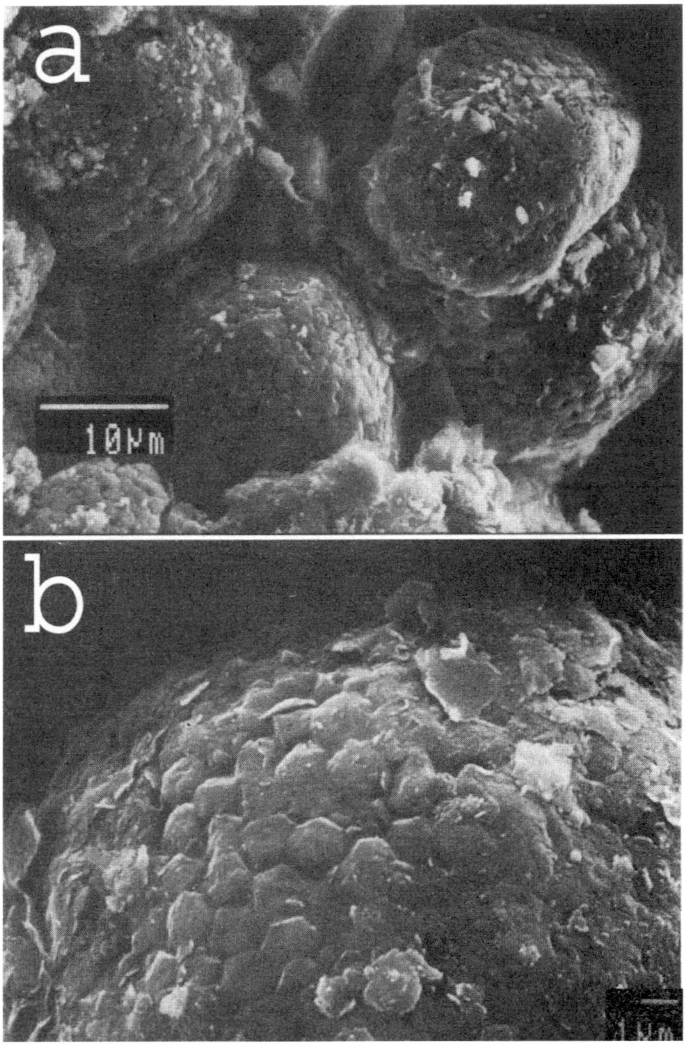

Figure 6. Scanning electron micrograph of tiny hexagonal plates of hematite crystals formed by recrystallization of large cubic pyrite crystals. a) The cubic morphology of pyrite was preserved. b) A close observation of tiny hematite crystallites.

yield a concentration of 0.06M. Fe oxidation was carried out by introducing air at flow rates of 25ml/l, which was monitored with a flow meter, and was kept stable during the 3h of synthesis. Buffering of the pH was obtained by adding a small amount of NaOH (1M). It appears that the pH has a major effect on the kinetics of recrystallization; hence, in samples that were prepared in a highly alkaline solution, transformation into magnetite was quicker, leading to preservation of the precursor. Usually, the recrystallization process yields a cubic morphology of magnetite but due to very fast oxidation and the recrystallization process, the crystals formed preserve the hexagonal morphology of green rust that was observed with a high resolution scanning electron microscope (Figure 9).

http://dx.doi.org/10.5772/53735

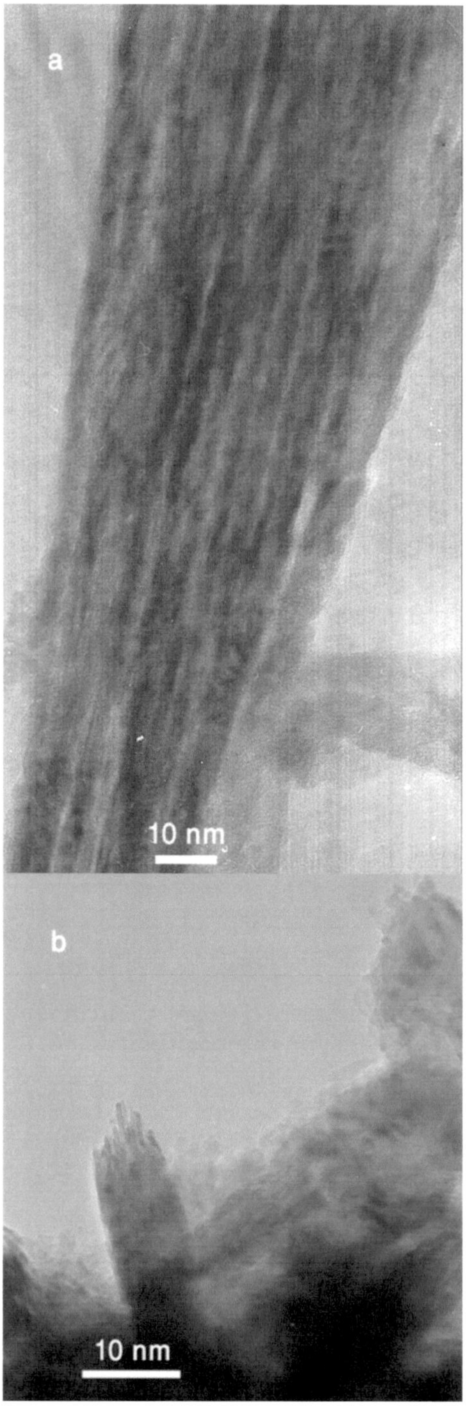

Figure 7. Iron oxides observed in the hyper-saline sediments of the Atlantis II Deep in the Red Sea. a) Multi-domainic goethite; b) Multi-domainic hematite that was formed in a recrystallization process probably from goethite.

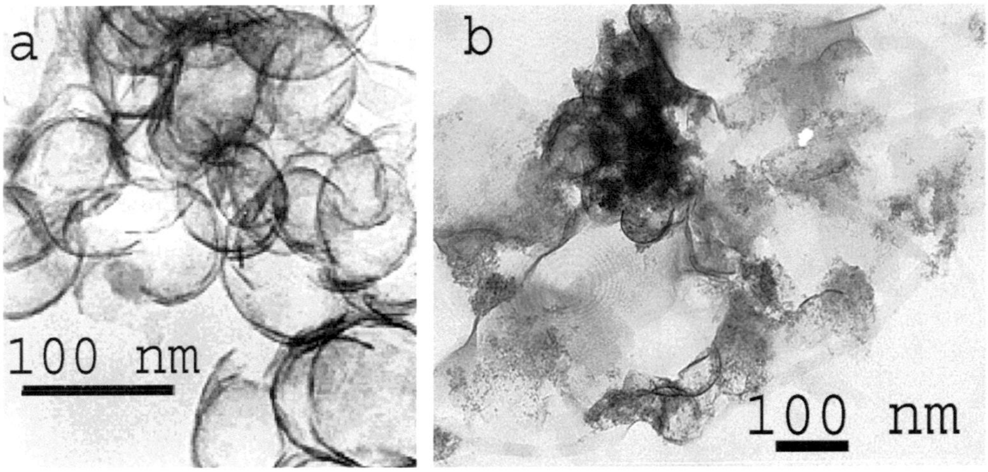

Figure 8. a) A cluster of rounded particles of a short range ordered Si-Fe phase (singerite); b) Disintegration and re-crystallization of singerite into iron-rich clay mineral (nontronite).

Figure 9. A high resolution scanning electron microscope image of synthesized magnetite in varying conditions. Plates appear in both images due to quick recrystallization from green rust or $Fe(OH)_2$ plates. Samples were synthesized at a) 70°C pH 9.4 and a solution of 4M NaCl b) 60°C pH 10.2 and in a 5M NaCl solution.

3. Conclusions

Iron oxides are not the only minerals formed by recrystallization processes. However, their abundance, small crystallite size, quick formation patterns that often involve preservation of their precursor enables their observation through various electron microscopes of the morphology or impurities from the precursor.

In this chapter it was shown that the initial morphology of ferrihydrite is preserved in recrystallization into more stable phases like goethite, akaganéite and hematite. It was also shown that hematite can be formed through several recrystallization processes from other iron oxides or iron bearing minerals.

In some cases, slow recrystallization process is captured within the sample leading to observation both the decomposing precursor and the newly formed product. For example transformation of singerite into nontronite clay mineral or ilmenite into hematite.

Author details

Nurit Taitel-Goldman

The Open University of Israel, Israel

References

[1] Cornell, R. M., & Giovanoli, R. (1986). Factors that govern the formation of multi-domainic goethite. Clays and Clay minerals, , 34, 557-564.

[2] Cornell, R. M., & Schwertmann, U. (2003). The iron oxides structure, properties reactions occurrences and uses. Wiley-VCH Verlag GmbH&Co. KGaA, Weinheim 664p.

[3] Saaroni, H., Halfon, N., Ziv, B., Alpert, P., & Kutiel, H. (2010). Links between the rainfall regime in Israel and location and intensity of Cyprus lows. International journal of climatology, , 30, 1014-1025.

[4] Taitel-Goldman, N., Singer, A., & Stoffers, P. (1999). A new short range ordered, Fe-Si phase in the Atlantis II Deep, Red Sea hydrothermal sediments. in: Proceedings 11th International Clay conference, Ottawa, Canada, 1997 (H. Kodama, A.R. Mermuth and J.K Torrance, editors), 697-705.

[5] Taitel-Goldman, N., & Singer, A. (2002). Metastable Si-Fe phases in hydrothermal sediments of Atlantis II Deep, Red Sea. Clays Minerals, 37 , 235-248.

[6] Taitel-Goldman, N. (2009). Nano-sized iron-oxides and clays of the Red Sea hydrothermal deeps: Characterization and formation processes. VDM Verlag Dr. Miller. 118p.

Section 4

Recrystallization in Ice

Chapter 7

Ice Recrystallization Inhibitors: From Biological Antifreezes to Small Molecules

Chantelle J. Capicciotti, Malay Doshi and
Robert N. Ben

Additional information is available at the end of the chapter

http://dx.doi.org/10.5772/54992

1. Introduction

Recrystallization is a phenomenon that is well documented in the geological and metallurgical literature. In metallurgy, the phenomenon can be formally defined as the process by which deformed grains are replaced by a new set of non-deformed grains that nucleate and grow until the original grains have been entirely consumed. A more precise definition is difficult as this process is quite complex. The phenomenon of recrystallization also occurs in ice, where it is similarly defined as the growth of large ice crystals (or grains) at the expense of small ones. Regardless of the definition or context in which recrystallization is discussed, it is a thermodynamically driven process which results in an overall reduction in the free energy of the system in which it is occurring.

While the exact mechanism(s) by which the phenomenon of recrystallization occurs remains controversial, the industrial significance and the benefits of preventing this process have been realized for hundreds of years. Within the context of ice, recrystallization has a direct impact on many areas such as glaciology, food preservation and cryo-medicine. However, it has been considerably less studied than the process of recrystallization in areas like metallurgy, materials and geology. This may not be entirely surprising as ice itself has very unique physical and chemical properties. While ice exists in several forms, ice I_h (pronounced "ice one h") is the most common form of ice found on Earth. The unique properties of ice and the complications these pose for the detailed study of ice will be described in this chapter with particular emphasis placed upon the efforts to identify and/or design inhibitors of the ice recrystallization process. While inhibitors of ice recrystallization have applications in preventing recrystallization processes in other substances, this review will focus on inhibiting ice recrystallization and its impact in cryopreservation.

As the phenomenon of recrystallization has origins in metallurgy, geology and materials a general discussion of this process with reference to these areas is necessary (Section 2.0), followed by a discussion on the structure and properties of ice and ice recrystallization (Section 3.0) and the importance of inhibiting ice recrystallization (Section 4.0). Finally, inhibitors of ice recrystallization and proposed mechanism(s) of action will be addressed, beginning with the first known inhibitors of ice recrystallization, biological antifreezes (Section 5.0), and concluding with novel synthetic peptides, glycopeptides, polymers and small molecules (Section 6.0). This chapter will conclude with a summary of the role of ice recrystallization in cryo-injury and a discussion on the cryoprotective ability of compounds that exhibit the ability to inhibit ice recrystallization, with the benefits and/or drawbacks of their use during cryopreservation (Section 7.0).

2. The phenomenon of recrystallization

As stated in the introduction, the process of recrystallization has been extensively studied and reviewed throughout the metallurgic literature. [1,2] While the mechanism is quite complex, it is generally defined as the thermally induced change in grain structure facilitated by the formation and/or migration of high angle grain boundaries and is driven by the stored energy of deformation. [1] A grain is defined as the microstructure that constitutes metals and alloys. In a metal, each grain consists of an ordered arrangement of atoms (depicted in Figure 1). [3,4] A grain boundary is the interface where two or more grains of different orientations meet and is considered a defect within the crystal structure. A grain boundary contains atoms that are not well aligned with neighboring grains, leading to less efficient packing and a less ordered structure within the grain boundary. [5] Thus, grain boundaries have a higher internal energy than ordered grains. [5,6] At elevated temperatures, atoms within grains are able to transfer between grain boundaries and neighboring grains. [3,4]

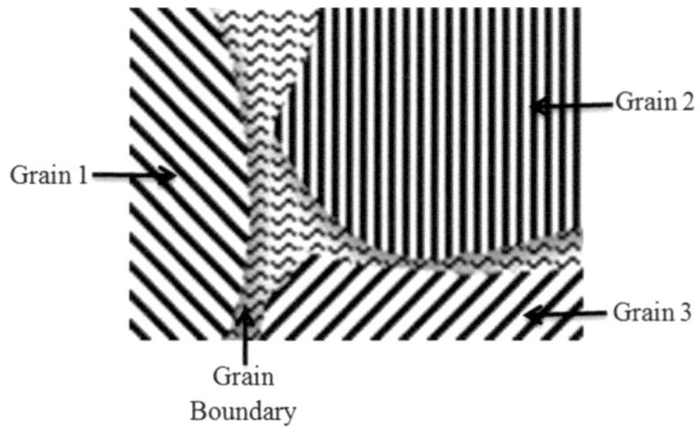

Figure 1. An illustration of grains and grain boundaries in polycrystalline metals and/or alloys.

The process of "plastic deformation" causes a permanent change in the shape of the metal or alloy. During this process, energy is stored mainly in the form of dislocations, ultimately changing the grain shape. [1,2,7] Dislocations are areas where atoms are out of position in the crystalline structure and are linear defects within the grain due to the misalignment of atoms. The amount of dislocations present after deformation is significantly greater than the amount of dislocations prior to deformation. [7] Consequently, the amount of stored energy and the amount of grain strain after deformation is also increased. Heating and annealing of the metal or alloy at or above the recrystallization temperature allows strain-free grains to nucleate and/or migrate within the polycrystalline lattice to minimize the amount of dislocations present within this new set of grains. Thus, the driving force of recrystallization in metals is to eliminate dislocations present in the material to reduce this amount of stored energy in the system. [2]

Recrystallization is an important step in the processing of metals and alloys and can be a desirable or undesirable effect. This is attributed to the fact that recrystallization in metals and alloys ultimately results in a decrease in the strength of the metal. Polycrystalline metals containing smaller grains and more dislocations are significantly stronger than those with larger grains according to the Hall-Petch relationship. [7-9] However, during recrystallization strain-free grains grow to reduce the amount of stored energy from dislocations. As such, the metal is softened and its ductility is increased due to the formation of larger strain-free grains. This process can be a significant problem in metals and/or alloys when these materials are used for structural support where a decrease in metal strength is often detrimental. In contrast, recrystallization can also be beneficial and purposely induced to soften and restore the ductility of metals and alloys that have been hardened by low temperature deformation or cold work, or to control the grain structure of the final metal or alloy product. [1,2,10] For example, metals and alloys that have been deformed by "cold working" (deformation below the recrystallization temperature of the metal or alloy) become stronger and more brittle. [7] Inducing recrystallization will anneal the material to allow it to be deformed further without the risk of cracking or breaking.

3. Recrystallization in ice

Ice has many different polymorphic forms. Individual water molecules in ice can possess different arrangements within three-dimensional space and this is dependent upon temperature and pressure. The most common form of ice below 0 °C and atmospheric pressure is the hexagonal ice I_h lattice unit. [11,12] It possesses a regular crystalline structure in which a single oxygen atom is hydrogen-bonded to two hydrogen atoms. The hexagonal ice I_h lattice unit is characterized by four axes, a_1, a_2, a_3 and c, and the surface of the hexagonal unit has eight faces. [11-14] Two of these faces are normal to the c-axis and are the basal faces, and the remaining six are prism faces. The structure of hexagonal ice is shown in Figure 2. The arrangement of intermolecular hydrogen bonds influences the properties and phases of ice. At 0 °C and atmospheric pressure ice grows most rapidly along the a-axis to give hexagonal shaped crystals which grow as sheets. [11-13,15]

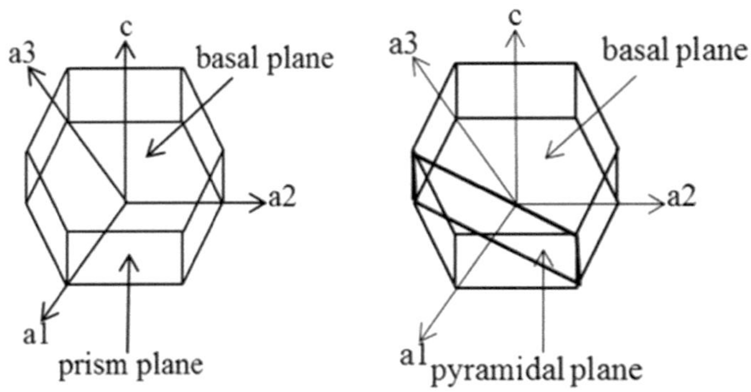

Figure 2. Schematic representation of the hexagonal ice I_h lattice unit illustrating the a_1, a_2, a_3 and c axes and the basal, prism and pyramidal planes.

When ice is in an aqueous solution, the interface between the ice lattice and bulk water is not an abrupt transition. Studies have indicated that a semi-ordered layer exists in between the highly ordered ice lattice and the less ordered bulk water surrounding ice crystals [14,16-22]. This layer has been named the quasi-liquid layer (QLL). While more than 150 years ago Michael Faraday proposed that the surface of ice when near the melting temperature is covered by a thin liquid layer, Fletcher was the first to propose a model for the existence of the QLL in 1962, which was subsequently revised in 1968. [16,17] Important insights on the properties of the QLL was described by Haymet where using molecular dynamic simulations and the TIP4P model of water, the structure and dynamics of the ice/water interface was studied. [18,19] Data from these simulations made it possible to calculate the density profile, molecular orientation and diffusion constants of water molecules in the QLL. The thickness of interface region between ice lattice and bulk water is approximately 10-15 Å thick, but this has been shown to be temperature dependent. [18,19,23] The average density profile, translational and orientational order and diffusion constants of water within the QLL interface also vary depending on the face of ice from which they are calculated. Studies have suggested that the QLL is thicker on the basal and prism faces than on the pyramidal and secondary prism planes. [14]

The exact molecular nature and thickness of the QLL interface has been debated throughout the literature and a wide variety of techniques have been used to study it including atomic force microscopy, [24] X-ray diffraction, [25] infrared spectroscopy, [26] proton-backscattering, [27] Raman spectroscopy, [28] quartz-crystal microbalance measurements, [29] light scattering techniques, [30-32] photoelectron spectroscopy, [33] optical ellipsometry, [22,34,35] optical reflection [36] and mechanical measurements. [37] Ellipsometric studies measuring the refractive index on the basal and prism faces of ice have suggested that the interface is more water-like in nature, rather than ice-like. [20-22,24,33,34] In contrast, other studies have suggested that the orientation and motion of water molecules in the QLL closely resembles that of ice. [25,27,36] The thickness of the QLL has been shown to be temperature dependent, [29,33] such that at temperatures approaching the melting point of ice (at -0.03 °C) the thickness was 15 nm, corresponding to approximately 40 monolayers water. [26] However, below -10 °C the thickness was less than 0.3 nm,

http://dx.doi.org/10.5772/54992

approximately one monolayer of water. The effect of temperature and thickness also varies depending on the face of ice (prism or basal) from which it is calculated, [20,22] and studies have also reported that there is twice as much anisotropy of the water molecules in the QLL for the prism face than the basal face. [34] Light scattering techniques have shown that ice crystals grow into the QLL and not into the bulk water layer. [38,39]

The recrystallization of ice in polycrystalline aqueous solutions is believed to occur through either grain boundary migration or Ostwald ripening. Grain boundary migration in ice is similar to grain boundary migration in metals and alloys where large ice grains grow larger at the expense of small ice grains. In metallurgy a grain consists of an ordered arrangement of atoms and a grain boundary is the interface where two (or more) grains meet. However, in ice a grain consists of the crystallographic orientation of the water molecules commonly observed in ice I_h (Figure 2). Grain boundaries are therefore the interfaces between different oriented ice grains. [40,41] Grain boundary migration occurs as individual molecules transfer from unfavorably oriented ice grains to favorably oriented ice grains. The boundaries of individual ice grains tend to be curved and the degree of curvature is proportional to the size of the grain. Boundaries of small ice crystals have a higher degree of curvature making them more convex (bulge outwards) and thus have a higher amount of surface energy. Large ice crystals have more concave grain boundaries and have a lower amount of surface energy. Grain boundaries migrate towards their center of curvature to reduce the overall degree of curvature, resulting in ice grains with concave boundaries (larger crystals) growing larger while those with convex boundaries (smaller crystals) decrease in size (depicted in Figure 3). [42,43] Thus, the driving force of grain boundary migration in ice arises from a reduction in grain boundary curvature, which results in an overall reduction in the energy of the system.

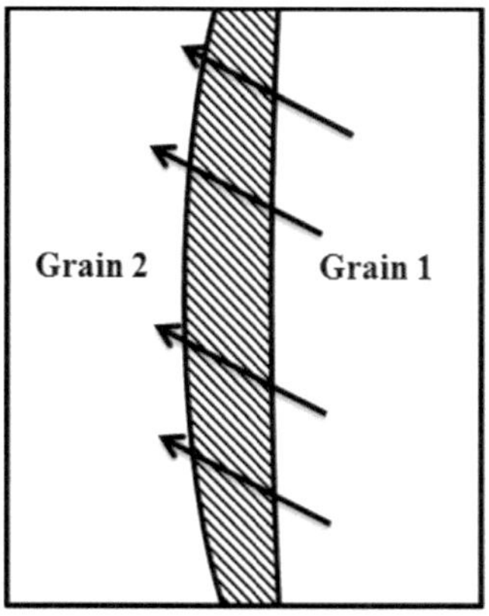

Figure 3. Representation of a liquid-layer (shaded) in a curved boundary between two ice grains. Large ice grains with concave boundaries (grain 2) grow larger while small grains with convex boundaries (grain 1) decrease in size to reduce the overall degree of grain boundary curvature. Arrows indicate the direction of boundary migration.

Grain boundary migration of polycrystalline ice assumes that water molecules are transferred directly from the shrinking ice grain to the growing grain. This assumption often neglects the presence of bulk-water or the QLL in between individual ice grains as the system is treated below -10 °C. [42] However, Ostwald ripening of polycrystalline ice in an aqueous solution considers the whole ice crystal/liquid water system and thus accounts for the presence of bulk-water and the QLL. In ice, Ostwald ripening is the thermodynamically driven process whereby large ice crystals grow larger at the expense of small crystals, resulting in an overall reduction in energy of the ice crystal/bulk-water interface. [44-46] Throughout the Ostwald ripening process a constant ice volume is maintained. Smaller ice crystals have a higher surface area to volume ratio, giving them higher surface free energy since water molecules on the surface are less stable than the water molecules within the ice crystal. [44,45] However, larger ice crystals have a greater volume to surface area ratio and thus are thermodynamically more stable than small ice crystals. As the total overall volume of ice remains constant during the Ostwald ripening process, water molecules transfer from the surface of smaller ice crystals to bulk-water and then are transferred onto the surface of larger ice crystals. The net result is an increase in the average ice crystal size and a decrease in the total number of ice crystals at a constant total ice volume, resulting in an overall reduction in the free energy of the system. [46]

4. Impact of recrystallization

Ice recrystallization is particularly problematic in the areas of frozen foods and cryopreservation of biological samples (cells, tissues etc.). Freezing of foods is a well-established process as it helps decrease the rates of deterioration. In the last 30 years, the frozen food industry has taken significant steps to improve the freezing and storage process of various food products, recognizing that all frozen food products have a finite shelf. [47] Changes in texture, taste and overall quality of a frozen food product are a direct result of the ice recrystallization process. It is well established that ice morphology is an important factor in determining food texture and quality. For example, ice cream containing small ice crystals has better texture and taste. [48]

In medicine, cryostorage is an important process to preserve biological materials or precious cell types such as stem cells (or other progenitors) as well as red blood cells. However, as with any cold storage practice, ice recrystallization remains a major problem and is a significant cause of cellular damage and cell death. [49,50] Section 7.1 of this chapter provides a detailed discussion on the role of ice recrystallization in cryo-injury however, to address these problems effective inhibitors of ice recrystallization are urgently required. Naturally occurring biological antifreezes are very effective inhibitors of ice recrystallization. Biological antifreezes (BAs) are peptides or glycopeptides typically found in organisms inhabiting sub-zero environments. The biological purpose of these compounds is to prevent the seeding of ice crystals *in vivo* and prevent cryoinjury and death.

5. Biological antifreezes as inhibitors of ice recrystallization

The first biological antifreezes were reported in the late 1950s. [51,52] Given their ability to prevent cryoinjury upon exposure to cold temperatures, [53-55] they have attracted a great deal of interest in the scientific and industrial communities. [56] Biological antifreezes are a complex class of compounds with dramatically different structures, making it difficult to understand how they inhibit ice recrystallization. Nevertheless, this important class of compounds is the foundation upon which all "rationally designed" novel ice recrystallization inhibitors are based, including the more recently reported small molecule inhibitors of ice recrystallization. [57-62]

5.1. Structures of Biological Antifreezes (BAs)

In the late 1950s and early 1960s it was observed by Scholander and colleagues that marine teleost fish did not freeze during the winter despite the water temperature being -1.9 °C, over a degree below the freezing point of their blood serum. [51,52] DeVries and Wohlschlag later attributed their survival to the presence of circulating proteins and glycoproteins. [53-55] These proteins later became known as biological antifreezes, specifically antifreeze proteins (AFPs) and antifreeze glycoproteins (AFGPs). A variety of AFPs and AFGPs have since been identified in a number of different fish, insects, plants and bacteria.

There are four classes of structurally diverse fish AFPs that have been identified. These are type I, [13,63-72] type II, [73-81] type III, [73-77,82-87] and type IV AFPs. [88-90] The four types of fish AFPs have a wide variation in their size, which can range from 3-12 kDa, and in their secondary structures, which can be α-helices, β-rolls, random coils and globular structures. AFGPs are also present in fish, and are comprised of a tripeptide repeat of (Thr-Ala-Ala)$_n$, in which the secondary hydroxyl group of threonine is glycosylated with the disaccharide β-D-galactosyl-(1-3)-α-*N*-acetyl-D- galactosamine (structure shown in Figure 4). [15,55,60,61,91-95] In general, AFGPs have a homologous structure and have been separated into eight subclasses, AFGP 1-8, based on their molecular masses which range from 2.6 kDa (n = 4) to 33.7 kDa (n = 50). [55] Minor sequence variations have been identified in AFGPs where the first alanine residue is replaced by proline, or where the glycosylated threonine residue is occasionally replace by arginine. [96-101] The solution structure of AFGPs has been debated in the literature. Early circular dichroism (CD) and nuclear magnetic resonance (NMR) studies suggested AFGPs adopt an extended random coil structure. [102-107] However, studies have also suggested that they adopt an ordered helix similar to a PPII type II helix. [106,108-110] It has also suggested that they adopt an amphipathic helical structure, with a hydrophilic face containing the exposed hydroxyl groups of the disaccharide moiety and a hydrophobic face containing the exposed methyl groups of the amino acid residues. [72] However, the most recent studies have indicated that AFGP 1-5 possess no form of long-range order and that AFGP-8 is predominantly random coil with short segments of localized order. [106-108] A brief summary of the key structural differences between AFPs and AFGPs is provided in Figure 4.

A number of other AFPs have been identified in other organisms. Various insect AFPs have been identified such as those from the spruce budworm moth (*Choristoneura fumiferana*, *Cf*AFP), [111,112] the yellow mealworm beetle (*Tenebrio molitor*, *Tm*AFP), [113,114] the fire-coloured beetle (*Dendroides canadensis*, *Dc*AFP), [115] and the snow flea (sfAFP). [116] Plant AFPs have also been identified from carrot (*Daucus carrota*), [117] bittersweet nightshade (*Solanum dulcamara*), [118] perennial ryegrass (*Lolium perenne*), [119-121] Antarctic hair grass

Characteristic	AFGP	Type I AFP	Type II AFP	Type III AFP	Type IV AFP
Mass (kDa)	2.6 - 33	3.3 – 4.5	11 – 24	6.5	12
Key Properties	AAT repeat; disaccharide	Alanine-rich α-helix	Disulfide bonded	β-sandwich	Alanine rich; helical bundle
Representative Structure	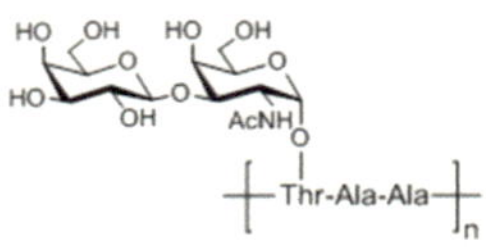				
Natural Source	Antarctic Notothenioids; northern cods	Right-eyed flounders; sculpins	Sea raven; smelt; herring	Ocean pout; wolfish; eel pout	Longhorn sculpin

Figure 4. Classification and structural differences between fish antifreeze proteins (AFPs) and antifreeze glycoproteins (AFGPs).

(*Deschampsia antartica*), [122] and several other species. [123,124] Additionally, AFPs have been identified in fungi and bacteria. [125-130] The secondary structures of the various AFPs from plants and insects are also diverse. [131,132] Regardless of where the AFPs are found or their secondary structure, they are all ice-binding proteins that are crucial for the species survival in the harsh cold environments to which they are exposed.

5.2. "Antifreeze" activities of biological antifreezes: Thermal Hysteresis (TH) and Ice Recrystallization Inhibition (IRI) activity

Biological antifreezes exhibit two types of antifreeze activities. The first and the most studied is thermal hysteresis (TH). This is defined as a selective depression of the freezing point of a solution relative to the melting point. [133-135] TH activity is the direct result of the binding of a BA to the surface of a seeded ice crystal. [136,137] The binding of the BA to the surface of ice facilitates a localized freezing point depression and induces a change in the ice crystal habit. This change in ice crystal habit is referred to as dynamic ice shaping (DIS) and is illustrated in Figure 5A. A more detailed description of this process is described in Section 5.3. The standard assay used to measure TH activity is nanolitre osmometry. [138] In this assay, a single ice crystal in an aqueous solution of the biological antifreeze is obtained, and the growth and behavior of the crystal upon increasing/decreasing the temperature can be observed. TH activity is reported as the difference between the observed freezing and melting points in Kelvin or degrees Celsius.

The second type of antifreeze activity exhibited by biological antifreezes is their ability to inhibit ice recrystallization (referred to as ice recrystallization inhibition (IRI) activity). [41,139]

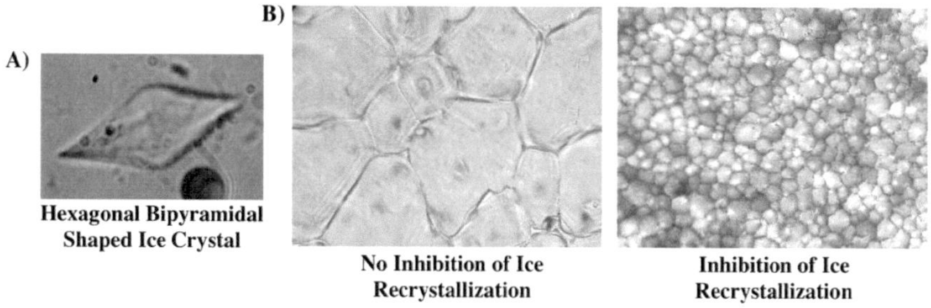

Figure 5. Photographs illustrating dynamic ice shaping (DIS) and ice recrystallization inhibition (IRI) activity. **A)** Ice crystal habit in the presence of 10 mg/mL AFGP-8. The binding of AFGP-8 to the surface of ice crystals induces a change in ice crystal habit, resulting in hexagonal bipyramidal (or spicule) ice crystal shapes. **B)** Photographs of annealed ice grains obtained from a splat-cooling assay. A compound that can inhibit ice recrystallization is able to maintain small ice crystal sizes within a frozen solution.

An illustration of this process is shown in Figure 5B. Inhibiting ice recrystallization results in very small ice crystals within a frozen sample. The ability to maintain small ice crystal size within a frozen solution is a highly desirable property and compounds exhibiting this property have tremendous medical, commercial and industrial applications.

While there are various methods for assessing IRI activity such as the capillary method assay [140,141] or the use of wide-angle X-ray scattering (WAXS) and differential scanning calorimetry (DSC), [142-144] the most commonly used is the splat-cooling assay. [139] In the splat-cooling assay recrystallization can be observed by the change in size of individual ice grains. Briefly, the sample solution is frozen as a thin circular wafer by either dropping a small aliquot onto a precooled (-80 °C) polished aluminum block from a height of approximately 2 meters, [139] or by pressing the solution between two coverslips and freezing. [117] The samples are then annealed at a temperature below 0 °C and the ice crystal size distribution of the sample after a given time is observed. Ice crystal size can be quantified by measuring the mean largest ice grain dimension along any axis [59,145] or by measuring the mean ice grain area. [46,146] Thus, smaller ice crystal sizes represent greater IRI activity. Commonly, analytes are assayed in a salt solution (NaCl, $CaCl_2$ or phosphate buffered saline (PBS)) or a 30-45% sucrose solution, and the solutions without analyte are used as positive controls for ice recrystallization for comparison. The presence of salt or other small solutes is very important as it ensures that liquid is present between ice crystal boundaries and the presence of these solutes negates non-specific IRI effects that can be observed in pure water. [41] While the original version of this assay was subjective in nature, it has recently been improved using Domain Recognition Software (DRS). [146] IRI can now be reliably quantified, providing accurate comparisons between samples and information on small and subtle changes in IRI activity within a series of analogues.

5.3. Biological antifreezes - Mechanisms of action for Thermal Hysteresis (TH) activity

The most widely accepted mechanism for thermal hysteresis (TH) involves an irreversible adsorption-inhibition process. [133-137] In this mechanism, BAs irreversibly bind to specific planes of a growing ice crystal. Preferential binding occurs on the prism faces of ice, thus inhibiting ice growth along the *a*-axis. [93,147-149] Ice crystal growth continues as the temperature of the solution is decreased below the hysteresis freezing point, however it occurs along the *c*-axis, giving rise to the characteristic hexagonal bipyramidal (or spicule) crystal shapes (illustrated in Figure 6). [133,150] The faces that BAs bind to can be determined experimentally by ice hemisphere etching. [136] In this experiment, a single ice crystal in a dilute solution of the BA is grown into a hemisphere such that all interfacial orientations are present during growth. As adsorption of the BA to ice is irreversible, the BA is incorporated into the crystal during growth. Sublimation of the ice crystal then results in visibly etched regions on the ice surface where the BA adsorbed and the orientation of these regions can be observed. While it has been determined that BAs adsorb preferentially to the prism planes of a seeded ice crystal, various insect and plant AFPs adsorb to the basal planes, and it is postulated this results in the superior TH activity exhibited by these proteins. [131,132,151]

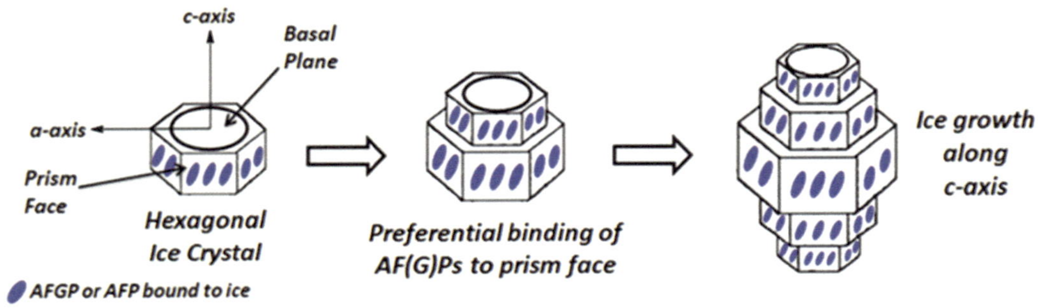

Figure 6. Formation of hexagonal bipyramidal ice crystals by inhibition of growth on the prism faces due to adsorption of BAs.

The irreversible binding of a BA to the surface of ice crystals results in a localized freezing point depression. This occurs *via* the Kelvin (or Gibbs-Thomson) Effect. [135] Given that ice growth cannot occur where the BA has adsorbed, growth occurs on the ice surfaces between adjacent BA molecules, resulting in curved ice surfaces (shown in Figure 7). The energetic cost of adding a water molecule (freezing) to this curved surface is high and it becomes unfavorable for more water molecules to add to this surface, thus a localized freezing point depression is observed. This process does not affect the energetics of the melting process, hence only the freezing point is depressed while the melting point remains constant, resulting in a thermal hysteresis gap (Figure 7A). [135,149,152]

There are two models that described how BAs inhibit ice growth within the thermal hysteretic gap. The first (illustrated in Figure 7B) was proposed by Raymond and DeVries and is known as the step pinning model. In this model, the growth of a step is inhibited by the BA which has pinned ice growth across the ice surface. [133] However, this model assumes that ice crystal growth occurs in steps advancing across the plane that the BA is adsorbed. The second model (illustrated in Figure 7C) is a three-dimensional model known as the mattress model and was

http://dx.doi.org/10.5772/54992

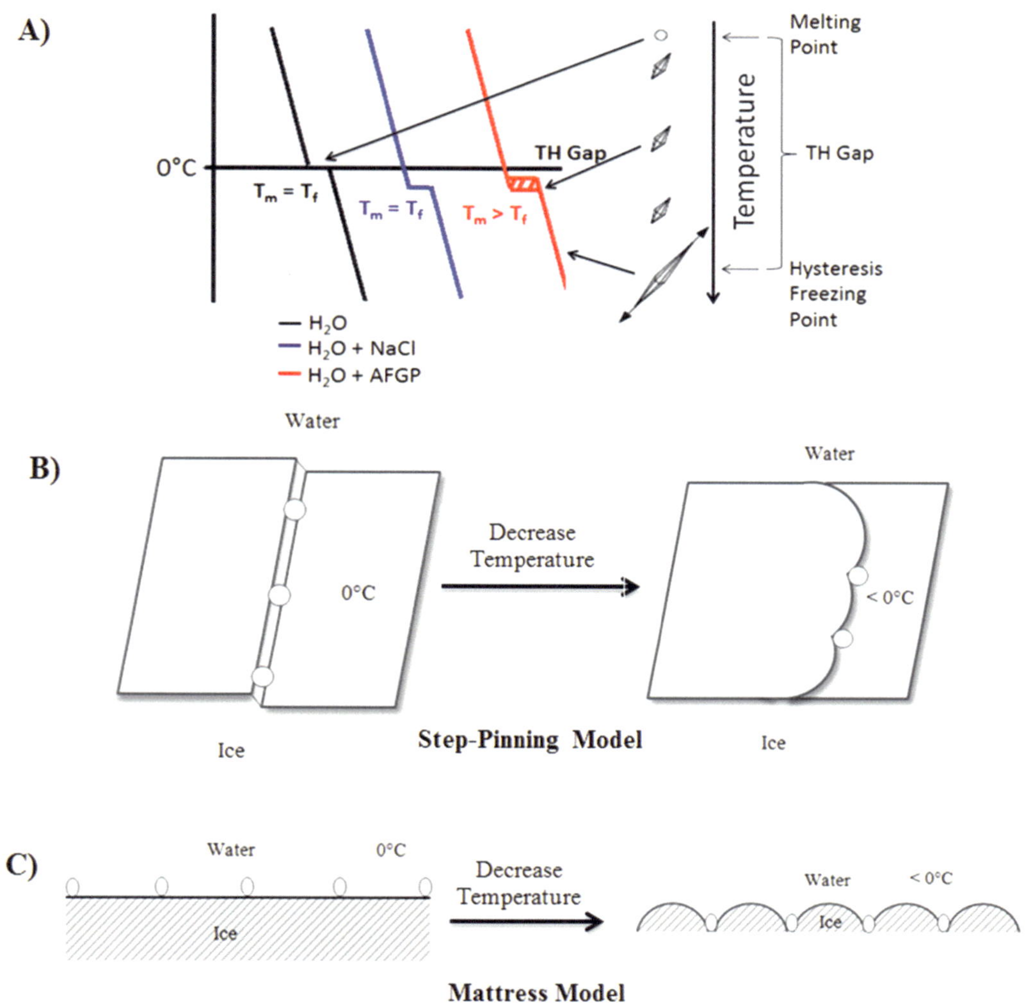

Figure 7. Illustrations of thermal hysteresis (TH) activity and the two models of ice growth inhibition. **A)** BAs have the ability to depress the freezing point of ice crystals relative to the melting point, resulting in a thermal hysteresis gap. **B)** Step-pinning model and **C)** mattress model depicting the irreversible adsorption-inhibition mechanism of BAs.

proposed by Knight and DeVries. In this model, the adsorbed BA molecules exhibit inhibition by pinning ice growth normal (perpendicular) to the ice surface. [136]

Both of these models assume an irreversible adsorption of the BA onto the surface of ice. However, there have been reports suggesting that the adsorption is reversible. The main argument in favour of this is that if adsorption were truly irreversible then significant levels of adsorption would be observed in the presence of very low concentrations of BAs, [67] however this has not been definitively observed. Furthermore, a large free energy of adsorption of BAs would be expected, but it has been observed that the free energy of adsorption is close to zero. [153] Consequently, alternative mechanisms have been proposed describing ice growth inhibition of BAs. [153-156] Regardless of these alternate mechanisms, sufficient data exists to suggest an irreversible adsorption-inhibition mechanism, and consequently this model is the generally accepted mechanism by which BAs exhibit TH activity.

It should be emphasized that the ability to bind to ice is believed to be a property unique to BAs. However, it has been reported that polyvinyl alcohol (PVA) can bind to ice and exhibit a small degree of thermal hysteresis. [157] It was originally proposed that adsorption of BAs to the surface of ice occurred through the hydrogen bonding of hydrophilic groups to the oxygen atoms in the ice lattice. [12,158] However, this is contradictory to the current mechanism of action for AFPs where the importance of hydrogen bonding between polar residues and ice has been questioned. Alternatively, it has been demonstrated that entropic and enthalpic contributions from hydrophobic residues are crucial for ice binding. [159-161] The importance of hydrophobic residues has been validated with a number of different AFPs through site-specific mutagenesis studies, [82,159,162,163] and in general it is believed that the ice-binding site of these AFPs is hydrophobic and has a discrete complementarity with the planes of ice to which it binds. [82,162-165]

In contrast to AFPs, the current hypothesis of how AFGPs bind to ice involves hydrogen bonding between the hydroxyl groups of the sugars and the ice lattice. [137] A landmark study conducted by Nishimura and co-workers investigated the key structural features of AFGPs that were crucial for ice binding and TH activity. [166] In this study it was reported that three key motifs were required for TH activity (shown in Figure 8): 1) the *N*-acetyl group at the C2 position of the galactosamine; 2) the α-configuration of the *O*-glycosidic linkage between the disaccharide and the peptide chain; 3) the γ-methyl group of the threonyl residue. In addition, the TH activity of homogenous AFGPs is dependent upon the length of the glycoprotein segment. [166,167]

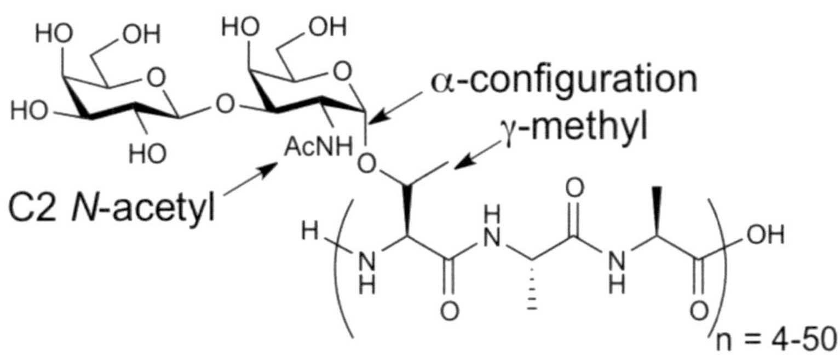

Figure 8. Important structural motifs on AFGPs for TH activity as determined by Nishimura and co-workers. [166]

Despite the tremendous number of structure-function studies conducted on AFPs and AFGPs over the last three decades, in all cases only TH activity has been assessed and correlated to structural modifications. The ability of these analogues to inhibit ice recrystallization has not been assessed, and consequently the structural features necessary for potent ice recrystallization inhibition (IRI) activity are not known. This is unfortunate as IRI activity is a highly desirable property for a compound to exhibit due to the many potential medical and industrial applications. Furthermore, while BAs do possess potent IRI activity, they cannot be used effectively as cryoprotectants. The ice binding ability associated with the TH activity of BAs alters the habit of ice crystals, and since the temperatures employed during cryopreservation are outside of the TH gap, this exacerbates cellular damage. [168-170] However, during the last several years considerable amount of progress has been made in discovering novel ice recrystallization inhibitors, some of which are synthetic analogues of AFGPs, and the work that has been conducted in this area will be the focus of the next section.

6. Inhibitors of ice recrystallization

Biological antifreezes are excellent inhibitors of ice recrystallization. However, as stated in the previous section, the dynamic ice shaping (DIS) capabilities prohibits their use in applications where ice recrystallization inhibition (IRI) activity is highly desirable. Thus, the purpose of the following section will be to summarize the progress towards designing molecules that exhibit the ability to inhibit ice recrystallization without the ability to bind to ice, and on understanding the key structural features that are important for the IRI activity exhibited by these molecules.

6.1. Peptide and glycopeptide analogues of biological antifreezes as ice recrystallization inhibitors

One of the first studies that examined ice recrystallization inhibition (IRI) activity of peptides and conventional polymers was conducted by Knight *et al.* in 1995. [41] In this study, a type I winter flounder antifreeze protein and six analogues of this protein were investigated for their ability to inhibit ice recrystallization, along with four polypeptides and three polymers including polyvinyl alcohol (PVA). One of the conclusions from this study was that all analogues of the antifreeze protein were completely IRI inactive in 0.1% and 0.5% NaCl solutions, a result that correlated with the reduced TH activity in comparison to the native AFP exhibited by these analogues. [41,171] It was also reported that poly-L-histidine, poly-L-hydroxyproline and PVA exhibited IRI activity at concentrations less than 1 mg/mL in pure water, whereas poly-L-aspartic acid, poly-L-asparagine, polyacrylic acid and polyvinylpyrrolidone were inactive. These polypeptides and polymers were not assessed for IRI activity in NaCl solutions.

This study ultimately suggested there was a correlation between TH and IRI activity in the type I AFP. [41,171] While it is well known that biological antifreezes exhibit both types of antifreeze activity, the relationship between TH and IRI has been debated throughout the literature. It was previously suggested that these two properties were directly correlated and derived from the ability of BAs to bind to ice. [139,153] In contrast, it has been suggested there is little or no correlation between TH and IRI as some plant AFPs typically exhibit a low degree of TH activity but a high degree of IRI activity. [119,120] Furthermore, the elevated TH activity exhibited by hyperactive insect AFPs is often not accompanied by highly potent IRI activity. [141] To date, few studies have emerged examining the relationship between TH and IRI activity in native BAs, and those that have, report IRI activity using methods other than the traditional splat-cooling assay, [141] making it difficult to ascertain definitive conclusions about the correlation between TH and IRI.

Payne and co-workers recently published a study in 2012 examining the correlation of glycopeptide/glycoprotein mass on both TH and IRI activity for a range of homogeneous synthetic AFGPs (synAFGPs). [167] A native chemical ligation-desulfurization approach was used for the first convergent synthesis of homogenous synAFGPs that ranged in molecular mass from 1.2 – 19.5 kDa (compounds **1-6**, Figure 9). Increasing the length of the glycopeptide to eight and twelve tripeptide repeats (synAFGP$_8$ and synAFGP$_{12}$, **3** and **4**) increased TH and IRI activity. However, increasing the number of tripeptide repeats to 16 (synAFGP$_{16}$, **5**) led to reduced TH and IRI activity. Additional elongation of the glycopeptide to 32 tripeptide repeats (synAFGP$_{32}$, **6**) restored the potent TH and IRI activities exhibited by these glycopeptides. Interestingly, while synAFGP$_{16}$ (**5**) exhibited less TH activity than synAFGP$_8$ (**3**), both had similar IRI activities. Furthermore, while synAFGP$_{12}$ (**4**) and synAFGP$_{32}$ (**6**) exhibited similar TH and IRI activities and were three times more IRI active than synAFGP$_8$ (**3**) and synAFGP$_{16}$ (**5**), they twice as TH active than synAFGP$_8$ and four times as TH active as synAFGP$_{16}$. These results support the hypothesis that the two types of antifreeze activities may not be as closely correlated as previously thought as the magnitude of change in TH activity was not reflected in IRI activity with these homogenous synAFGPs. While further work is still required in this area to verify this hypothesis, studies on synthetic structural analogues of AFGPs have shown it is possible to decouple the two types of antifreeze activities from each other, resulting in compounds that exhibit "custom-tailored" antifreeze activity and are only IRI active and not TH active. [57,58]

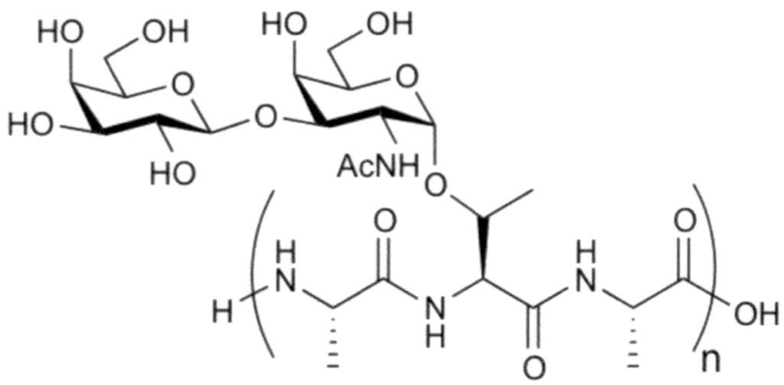

1, n = 2, synAFGP$_2$
2, n = 4, synAFGP$_4$
3, n = 8, synAFGP$_8$
4, n = 12, synAFGP$_{12}$
5, n = 16, synAFGP$_{16}$
6, n = 32, synAFGP$_{32}$

Figure 9. Structures of homogeneous synthetic AFGPs (synAFGPs) reported by Payne and co-workers. [167]

Most of the peptide and glycopeptides that have been assessed for IRI activity have been synthetic structural analogues of AFGPs. The Ben laboratory published the first series of analogues with dramatic structural modifications relative to the AFGP structure, and these analogues maintained the potent IRI activity exhibited by AFGP-8 at equimolar concentrations but did not exhibit TH activity. These analogues were carbon-linked or *C*-linked analogues, and consequently did not possess the *O*-glycosidic linkage found in AFGPs which is susceptible to hydrolysis under basic or acidic conditions. The first of these analogues was reported in 2003 (shown in Figure 10). [172] In comparison to AFGPs, the terminal galactose unit and the *N*-acetyl group were removed leaving only an α-D- galactosyl unit that was conjugated to lysine residues. Lysine was used due to its structural similarity to an arginine residue, which was occasionally found in native AFGPs (see section 5.1). In addition, the alanine residues present in AFGPs were substituted with glycine residues to avoid racemization encountered during solid-phase synthesis. [172,173] The monomer tripeptide unit (**7**) and the analogue with three repeating tripeptide units (**8**) did not exhibit IRI activity. However, the analogues with six and nine repeating tripeptide units (derivatives **9** and **10**, respectively) were both moderately IRI active. Derivatives **9** and **10** were also assessed for TH activity and both exhibited a small TH gap of 0.06 °C and induced the formation of hexagonal shaped ice crystals.

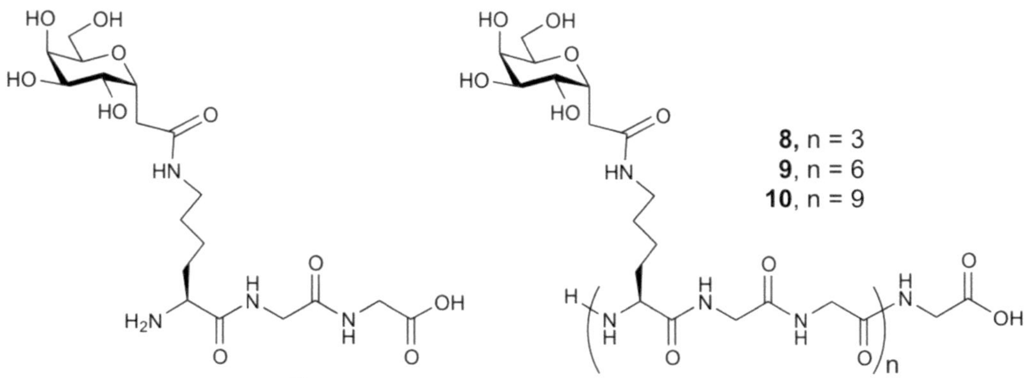

7

Figure 10. Structure of first-generation lysine-based *C*-linked AFGP analogues reported by Ben. [172]

The Ben laboratory has published two other *C*-linked AFGP analogues that exhibit potent IRI activity. These are derivatives **11** and **12** (Figure 11). Derivative **11** contains four tripeptide repeats, in which a *C*-linked galactosyl unit is incorporated. [57] Derivative **12** also contains four tripeptide repeats, and is structurally similar to lysine derivatives **8-10**, however the *C*-linked α-D- galactosyl unit is conjugated to an ornithine residue. [58] Both of these derivatives exhibited potent IRI activity at 5.5 μM and their activity was similar to that exhibited by AFGP-8 at 5.5 μM. Unlike AFGP-8, neither of these derivatives exhibited TH activity and while **12** exhibited very weak dynamic ice shaping, [58] **11** did not exhibit any ice shaping capabilities. [57] This suggested that the exhibited IRI activity was not likely due to ice binding. These analogues were the first examples where the two properties of biological antifreezes, TH and IRI activity, were decoupled from each other. Additionally, these *C*-linked AFGP analogues were the first compounds that possessed "custom-tailored" antifreeze activity, meaning they exhibited potent IRI activity with little or no measureable TH activity.

Following the discovery of the two novel synthetic ice recrystallization inhibitors **11** and **12**, two studies have been reported that identify the structural features necessary for the potent IRI activity of these *C*-linked analogues. The first structure-function study was conducted on

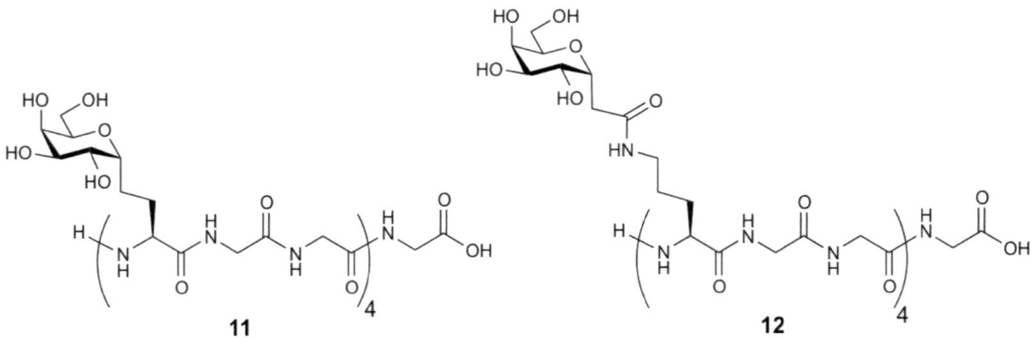

Figure 11. Structures of potently IRI active *C*-linked AFGP analogues **11** and **12** reported by the Ben laboratory. [57,58] Analogues **11** and **12** are the first compounds reported to exhibit "custom-tailored" antifreeze activity, meaning they exhibit potent IRI activity but not TH activity.

C-linked AFGP analogue **12** and examined the importance of the carbohydrate moiety. [58] The galactosyl moiety of **12** was substituted with three other monosaccharides: glucose, mannose and talose (analogues **13**, **14** and **15**, respectively, Figure 12). It was found that replacing the galactosyl unit with other monosaccharides was highly detrimental for IRI activity. The glucose analogue **13** exhibited weak activity, whereas the mannose and talose analogues (**14** and **15**) were inactive. The results showed that the stereochemical relationship of the hydroxyl groups on the carbohydrate moiety on the polypeptide has a direct affect upon IRI activity. The stereochemical relationship of the hydroxyl groups on simple carbohydrates (mono- and disaccharides) is known to influence the hydration of carbohydrates. [174-176] This lead to the observation that carbohydrate hydration was important for IRI activity. [58] A more detailed discussion of carbohydrate hydration and its influence on IRI activity is provided in section 6.3 of this chapter. Briefly, carbohydrate hydration influences IRI activity by altering the ordering of bulk-water based on the compatibility of the carbohydrate within the three-dimensional hydrogen-bonded network of water. [58,177] The hydration of a carbohydrate is related to the compatibility of the sugar with the three-dimensional hydrogen-bond network of water. [174-176] Of the monosaccharides assessed, talose is the most compatible and is thought to have the best "fit" into this hydrogen-bond network, whereas galactose is the least compatible and has the worse "fit". It was hypothesized that a poorer "fit" of the carbohydrate into the hydrogen-bond network of bulk water resulted in a more disordered bulk water layer between the semi-ordered quasi-liquid layer and ordered ice crystal layer. Consequently, transferring water molecules from a more disordered bulk water layer to an ordered layer was energetically unfavorable. Thus, carbohydrates that are highly hydrated resulted in greater IRI activity. [177] While the overall hydration of the *C*-linked glycoconjugates **12-15** is not known, having a more highly hydrated carbohydrate moiety conjugated on the glycopeptide (ie. galactose) was significantly better for IRI activity than a less hydrated carbohydrate moiety. [58]

The second structure-function study examined how the distance between the galactosyl moiety and the polypeptide backbone influenced IRI activity. In this study, the distance

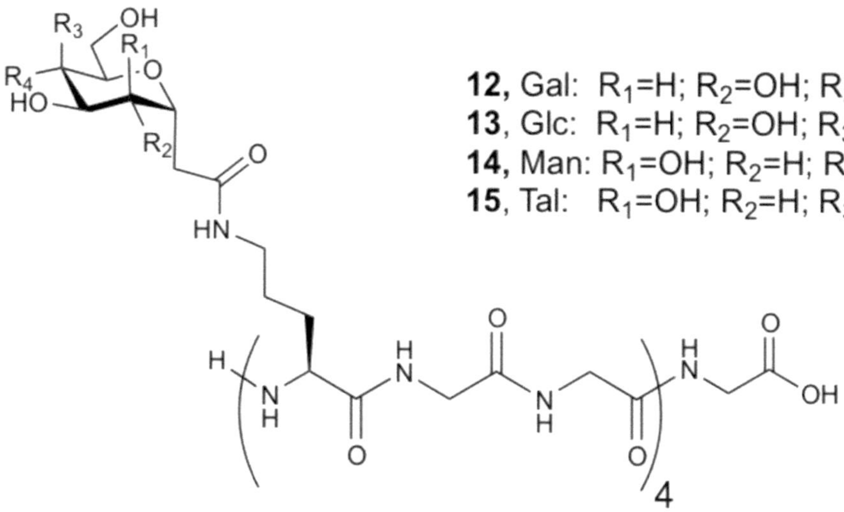

12, Gal: R_1=H; R_2=OH; R_3=OH; R_4=H,
13, Glc: R_1=H; R_2=OH; R_3=H; R_4=OH,
14, Man: R_1=OH; R_2=H; R_3=H; R_4=OH,
15, Tal: R_1=OH; R_2=H; R_3=OH; R_4=H

Figure 12. Structures of C-linked AFGP analogues containing various monosaccharide moieties. [58]

between the carbohydrate and peptide backbone of derivative **11** was increased such that the side chain linking the carbohydrate to the backbone was two, three or four carbons in length (analogues **11**, **16** and **17**, respectively, Figure 13). [57] The distance between the carbohydrate and peptide backbone of derivative **12** was both increased and decreased such that the side chain linking the carbohydrate to the backbone was a total of four, five, six or seven atoms in length (analogues **18**, **19**, **12** and **20**, respectively, Figure 13). [58,178] All of the analogues in which the side chain lengths were modified failed to exhibit IRI activity. These results indicated that the optimal length of linker between the carbohydrate and peptide backbone is two carbons for analogue **11** and six atoms for analogue **12**. [57,178] Molecular dynamic simulations indicated that **12** adopted a unique conformation in solution that was distinctly different than analogues **18-20**. [178] While **18-20** were found to adopt a conformation in which the carbohydrate moiety was extended away from the polypeptide backbone, the side chain of **12** was folded back on itself. It was speculated this fold formed a hydrophobic "pocket" between the carbohydrate and the peptide, resulting in potent IRI activity.

In addition to C-linked AFGP analogues, other synthetic variants of AFGPs have recently been assessed for their ability to inhibit ice recrystallization. In 2010, Sewald and co-workers synthesized analogues of AFGP-8 in which alanine residues were replaced with proline residues and the native disaccharide was replaced with the monosaccharide α-*N*-acetyl-D-galactosamine (Figure 14). [179] It was reported that the glycopeptide analogues containing tripeptide repeats of (Ala-Ala-Thr(GalNHAc))$_n$ were found to exhibit IRI activity (compounds **21-23**). This activity was dependent upon the length of the glycopeptide, and the compound with five tripeptide repeats (**23**) was found to be the most active at a lower concentration (12.5 μM) in comparison to the compound with three tripeptide repeats (**21**) which was active at a much higher concentration (0.8 mM). AFGP analogues **21-23** were found to induce hexagonal ice crystal shaping, suggesting that they are interacting with the ice lattice, however the TH

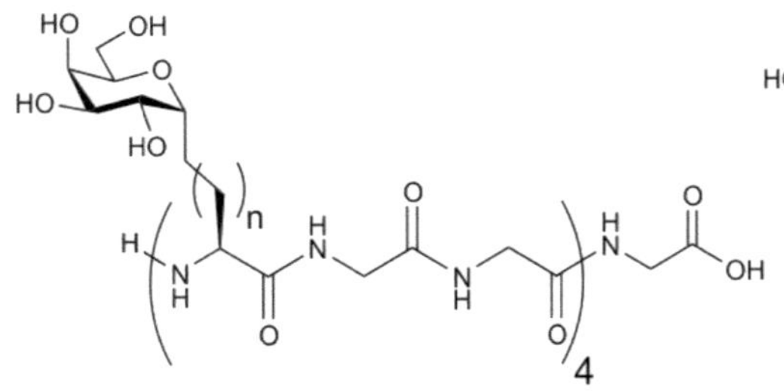

11, n = 1
16, n = 2
17, n = 3

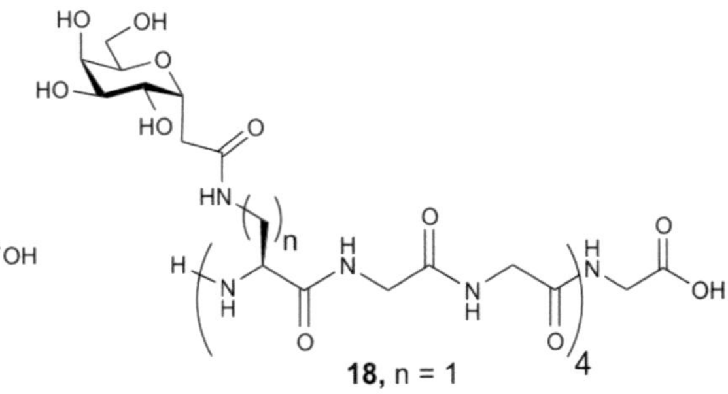

18, n = 1
19, n = 2
12, n = 3
20, n = 4

Figure 13. Structures of *C*-linked AFGP analogues containing different side chain lengths between the carbohydrate moiety and the polypeptide backbone, reported by the Ben laboratory. [57,178]

activity of these compounds was not assessed. Irregular incorporation of proline into these derivatives was detrimental to IRI activity as analogues **24-26** were only slightly active at a much higher concentration than the alanine-containing derivatives. However, incorporation of proline into a glycopeptide possessing four tripeptide repeats of (Pro-Ala-Thr(GalN-HAc))$_n$ (**27**) resulted in similar IRI activity as the analogue containing four tripeptide repeats of (Ala-Ala-Thr(GalNHAc))$_n$ (**22**).

Three studies have been reported where AFGP analogues containing triazole rings have been synthesized and assessed for their ability to inhibit ice recrystallization. The triazole ring was incorporated to provide a convergent synthetic approach to these analogues and to overcome the low yields often associated with glycosylation. The key step in the synthesis of these analogues was the Cu(I)-catalyzed Huisgen azide-alkyne cycloaddition (or "click" chemistry). [180-182] In 2009, the Brimble group described the synthesis of two AFGP derivatives in which a furanose carbohydrate moiety was conjugated to a polypeptide backbone with a triazole-linker (Figure 15, compounds **28** and **29**). [61,180] The IRI activity of these derivatives was not assessed, however neither compound exhibited thermal hysteresis or induced dynamic ice shaping. [61] Sewald and co-workers have also reported the synthesis of a number of triazole-containing AFGP peptoid analogues, three of which were assessed for IRI activity (**30-32**, Figure 15), but these analogues failed to inhibit ice recrystallization. [181] Finally, in 2011 the Ben laboratory reported the synthesis of *C*-linked triazole-containing AFGP derivatives **33-36** (Figure 15) that were structurally similar to one of their more IRI active glycopeptides reported previously (analogue **12**, Figure 11). [182] While analogues **33-36** only exhibited weak IRI activity, this study highlighted the importance of the amide-bond present in the side chain of **12** (Figure 11) and identified this structural feature as crucial for potent IRI activity. Collectively, the result from these three studies suggest that while utilizing "click" chemistry to conjugate the carbohydrate moiety to a polypeptide backbone may offer advantages synthetically, the triazole-linker is detrimental for IRI activity.

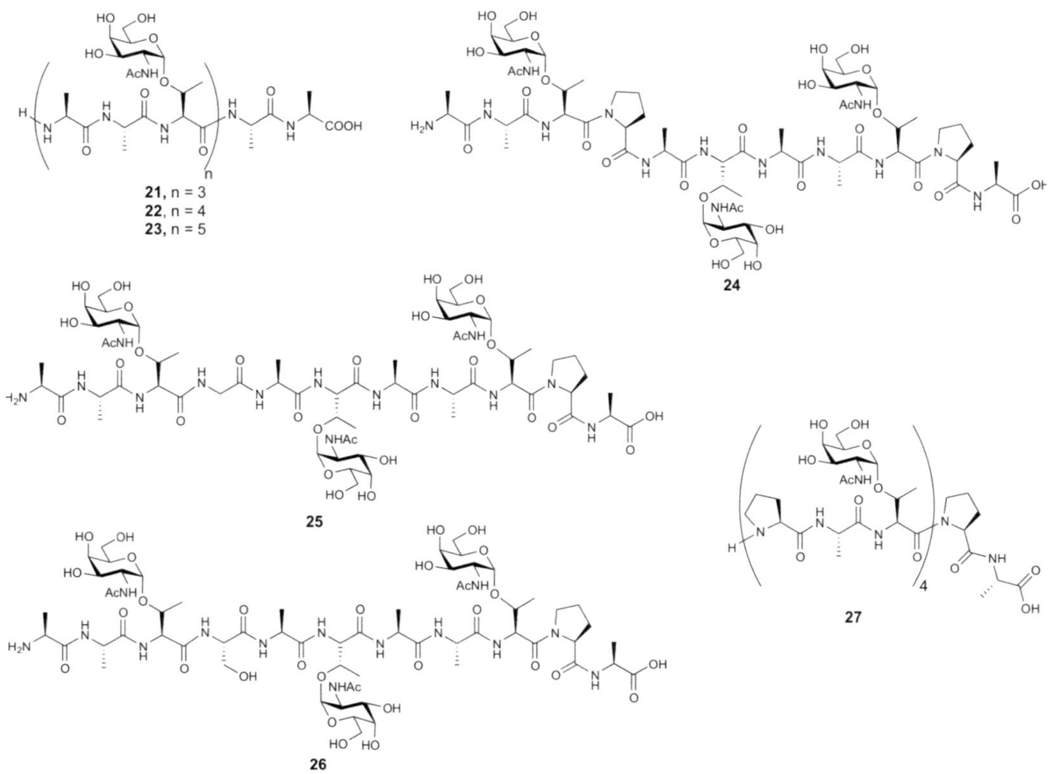

Figure 14. Structures of AFGP analogues reported by Sewald and co-workers. [179]

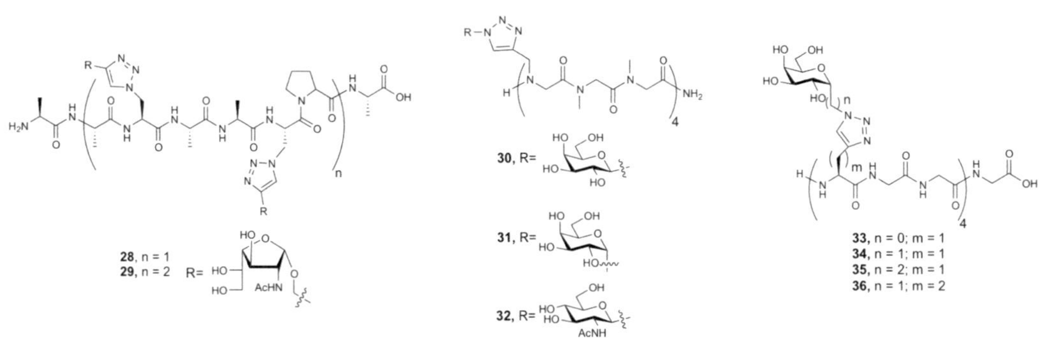

Figure 15. Structures of triazole-containing AFGP analogues reported by the Brimble (**28**-**29**), [180] Sewald (**30**-**32**) [181] and Ben (**33**-**36**) laboratories. [182]

6.2. Synthetic polymers as ice recrystallization inhibitors

All of the compounds discussed thus far that have exhibited the ability to inhibit ice recrystallization have been peptide or glycopeptide-based molecules. While some of these derivatives show great promise for the many applications of ice recrystallization inhibitors, the main limitation is that large-scale preparation of these compounds for *in vitro* or *in vivo* applications is problematic. Thus, interest has arisen in small molecules (section 6.3) and synthetic polymers (described below) that can inhibit ice recrystallization. Such compounds can be more efficiently synthesized. Knight *et al.* in 1995 made the first observation that synthetic polymers could inhibit ice recrystallization. [41] In this study it was found that poly-L-histidine, poly-L-hydroxyproline and polyvinyl alcohol (PVA) exhibited IRI activity at concentrations less than 1 mg/mL in pure water, whereas poly-L-aspartic acid, poly-L-asparagine, polyacrylic acid and polyvinylpyrrolidone were inactive. With the exception of PVA, which retained its IRI activity in a NaCl solution, these polypeptides and polymers were not assessed for IRI activity in a salt solution to negate false positive effects. [41] Following this study, the activity of PVA has been further investigated and various synthetic polymers have been examined for their ability to inhibit ice recrystallization.

In 2003 Inada *et al.* reported an extensive study on the IRI activity of PVA. The activity of PVA was found to be dependent on its molecular mass, with an increase in activity observed with higher molecular weight polymers of PVA. [145] Polymers with an average molecular weight of ~90 000 g/mol were found to exhibit comparable activity to a type I AFP from winter flounder at similar concentrations. However, due to the large difference in molecular weights between PVA and the AFP, the quantity of PVA required to exhibit this activity was significantly higher than that of the AFP. In 2009, Gibson *et al.* re-examined the molecular weight dependence of PVA and showed that PVA with an average molecular weight of ~115 500 has potent IRI activity at a concentration of 5 mg/mL. [59] It was suggested that the ability of PVA to inhibit ice recrystallization is attributed to its ability to interact with the ice crystal lattice. Budke and Koop reported that PVA induces dynamic ice shaping capabilities and suggested this is occurring as the spacing of the PVA hydroxyl groups are closely matched to that of the prism planes of ice, allowing adsorption to these planes. [183] Furthermore, Inada and Lu have shown that PVA exhibits a small TH gap of 0.037 °C at 50 mg/mL, suggesting that an adsorption to ice is occurring. [157]

In addition to PVA, a number of other water-soluble polymers have also been investigated for their ability to inhibit ice recrystallization. [95] In 2009, Gibson *et al.* reported the IRI activity of various structurally diverse polymers (Figure 16). [59] Polyacrylic acid (PAA, **37**), poly(2-aminoethyl methacrylate) (**38**), polyethylene glycol (PEG, **39**), poly-L-Lysine (**40**) and poly-L-glutamic acid (**41**) exhibited only weak IRI activity, and an increase in concentration did not improve activity for any of these polymers. Poly-L-hydroxyproline (**42**) was found to exhibit IRI activity and this activity was dependent on polymer concentration. Poly-L-hydroxyproline has a PPII helical secondary structure [184] similar to the structure AFGPs are suggested to adopt. However, it was suggested this secondary structure is not required for IRI activity as PVA and poly-L-hydroxyproline exhibited similar IRI activities, but PVA is largely unstructured in solution. [59] Two vinyl-derived glycopolymers were also assessed for their ability to inhibit ice recrystallization (**43** and **44**, Figure 16). The highest molecular weight glycopolymer with a glucose residue (**43**, at ~105000 g/mol) did exhibit a moderate ability to inhibit ice recrystallization. However, incorporating a different carbohydrate residue (**44**) failed to increase IRI activity. [59,95]

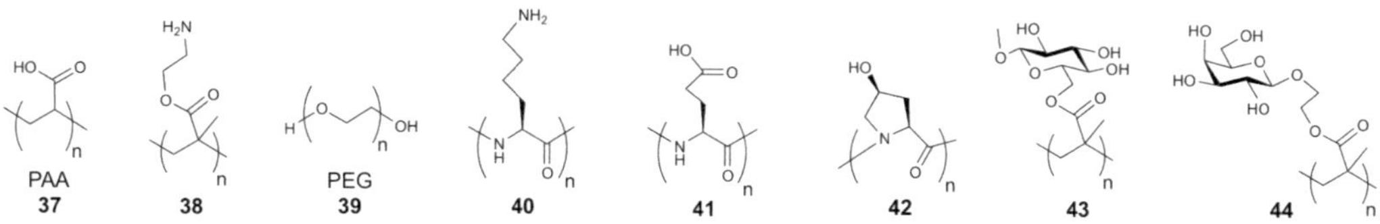

Figure 16. Structures of synthetic polymers assessed for IRI activity. [41,59,95,145]

6.3. Small molecules as ice recrystallization inhibitors

The Ben laboratory was the first group to report that small molecules, which were not peptide or polymer-based, could inhibit ice recrystallization. In 2008, Tam *et al.* reported a study examining the correlation between carbohydrate hydration and ice recrystallization inhibition. [177] This study arose from the observation that having a more hydrated carbohydrate moiety on one of their most active *C*-linked AFGP analogues (**12**) was a contributing factor to its exhibited IRI activity (see section 6.1, Figure 12). [58] Consequently, four monosaccharides and five disaccharides with known hydration parameters [174-176] were assessed for their ability to inhibit ice recrystallization. The structures of the mono- and disaccharides along with corresponding hydration numbers, isentropic molar compressibility and partial molar volume values are shown in Table 1. At a concentration of 22 mM, D- galactose exhibited moderate IRI activity, D- glucose and D- mannose had weak activity while D- talose was inactive. [177] These results showed a strong linear correlation between the hydration number of the monosaccharides and their respective IRI activity. The disaccharides examined also showed this strong linear correlation of their hydration number to IRI activity. Melibiose exhibited moderate IRI activity, while lactose and trehalose showed weak activity and maltose and sucrose were inactive.

The hydration layer or hydration shell of a carbohydrate can be defined as the number of tightly bound water molecules that surround the carbohydrate in aqueous solution. The hydration of carbohydrates has been the focus of many studies, and hypotheses for rationalizing observed hydration characteristics include hydration numbers, [185-188] anomeric effect, [189] hydrophilic volume, [190] hydrophobic index, [191] the ratio of axial versus equatorial hydroxyl groups [192,193] and the compatibility with bulk-water based upon the position of the next-nearest-neighbor hydroxyl group. [194,195] In the early 1990s, Galema *et al.* studied key parameters thought to dictate hydration characteristics and these were correlated to carbohydrate stereochemistry. Using kinetic experiments and density ultrasound measurements, the partial molar volumes, isentropic partial molar compressibilities and hydration numbers were determined for many commercially available mono- and disaccharides. [174-176] The isentropic partial molar compressibility and partial molar volume values of the carbohydrates quantify their "compatibility" with the three-dimensional hydrogen-bond network of bulk-

Carbohydrate	Molar Compressibility (K₂°(s)x10⁴,cm³ mol⁻¹ bar⁻¹)	Hydration Number	Carbohydrate	Molar Compressibility (K₂°(s)x10⁴,cm³ mol⁻¹ bar⁻¹)	Hydration Number
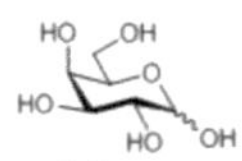 D-Galactose	-20.8 (0.5) -20.4 (0.4)	8.7 (0.2)	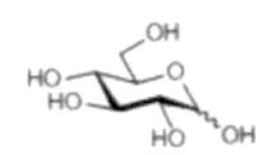 D-Glucose	-17.6 (0.3)	8.4 (0.2)
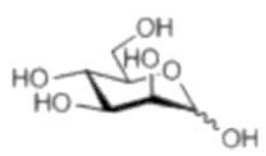 D-Mannose	-16.0 (0.5)	8.1 (0.2)	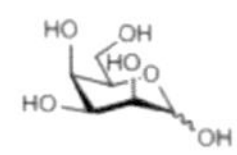 D-Talose	-11.9 (0.3)	7.7 (0.2)
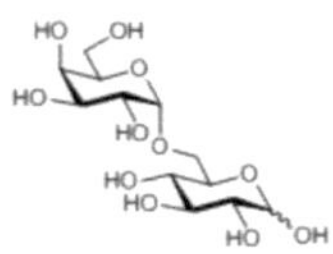 Melibiose	-31.2 (1.0)	15.5 (0.3)	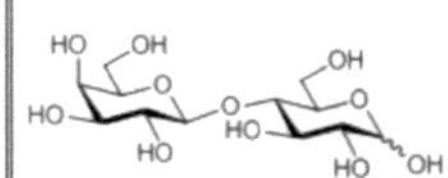 Lactose	-31.1 (0.2) -30.4 (1.0)	15.3 (0.3)
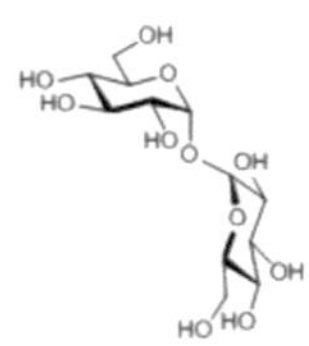 Trehalose	-30.2 (0.3)	15.3 (0.3)	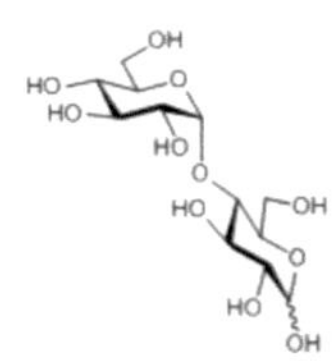 Maltose	-23.7 (1.0)	14.5 (0.3)
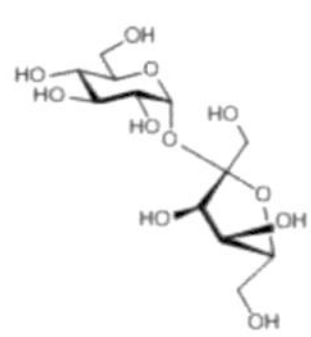 Sucrose	-17.8 (0.5)	13.9 (0.3)			

Table 1. Isentropic molar compressibilities ($10^4K_2°(s)$, cm^3 mol^{-1} bar^{-1}) and hydration numbers of various monosaccharides and disaccharides. Errors of molar compressibility values and hydration numbers are shown in parentheses. [175,177]

water as they are related to the size or volume the carbohydrate occupies upon hydration by water. Hydration numbers are calculated using isentropic coefficients of compressibility and they predict the number of water molecules that are hydrogen-bonded to the carbohydrate. In this study, it was observed that the compatibility of the carbohydrate with the three-dimensional hydrogen-bond network of bulk-water was directly related to the stereochemical relationship of the hydroxyl groups on the carbohydrate. D- Talose, with axial hydroxyl groups on C2 and C4, had a higher isentropic molar compressibility value and a lower hydration number, and fit well into the three-dimensional hydrogen-bonded network of bulk-water. In contrast, D- galactose, with an axial hydroxyl group on C4 and equatorial hydroxyl group on C2, had a lower isentropic molar compressibility value and a higher hydration number, and had a poor fit into the three-dimensional hydrogen-bonded network of bulk-water. Thus, D-talose was the most compatible and caused the least disturbance on the hydrogen-bonded network of bulk-water, whereas D- galactose was the least compatible and caused a greater disturbance on the hydrogen-bonded network of bulk-water. The carbohydrates with an equatorial C4 hydroxyl and either an equatorial or axial C2 hydroxyl group (ie. D- glucose and D- mannose) had a moderate fit and caused a moderate disturbance of the three-dimensional hydrogen-bonded network of bulk-water.

In the study conducted by Tam *et. al* which investigated the IRI activity of several mono- and disaccharides, a correlation was observed between IRI activity and carbohydrate hydration. [177] As none of the carbohydrates exhibited thermal hysteresis or dynamic ice shaping, it was unlikely that the IRI activity exhibited by the carbohydrates was due to an interaction with the ice lattice. This lead to an alternative proposed mechanism for the inhibition of ice recrystallization based upon the compatibility of a solute with bulk-water. As described in detail in section 2.0 of this chapter, a semi-ordered quasi-liquid layer (QLL) exists between the highly ordered ice lattice and bulk-water. For ice recrystallization to occur, bulk-water molecules transfer to the QLL, then subsequently from the QLL to the growing ice lattice. [38,39] Tam *et al.* have suggested that the carbohydrates are concentrated at the bulk-water-QLL interface. [177] A carbohydrate that had a poor fit into bulk-water will cause a greater disturbance to its three-dimensional hydrogen-bonded network, increasing the energy associated with the transfer of bulk-water to the QLL. It was therefore hypothesized that the inhibition of ice recrystallization observed with carbohydrates occurred at the bulk-water-QLL interface as more highly hydrated carbohydrates, such as D- galactose, disrupted the pre-ordering of bulk-water making it energetically unfavorable for water molecules to transfer to the QLL. Less hydrated carbohydrates, such as D- talose, fit well into bulk-water and caused less of a disturbance to the pre-ordering of bulk-water, thus inhibition of ice recrystallization was not observed.

The disaccharides assessed in this study also showed a strong linear correlation of their IRI activity to their hydration numbers (values give in table 1). [177] However, the increase in hydration numbers for disaccharides relative to monosaccharides was not reflected with an increase in IRI activity. For instance, melibiose has a hydration number of 15.5, yet it exhibited similar IRI activity to D- galactose, which has a hydration number of 8.7. Furthermore, D-galactose was significantly more IRI active than maltose, despite maltose having a much larger hydration number (8.7 for D- galactose and 14.5 for maltose). This was attributed to a difference in total steric volume between the monosaccharides (containing one carbohydrate unit) and disaccharides (containing two carbohydrate units). By dividing the carbohydrate hydration number by their partial molar volumes an indication of the degree of hydration per molar volume of carbohydrate was obtained. This value was referred to as the hydration index (HI) and it provided the degree of hydration of the substrate as a function of its size or volume. This metric was useful in justifying why highly hydrated monosaccharides exhibited similar IRI activity as highly hydrated disaccharides at 22 mM, despite hydration numbers for monosaccharides being almost half the value of disaccharides. [177] However, at higher carbohydrate concentrations, such as 220 mM, the disaccharides were twice as IRI active as the monosaccharides. [196] Thus, hydration numbers, not hydration indices, were better predictors of IRI activity at this concentration, but ultimately IRI activity still correlated with carbohydrate hydration.

Following the report that simple commercially available carbohydrates exhibit moderate IRI activity, the Ben laboratory has reported the ability of various other carbohydrate derivatives to inhibit ice recrystallization. Most of these compounds have been derivatives of D- galactose. [177] *C*-allylated derivatives of galactose (**45** and **49**), glucose (**46** and **50**), mannose (**47**) and talose (**48**) were assessed for ice recrystallization inhibition activity (Figure 17) to investigate the influence of a carbon substituent at the C1 position as their most IRI active AFGP analogues were *C*-linked glycoconjugates (see section 6.1). [177] The α-*C*-allyl-glycosides (**45-48**) had similar activities as the native monosaccharide units (ie. D- galactose and α-*C*-allyl-galacto-pyranoside exhibited similar IRI activities), and the trend of activity for these *C*-linked derivatives was identical to the trend observed with the corresponding native monosaccharides (ie. galactose was most active and talose was least active). However, the β-*C*-allyl-glycosides (**49-50**) showed a significant decrease in activity in comparison to the native monosaccharides (D- galactose and D- glucose) and the α-linked derivatives. Other D- galactose derivatives have been assessed for their ability to inhibit ice recrystallization, including compounds **51-57** (Figure 17). [197,198] All of these derivatives had weak to poor IRI activity, and were less active than native D- galactose.

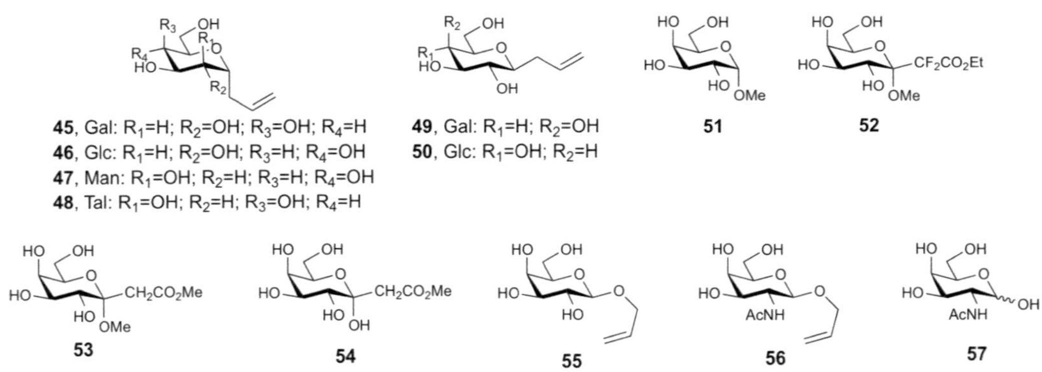

Figure 17. Structures of D- galactose-based analogues assessed for IRI activity by the Ben laboratory. [197,198]

In addition to monosaccharide derivatives, structural analogues of the disaccharide β-D-galactosyl-(1-3)-α-*N*-acetyl-D- galactosamine found in native AFGPs were investigated for IRI activity. These include disaccharide **58** (Figure 18), a close analogue of the disaccharide found in native AFGPs, regioisomers of **58** where the terminal β–D- galactosyl unit was linked to the C4 or C6 hydroxyl group of the *N*-acetyl-D- galactosamine moiety (**60** and **61**, respectively), and disaccharide **59**, in which the C2 *N*-acetyl group was replaced with a hydroxyl group. [197] These four disaccharides were assessed for IRI activity at 22 mM, and interestingly the most active disaccharide was not the analogue of the disaccharide found in native AFGPs. The β-(1,4)-linked disaccharide **60** was the most active disaccharide analogue assessed. The β-(1,6)-linked disaccharide **61** and both β-(1,3)-linked disaccharides, **58-59**, exhibited similar IRI activity and were less active than the β-(1,4)-linked analogue. These disaccharides were not conjugated to the native polypeptide backbone (Ala-Ala-Thr) found in AFGPs to investigate if the same trend was observed with the glycoconjugates. However, this study highlighted how the structural features necessary for TH and IRI activity may be different as the functional groups which were required for the TH activity of AFGPs (see section 5.3, Figure 8) [166] were not required for the IRI activity of the disaccharide analogues. [197]

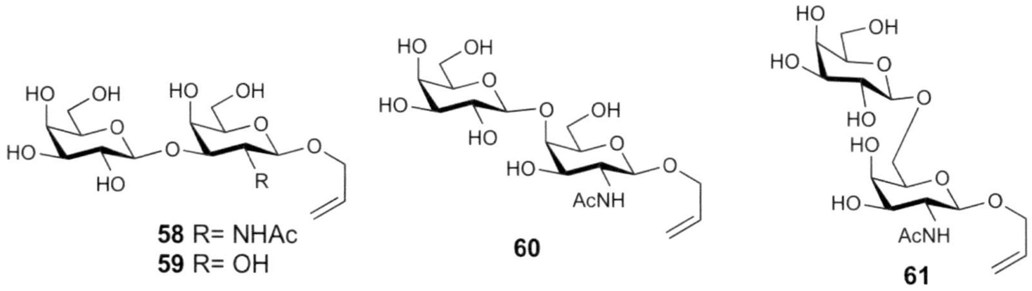

Figure 18. Structural disaccharide analogues of the native β-D- galactosyl-(1-3)-*N*-acetyl-D- galactosamine disaccharide found in AFGPs. [197]

While the small molecules described above had the ability to inhibit ice recrystallization, all exhibited only weak to moderate activity at much higher concentrations than those of the potently IRI active glycoconjugates. However, in 2012 the Ben laboratory reported the first examples of small carbohydrate-based molecules that were extremely potent inhibitors ice recrystallization, some that were highly IRI active at concentrations much lower than 22 mM. To date, these are the most potent IRI active small molecules. The molecules investigated were carbohydrate-based surfactants and hydrogelators (structures shown in Figures 19), two of which were found to exhibit potent IRI activity. [62] The carbohydrate-based non-ionic surfactant β-octyl-D- galactopyranoside (**62**) was highly IRI active, with potent activity reported at 11 mM. In contrast, carbohydrate-based non-ionic surfactant β-octyl-D- glucopyranoside (**63**) was only weakly active even at 44 mM. These results were in agreement with previous studies were that derivatives of the more highly hydrated D- galactose were significantly better inhibitors of ice recrystallization than derivatives of the less hydrated D- glucose. [58,177] While these carbohydrate-based surfactants were known to form micelles in solution, it was concluded that micelle formation was unrelated to IRI activity. β-octyl-D- galactopyranoside (**62**) was highly active at a concentration well below its critical micelle concentration (CMC) of 30 mM, where as β-octyl-D- glucopyranoside (**63**) did not exhibit an ability to inhibit ice recrystallization even well above its CMC value of 22 mM. [62] Furthermore, other structurally different non-ionic and anionic surfactants exhibited weak to moderate activity at concentrations well above their respective CMC values. None of the non-ionic carbohydrate-based surfactants assessed in this study possessed TH activity or dynamic ice shaping abilities, suggesting that the activity exhibited by these compounds was not due to an interaction with the ice lattice.

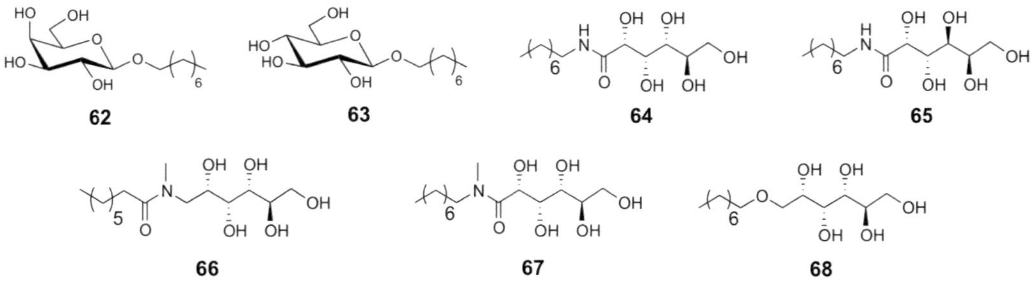

Figure 19. Structures of carbohydrate-based non-ionic surfactants and hydrogelators assessed for IRI activity by the Ben laboratory. [62] β-octyl-D- galactopyranoside (**62**) and *N*-octyl-D- gluconamide (**64**) are the first report of potent small molecule ice recrystallization inhibitors.

The second class of compounds investigated were carbohydrate-based hydrogelators, as in aqueous solution they were known to aggregate and sequester bulk-water forming fibres and hydrogels. D- glucose hydrogelator derivative *N*-octyl-D- gluconamide (**64**) was found to be a potent inhibitor of ice recrystallization at 0.5 mM, a concentration much lower than that of other reported carbohydrate derivatives. [62] However, the D- galactose hydrogelator derivative *N*-octyl-D- galactonamide (**65**) was only weakly IRI active at this same concentration. *N*-octyl-D- gluconamide (**64**) is the first example of a small molecule exhibiting potent activity at a concentration much lower than 22 mM, and it was also the first example of a glucose-based derivative exhibiting better activity than a galactose-based derivative. Structure-function work conducted in this study suggested that the amide bond in **64** is an essential structural feature for its activity as **66-68** (Figure 19) were significantly less active at much higher concentrations. While *N*-octyl-D- gluconamide (**64**) was able to form hydrogels in solution, it was concluded using solid-state NMR studies and characterization of the hydrogels that the ability to form a hydrogel was not a prerequisite for potent IRI activity. This conclusion was further supported by the fact that *N*-octyl-D- galactonamide (**65**) also formed hydrogels in solution, yet it did not possess IRI activity. Finally, these studies also suggested that ice binding was not a prerequisite for potent activity as solid-state NMR studies and TH measurements failed to indicate an interaction with the ice lattice. To date, the report that small molecules can exhibit potent IRI activity remains a significant discovery that will facilitate the rational design of small molecule ice recrystallization inhibitors suitable for medical, commercial and industrial applications.

7. Cryopreservation

Cryopreservation is a very attractive process for the preservation of biological materials. While vitrification and hypothermic storage each offer their own unique advantages and their own limitations, cryopreservation has a major advantage. At the temperatures associated with cryopreservation (typically -190 °C) all biochemical processes are effectively stopped. However, cryopreservation is a complex process during which careful attention to sample volume, cooling rates and cryoprotectants (dimethyl sulfoxide and glycerol) are extremely important to ensure cells survive the process. Unfortunately, all cryoprotectants exhibit cytoxicity and this complicates the cryopreservation process as the cryoprotectant must often be removed during the thawing cycle. Indeed there is a common myth that cooling rates of 1 °C/min with 10% dimethyl sulfoxide (DMSO) is sufficient for all cryopreservation applications. Unfortunately, this is incorrect and there is an urgent need for novel cryoprotectants, especially in light of the recent developments in the field of regenerative medicine where the supply of various progenitor cells is problematic for the many clinical applications. To highlight the complexity of this process and the need for new and improved cryoprotectants a brief description of cellular injury during cryopreservation will be presented in the following section.

7.1. The complex mechanisms of cryoinjury

Traditionally, there exist three characterized mechanisms of cell death that occur during cryopreservation. These are cell rupture due to damage to the external cell membrane, necrosis and cold induced apoptosis. Cell rupture is usually the result of osmotic imbalance causing a loss in membrane integrity. [50] Cell necrosis is characterized by cellular swelling (due to an increase in immune response), compromised cell membrane integrity, random DNA fragmentation by cellular endonucleases, cell lysis and the release of cytokines. Apoptosis (programmed cell death) is a highly complex and closely regulated biochemical pathway (the details of which will not be covered in this chapter). It may appear at first that cell death due to apoptosis is not related to cryopreservation however, it has been demonstrated that cold-induced apoptosis is common in cryopreserved cells. [50,199]

The formation of ice under typical cryopreservation conditions is inevitable, but cooling rates become extremely important in mitigating the damage associated with ice formation. For every cell type there is an optimal cooling and warming rate that is determined by the permeability of the cell membrane to water and the cryoprotectant. Hence, cryopreservation is performed with either slow or fast cooling rates depending on cell type. In most instances, ice will prefer to form outside of the cell. [200] Formation of extracellular ice creates an increased osmotic pressure across the cell membrane. This "osmotic flux" intensifies as ice growth continues after the nucleation event. As the ice crystal grows all solutes are excluded from the ice lattice [201] and are concentrated in the extracellular medium. Cells with less permeable membranes will rupture with increasing osmotic pressure if they cannot dehydrate fast enough.

The process of dehydration during freezing is somewhat of a "double-edged sword". In one instance, the amount of intracellular water decreases, reducing the chance for intracellular ice formation – a lethal process. However, it has been shown that dehydration and exposure to excessively high concentrations of electrolytes is also lethal to the cell. [202] This is referred to as solute damage or the "solute effect" and it facilitates damage to the cell membrane that is irreparable. [202] Conversely, when cells are frozen very slowly, dehydration and excessive cell shrinkage facilitates cell death. Excessive dehydration can be prevented using cryoprotectants. Two classes of cryoprotectant are commonly employed. Non-penetrating cryoprotectants do not cross the cell membrane and hence remain outside the cell, thereby increasing the osmolality of the extracellular solution, facilitating dehydration of the cell prior to freezing and preventing formation of intracellular ice. Penetrating cryoprotectants, such as DMSO and glycerol, readily cross the cell membrane and decrease the concentration of intracellular electrolytes while maintaining greater cell volumes. The major problem with penetrating cryoprotectants is cytotoxicity due to the disruption of intracellular signaling. [203] In summary, cryopreservation of cells using slow-freezing results in dehydration of the cell in response to increasing osmotic pressures as electrolytes are concentrated outside the cell during extracellular ice growth. While dehydration of the cells helps to prevent intracellular ice growth, it is also detrimental to cell survival.

Cryopreservation using high cooling rates traps water inside the cell promoting the formation intracellular ice. [204] The exact mechanism by which this occurs is not clear [205] however, most cryobiologists believe that intracellular ice formation results in cell death. Hence, practical fast-freezing protocols must dehydrate cells prior to freezing in order to mitigate intracellular ice formation. [206] Of course cryoprotectants are necessary to accomplish this, but the role of the cryoprotectant during fast cooling is different than during slow cooling. Non-penetrating cryoprotectants are employed in an effort to dehydrate the cell and minimize the chance of intracellular ice formation. Interestingly, while the correlation between intracellular ice formation and cell death has been recognized, there is evidence to suggest that formation of intracellular ice does not directly kill cells. [200] Studies have shown that survival of cells post-cryopreservation is dependent upon the rate at which the cells are warmed during thawing and that cell death associated with intracellular ice formation is not caused by the initial nucleation of ice but by an alternate process during warming. [207,208] Possible mechanisms by which intracellular ice damages cells have been reviewed extensively in the literature and it has been concluded that cell death is occurring as a result of ice recrystallization. [202,209] This hypothesis is supported by the fact that may freeze-tolerant organisms inhabiting sub-zero environments produce large quantities of recrystallization-inhibitors *in vivo* to ensure survival. [139,210] In addition, mechanical damage to cell membranes from ice recrystallization has been identified as a primary cause of cell injury during cryopreservation. [50]

7.2. Preservation of biological materials using biological antifreezes and their analogues

Cellular damage due to ice recrystallization occurs during the storage and thawing cycles of cryopreservation and, given the cryoprotective nature of BAs, it is not surprising that they have been investigated as cryoprotectants to increase cell viability post-thaw. In principle, BAs have the advantage of being relatively non-toxic compared to common cryoprotectants such as DMSO and glycerol. While BAs seem like ideal cryoprotectants, they have not been very effective and often fail to protect mammalian cells from cryoinjury at temperatures outside of the TH gap. This section will discuss specific examples where BAs were used to cryopreserve biological materials, including the benefits and problems associated with their use.

BAs have been examined as protective agents for the hypothermic storage and cryopreservation of various biological materials. AFPs have been reported to protect cell membranes during hypothermic storage. For instance, Rubinsky and co-workers demonstrated that AFPs [211] and AFGPs [212] of various molecular weights and in concentrations ranging from 1-40 mg/mL can successfully preserve the structural integrity of pig oolemma and bovine immature oocytes. Furthermore, these oolemma and oocytes underwent successful *in vitro* maturation and fertilization. [211,212] In addition, it has been shown that AFPs can stabilize plasma membranes. [213] Crowe and co-workers demonstrated that while a 1 mg/mL solution of AFGP prevented cold-induced activation of human blood platelets following hypothermic storage, a type I AFP had no effect. [214,215] Despite these promising examples, toxic effects during hypothermic storage from the BAs during hypothermic storage have also been reported. Both AFPs and AFGPs have exhibited significant toxic effects and have compromised cell viabilities in spinach thylakoids, [216] ram spermatozoa [217] and chimpanzee spermatozoa. [218]

In addition to hypotherminc storage, BAs have also been utilized for cryostorage of biological materials. Several studies have reported benefits of using AFPs and AFGPs as cryoprotectants. Rubinsky and co-workers observed dramatically improved morphological integrity of immature oocytes and two-cell-stage embroys of mice and pigs that were subjected to vitrification in the presence of 40 mg/mL AFGPs. [219,220] Similar results were observed with mature mouse oocytes [221], bovine and ovine embryos at the morula/blastocyst stage, [222] ram spermatozoa, [217] chimpanzee spermatozoa [218] and porcine oocytes. [223] While post-thaw viabilities were increased in the presence of BAs with ram and chimpanzee spermatozoa and porcine oocytes, cytotoxic effects during cooling were also observed. [217,218,223]

In contrast, other investigations have reported that BAs fail to protect cells during cryopreservation and actually facilitate cellular damage during cryopreservation. For instance, no specific benefits were observed in survival rates of vitrified bovine blastocysts, [224] two-step-cryopreserved oyster oocytes [225] and equine embroys using various AFPs. [226] Freezing of red blood cells in the presence of glycerol with AFPs (at concentrations between 25 and 1000 μg/mL) [227] and AFGPs (at 40 μg/mL) has been reported to damage cells during cryopreservation. [228] A similar result was also observed during the cryopreservation of hematopoietic cells with AFPs in DMSO. [229] Additionally, this cellular damage during cryopreservation with BAs has also been reported with spinach thylakoids, [216] intact rat heart (from cardiac explant) [230] and cardiomyocytes. [231] This damage has been attributed to the change in ice crystal morphology that is induced in the presence of BAs (dynamic ice shaping). [228,231] Furthermore, it has been suggested that BAs may also increase the incidence of intracellular ice formation, thereby decreasing cell viabilities post-thaw. [232] Finally, reports have demonstrated both beneficial and detrimental effects with BAs during cryopreservations, depending on AFP concentration and type. [233] At low concentrations AFPs were reported to increase the survival rate of red blood cells however, at higher concentrations where the ice recrystallization inhibition ability of the AFP was significantly enhanced, they decreased survival rates. [234,235]

In contrast to native biological antifreezes, the benefit of analogues possessing "custom-tailored" antifreeze activity for cryopreservation has been demonstrated. In 2011, the Ben laboratory demonstrated that C-linked AFGP analogues that exhibit potent IRI activity but not TH activity function as effective cryoprotectants. Using a human embryonic liver cell line, 1.0-1.5 mg/mL of *C*-linked AFGP analogues **11** or **12** doubled cell viability relative to the negative control (cell medium only). [236] The post-thaw viability was comparable to that obtained with a 2.5% DMSO solution. This effect was attributed to the IRI activity of these *C*-linked AFGP analogues. This conclusion was validated when it was demonstrated that IRI active carbohydrates exhibiting minimal cytotoxicity significantly increased cell viabilities post-thaw. [196] To date, these are the only examples where potent inhibitors of ice recrystallization not displaying thermal hysteresis activity or dynamic ice shaping capabilities have been successfully utilized as cryoprotectants.

8. Conclusions and outlook

While Nature has provided various organisms with peptides and glycopeptides to mitigate cellular damage during exposure to cold temperatures, these compounds have failed to be effective cryoprotectants in various medical and commercial applications. This is somewhat ironic as these compounds are potent inhibitors of ice recrystallization, a process that contributes significantly to cellular injury. The recent discovery that IRI activity can be selectively enhanced while suppressing TH activity in various analogues of biological antifreezes is a significant advancement towards the rational design of novel cryoprotectants. Some of these molecules have even demonstrated the ability to enhance cell viabilities post-thaw. While these compounds do not yet exhibit viabilities comparable to 10 % DMSO solutions, it is feasible that with a better understanding of the structural features necessary for potent IRI activity future analogues will be efficient cryoprotectants replacing conventional ones such as DMSO and glycerol. The recent discovery that small molecules are extremely potent inhibitors of ice recrystallization represents a "quantum leap" forward in this area. Further studies with these compounds *in vitro* and *in vivo* will elucidate their effectiveness as cryoprotectants while overcoming the problems of high cost and large-scale synthesis associated with the higher molecular weight analogues of biological antifreeze that exhibit the potent IRI activity, a property necessary for an effective cryoprotectant.

Author details

Chantelle J. Capicciotti, Malay Doshi and Robert N. Ben*

Department of Chemistry, D'Iorio Hall, University of Ottawa, Ottawa, ON, Canada

References

[1] Doherty, R.D., Hughes, D.A., Humphreys, F.J., Jonas, J.J., Jensen, D.J., Kassner, M.E., King, W.E., McNelley, T.R., McQueen, H.J., and Rollett, A.D., Current Issues in Recrystallization: A Review. Materials Science and Engineering: A 1997; A238(2) 219-274.

[2] Rios, P.R., Siciliano, F.J., Sandim, H.R.Z., Plaut, R.L., and Padilha, A.F., Nucleation and Growth During Recrystallization. Materials Research 2005; 8(3) 225-238.

[3] Gleiter, H., The Mechanism of Grain Boundary Migration. Acta Metallurgica 1969; 17(5) 565-573.

[4] Gleiter, H., Theory of Grain Boundary Migration Rate. Acta Metallurgica 1969; 17(7) 853-862.

[5] Humphreys, F.J. and Hatherly, M. Chapter 4 - The Structure and Energy of Grain Boundaries. In: D. Sleeman, editor. Recrystallization and Related Annealing Phenomena (2nd ed.). Oxford: Elsevier; 2004. p 91-119.

[6] Berdichevsky, V.L., Thermodynamics of Microstructure Evolution: Grain Growth. International Journal of Engineering Science 2012; 57 50-78.

[7] Gil Sevillano, J., van Houtte, P., and Aernoudt, E., Large Strain Work Hardening and Textures. Progress in Materials Science 1980; 25(2-4) 69-134.

[8] Hall, E.O., The Deformation and Ageing of Mild Steel: III Discussion of Results. Proceedings of the Physical Society. Section B 1951; 64(9) 747-753.

[9] Petch, N.J., The Cleavage Strength of Polycrystals. Journal of the Iron and Steel Institute, London 1953; 174 25-28.

[10] Dehghan-Manshadi, A. and Hodgson, P.D., Dependency of Recrystallization Mechanism to the Initial Grain Size. Metallurgical and Materials Transactions A 2008; 39A(12) 2830-2840.

[11] Fletcher, N.H. Chapter 2 - Structure and Energry of Ordinary Ice. In: editor. The Chemical Physics of Ice (1st ed.). London: Cambridge University Press; 1970. p 23-48.

[12] DeVries, A.L. and Lin, Y., Structure of a Peptide Antifreeze and Mechanism of Adsorption to Ice. Biochimica et Biophysica Acta (BBA) - Protein Structure 1977; 495(2) 388-392.

[13] Harding, M.M., Ward, L.G., and Haymet, A.D.J., Type I 'Antifreeze' Proteins. European Journal of Biochemistry 1999; 264(3) 653-665.

[14] Hayward, J.A. and Haymet, A.D.J., The Ice/Water Interface: Molecular Dynamics Simulations of the Basal, Prism, {2021}, and {2110} Interfaces of Ice Ih. The Journal of Chemical Physics 2001; 114(8) 3713-3726.

[15] Harding, M.M., Anderberg, P.I., and Haymet, A.D.J., 'Antifreeze' Glycoproteins from Polar Fish. European Journal of Biochemistry 2003; 270(7) 1381-1392.

[16] Fletcher, N.H., Surface Structure of Water and Ice. Philosophical Magazine 1962; 7(74) 255-269.

[17] Fletcher, N.H., Surface Structure of Water and Ice II. A Revised Model. Philosophical Magazine 1968; 18(156) 1287-1300.

[18] Karim, O.A. and Haymet, A.D.J., The Ice/Water Interface. Chemical Physics Letters 1987; 138(6) 531-534.

[19] Karim, O.A. and Haymet, A.D.J., The Ice/Water Interface: A Molecular Dynamics Simulation Study. Journal of Chemical Physics 1988; 89(11) 6889-6896.

[20] Furukawa, Y., Yamamoto, M., and Kuroda, T., Ellipsometric Study of the Transition Layer on the Surface of an Ice Crystal. Journal of Crystal Growth 1987; 82(4) 665-677.

[21] Furukawa, Y. and Ishikawa, I., Direct Evidence for Melting Transition at Interface between Ice Crystal and Glass Substrate. Journal of Crystal Growth 1993; 128(1–4, Part 2) 1137-1142.

[22] Beaglehole, D. and Nason, D., Transition Layer on the Surface on Ice. Surface Science 1980; 96(1-3) 357-363.

[23] Karim, O.A., Kay, P.A., and Haymet, A.D.J., The Ice/Water Interface: A Molecular Dynamics Simulation Using the Simple Point Charge Model. The Journal of Chemical Physics 1990; 92(7) 4634-4635.

[24] Döppenschmidt, A. and Butt, H.-J., Measuring the Thickness of the Liquid-Like Layer on Ice Surfaces with Atomic Force Microscopy. Langmuir 2000; 16(16) 6709-6714.

[25] Dosch, H., Lied, A., and Bilgram, J.H., Disruption of the Hydrogen-Bonding Network at the Surface of Ih Ice Near Surface Premelting. Surface Science 1996; 366(1) 43-50.

[26] Sadtchenko, V. and Ewing, G.E., Interfacial Melting of Thin Ice Films: An Infrared Study. Journal of Chemical Physics 2002; 116(11) 4686-4697.

[27] Golecki, I. and Jaccard, C., Intrinsic Surface Disorder in Ice Near the Melting Point. Journal of Physics C: Solid State Physics 1978; 11(20) 4229-4237.

[28] Kahan, T.F., Reid, J.P., and Donaldson, D.J., Spectroscopic Probes of the Quasi-Liquid Layer on Ice. Journal of Physical Chemsitry A 2007; 111(43) 11006-11012.

[29] Kaverin, A., Tsionsky, V., Zagidulin, D., Daikhin, L., Alengoz, E., and Gileadi, E., A Novel Approach for Direct Measurement of the Thickness of the Liquid-Like Layer at the Ice/Solid Interface. Journal Physical Chemistry B 2004; 108(26) 8759-8762.

[30] Güttinger, H., Bilgram, J.H., and Känzig, W., Dynamic Light Scattering at the Ice Water Interface During Freezing. Journal of Physics and Chemistry of Solids 1979; 40(1) 55-66.

[31] Brown, R.A., Keizer, J., Steiger, U., and Yeh, Y., Enhanced Light Scattering at the Ice-Water Interface During Freezing. The Journal of Physical Chemistry 1983; 87(21) 4135-4138.

[32] Bilgram, J.H., Dynamics at the Solid-Liquid Transition: Experiments at the Freezing Point. Physics Reports 1987; 153(1) 1-89.

[33] Bluhm, H., Ogletree, D.F., Fadley, C.S., Hussain, Z., and Salmeron, M., The Premelting of Ice Studied with Photoelectron Spectroscopy. Journal of Physics: Condensed Matter 2002; 14(8) L227-L233.

[34] Beaglehole, D. and Wilson, P., Thickness and Anisotropy of the Ice-Water Interface. The Journal of Physical Chemistry 1993; 97(42) 11053-11055.

[35] Beaglehole, D. and Wilson, P., Extrinsic Premelting at the Ice-Glass Interface. The Journal of Physical Chemistry 1994; 98(33) 8096-8100.

[36] Elbaum, M., Lipson, S.G., and Dash, J.G., Optical Study of Surface Melting on Ice. Journal of Crystal Growth 1993; 129(3-4) 491-505.

[37] Gilpin, R.R., Wire Regelation at Low Temperatures. Journal of Colloid and Interface Science 1980; 77(2) 435-448.

[38] Halter, P.U., Bilgram, J.H., and Känzig, W., Properties of the Solid-Liquid Interface Layer of Growing Ice Crystals: A Raman and Rayleigh Scattering Study. The Journal of Chemical Physics 1988; 89(5) 2622-2629.

[39] Bilgram, J.H., The Structure and Properties of Melt and Concentrated Solutions. Progress in Crystal Growth and Characterization of Materials 1993; 26 99-119.

[40] Knight, C.A., Grain Boundary Migration and Other Processes in the Formation of Ice Sheets on Water. Journal of Applied Physics 1966; 37(2) 568-574.

[41] Knight, C.A., Wen, D., and Laursen, R.A., Nonequilibrium Antifreeze Peptides and the Recrystallization of Ice. Cryobiology 1995; 32(1) 23-34.

[42] Alley, R.B., Perepezko, J.H., and Bentley, C.R., Grain Growth in Polar Ice: I. Theory. Journal of Glaciology 1986; 32(112) 415-424.

[43] Alley, R.B., Perepezko, J.H., and Bentley, C.R., Grain Growth in Polar Ice: II. Application. Journal of Glaciology 1986; 32(112) 425-433.

[44] Sutton, R.L., Lips, A., Piccirillo, G., and Sztehlo, A., Kinetics of Ice Recrystallization in Aqueous Fructose Solutions. Journal of Food Science 1996; 61(4) 741-745.

[45] Hagiwara, T., Hartel, R., and Matsukawa, S., Relationship between Recrystallization Rate of Ice Crystals in Sugar Solutions and Water Mobility in Freeze-Concentrated Matrix. Food Biophysics 2006; 1(2) 74-82.

[46] Budke, C., Heggemann, C., Koch, M., Sewald, N., and Koop, T., Ice Recrystallization Kinetics in the Presence of Synthetic Antifreeze Glycoprotein Analogues Using the Framework of LSW Theory. The Journal of Physical Chemistry B 2009; 113(9) 2865-2873.

[47] Goff, H.D., Measuring and Interpreting the Glass Transition in Frozen Foods and Model Systems. Food Research International 1994; 27(2) 187-189.

[48] Petzold, G. and Aguilera, J.M., Ice Morphology: Fundamentals and Technological Applications in Foods. Food Biophysics 2009; 4(4) 378-396.

[49] Baust, J.M., Van Buskirk, R., and Baust, J.G., Cell Viability Improves Following Inhibition of Cryopreservation-Induced Apoptosis. In Vitro Cellular & Developmental Biology - Animal 2000; 36(4) 262-270.

[50] Baust, J.M., Molecular Mechanisms of Cellular Demise Associated with Cryopreservation Failure. Cell Preservation Technology 2002; 1(1) 17-31.

[51] Scholander, P.F., van Dam, L., Kanwisher, J.W., Hammel, H.T., and Gordon, M.S., Supercooling and Osmoregulation in Arctic Fish. Journal of Cellular and Comparative Physiology 1957; 49(1) 5-24.

[52] Gordon, M.S., Amdur, B.H., and Scholander, P.F., Freezing Resistance in some Northern Fishes. The Biological Bulletin 1962; 122(1) 52-62.

[53] DeVries, A.L. and Wohlschlag, D.E., Freezing Resistance in Some Antarctic Fishes Science 1969; 163(3871) 1073-1075.

[54] DeVries, A.L., Komatsu, S.K., and Feeney, R.E., Chemical and Physical Properties of Freezing Point-Depressing Glycoproteins from Antarctic Fishes. The Journal of Biological Chemistry 1970; 245(11) 2901-2908.

[55] DeVries, A.L., Glycoproteins as Biological Antifreeze Agents in Antarctic Fishes. Science 1971; 172(3988) 1152-1155.

[56] Venketesh, S. and Dayananda, C., Properties, Potentials, and Prospects of Antifreeze Proteins. Critical Reviews in Biotechnology 2008; 28(1) 57-82.

[57] Liu, S. and Ben, R.N., *C*-Linked Galactosyl Serine AFGP Analogues as Potent Recrystallization Inhibitors. Organic Letters 2005; 7(12) 2385-2388.

[58] Czechura, P., Tam, R.Y., Dimitrijevic, E., Murphy, A.V., and Ben, R.N., The Importance of Hydration for Inhibiting Ice Recrystallization with *C*-Linked Antifreeze Glycoproteins. Journal of the American Chemical Society 2008; 130(10) 2928-2929.

[59] Gibson, M.I., Barker, C.A., Spain, S.G., Albertin, L., and Cameron, N.R., Inhibition of Ice Crystal Growth by Synthetic Glycopolymers: Implications for the Rational Design of Antifreeze Glycoprotein Mimics. Biomacromolecules 2009; 10(2) 328-333.

[60] Garner, J. and Harding, M.M., Design and Synthesis of Antifreeze Glycoproteins and Mimics. ChemBioChem 2010; 11(18) 2489-2498.

[61] Peltier, R., Brimble, M.A., Wojnar, J.M., Williams, D.E., Evans, C.W., and DeVries, A.L., Synthesis and Antifreeze Activity of Fish Antifreeze Glycoproteins and Their Analogues. Chemical Science 2010; 1(5) 538-551.

[62] Capicciotti, C.J., Leclère, M., Perras, F.A., Bryce, D.L., Paulin, H., Harden, J., Liu, Y., and Ben, R.N., Potent Inhibition of Ice Recrystallization by Low Molecular Weight

Carbohydrate-Based Surfactants and Hydrogelators. Chemical Science 2012; 3(5) 1408-1416.

[63] Ewart, K.V., Lin, Q., and Hew, C.L., Structure, Function and Evolution of Antifreeze Proteins. Cellular and Molecular Life Sciences 1999; 55(2) 271-283.

[64] Jia, Z. and Davies, P.L., Antifreeze Proteins: an Unusual Receptor-Ligand Interaction. Trends in Biochemical Sciences 2002; 27(2) 101-106.

[65] Fletcher, G.L., Hew, C.L., and Davies, P.L., Antifreeze Proteins of Teleost Fishes. Annual Review of Physiology 2001; 63 359-390.

[66] Davies, P.L. and Sykes, B.D., Antifreeze Proteins. Current Opinion in Structural Biology 1997; 7(6) 828-834.

[67] Hew, C.L. and Yang, D.S.C., Protein Interaction with Ice. European Journal of Biochemistry 1992; 203(1-2) 33-42.

[68] Duman, J.G. and DeVries, A.L., Isolation, Characterization, and Physical Properties of Protein Antifreezes from the Winter Flounder, *Pseudopleuronectes americanus*. Comparative Biochemistry and Physiology Part B: Comparative Biochemistry 1976; 54(3) 375-380.

[69] Duman, J.G. and DeVries, A.L., Freezing Resistance in Winter Flounder *Pseudopleuronectes americanus*. Nature 1974; 247 237-238.

[70] Ananthanarayanan, V.S. and Hew, C.L., Structural Studies on the Freezing-Point-Depressing Protein of the Winter Flounder *Pseudopleuronectes americanus*. Biochemical and Biophysical Research Communications 1977; 74(2) 685-689.

[71] Yang, D.S.C., Sax, M., Chakrabartty, A., and Hew, C.L., Crystal Structure of an Antifreeze Polypeptide and Its Mechanistic Implications. Nature 1988; 333 232-237.

[72] Raymond, J.A., Radding, W., and DeVries, A.L., Circular Dichroism of Protein and Glycoprotein Fish Antifreezes. Biopolymers 1977; 16(11) 2575-2578.

[73] Cheng, Y., Yang, Z., Tan, H., Liu, R., Chen, G., and Jia, Z., Analysis of Ice-Binding Sites in Fish Type II Antifreeze Protein by Quantum Mechanics. Biophysical Journal 2002; 83(4) 2202-2210.

[74] Baardsnes, J. and Davies, P.L., Contribution of Hydrophobic Residues to Ice Binding by Fish Type III Antifreeze Protein. Biochimica et Biophysica Acta (BBA) – Proteins and Proteomics 2002; 1601(1) 49-54.

[75] Sönnichsen, F.D., DeLuca, C.I., Davies, P.L., and Sykes, B.D., Refined Solution Structure of Type III Antifreeze Protein: Hydrophobic Groups May Be Involved in the Energetics of the Protein-Ice Interaction. Structure 1996; 4(11) 1325-1337.

[76] Madura, J.D., Taylor, M.S., Wierzbicki, A., Harrington, J.P., Sikes, C.S., and Sönnichsen, F., The Dynamics and Binding of a Type III Antifreeze Protein in Water and on Ice. Journal of Molecular Structure: THEOCHEM 1996; 388(11) 65-77.

[77] Miura, K., Ohgiya, S., Hoshino, T., Nemoto, N., Odaira, M., Nitta, R., and Tsuda, S., Determination of the Solution Structure of the N-Domain Plus Linker of Antarctic Eel Pout Antifreeze Protein RD3. The Journal of Biochemistry 1999; 126(2) 387-394.

[78] Slaughter, D., Fletcher, G.L., Ananthanarayanan, V.S., and Hew, C.L., Antifreeze Proteins from the Sea Raven, *Hemitripterus americanus*. Further Evidence for Diversity among Fish Polypeptide Antifreezes. The Journal of Biological Chemistry 1981; 256(4) 2022-2026.

[79] Sönnichsen, F.D., Sykes, B.D., and Davies, P.L., Comparative Modeling of the Three-Dimensional Structure of Type II Antifreeze Protein. Protein Science 1995; 4(3) 460-471.

[80] Ng, N.F., Trinh, K.Y., and Hew, C.L., Structure of an Antifreeze Polypeptide Precursor from the Sea Raven, *Hemitripterus americanus*. The Journal of Biological Chemistry 1986; 261(33) 15690-15695.

[81] Ng, N.F. and Hew, C.L., Structure of an Antifreeze Polypeptide from the Sea Raven. Disulfide Bonds and Similarity to Lectin-Binding Proteins. The Journal of Biological Chemistry 1992; 267(23) 16069-16075.

[82] DeLuca, C.I., Davies, P.L., Ye, Q., and Jia, Z., The Effects of Steric Mutations on the Structure of Type III Antifreeze Protein and Its Interaction with Ice. Journal of Molecular Biology 1998; 275(3) 515-525.

[83] Jia, Z., Deluca, C.I., and Davies, P.L., Crystallization and Preliminary X-Ray Crystallographic Studies on Type III Antifreeze Protein. Protein Science 1995; 4(6) 1236-1238.

[84] Hew, C.L., Slaughter, D., Joshi, S.B., Fletcher, G.L., and Ananthanarayanan, V.S., Antifreeze Polypeptides from the Newfoundland Ocean Pout, *Macrozoarces americanus*: Presence of Multiple and Compositionally Diverse Components. Journal of Comparative Physiology B: Biochemical, Systemic, and Environmental Physiology 1984; 155(1) 81-88.

[85] Schrag, J.D., Cheng, C.-H.C., Panico, M., Morris, H.R., and DeVries, A.L., Primary and Secondary Structure of Antifreeze Peptides from Arctic and Antartic Zoarcid Fishes. Biochimica et Biophysica Acta (BBA) - Protein Structure and Molecular Enzymology 1987; 915(3) 357-370.

[86] Sönnichsen, F.D., Sykes, B.D., Chao, H., and Davies, P.L., The Nonhelical Structure of Antifreeze Protein Type III. Science 1993; 259(5098) 1154-1157.

[87] Xue, Y.Q., Sicheri, F., Ala, P., Hew, C.L., and Yang, D.S.C., Single Crystals of a Type III Antifreeze Polypeptide from Ocean Pout. Journal of Molecular Biology 1994; 237(3) 351-352.

[88] Deng, G., Andrews, D.W., and Laursen, R.A., Amino Acid Sequence of a New Type of Antifreeze Protein, from the Longhorn Sculpin *Myoxocephalus octodecimspinosis*. FEBS letters 1997; 402(1) 17-20.

[89] Deng, G. and Laursen, R.A., Isolation and Characterization of an Antifreeze Protein from the Longhorn Sculpin, *Myoxocephalus octodecimspinosis*. Biochimica et Biophysica Acta (BBA) - Protein Structure and Molecular Enzymology 1998; 1388(2) 305-314.

[90] Cheng, C.-H.C., Evolution of the Diverse Antifreeze Proteins. Current Opinion in Genetics & Development 1998; 8(6) 715-720.

[91] Komatsu, S.K., DeVries, A.L., and Feeney, R.E., Studies of the Structure of Freezing Point-Depressing Glycoproteins from an Antarctic Fish. The Journal of Biological Chemistry 1970; 245(11) 2909-2913.

[92] DeVries, A.L., Vandenheede, J., and Feeney, R.E., Primary Structure of Freezing Point-Depressing Glycoproteins. The Journal of Biological Chemistry 1971; 246(2) 305-308.

[93] Feeney, R.E., Burcham, T.S., and Yeh, Y., Antifreeze Glycoproteins from Polar Fish Blood. Annual Review of Biophysics and Biophysical Chemistry 1986; 15 59-78.

[94] Bouvet, V. and Ben, R.N., Antifreeze Glycoproteins: Structure, Conformation, and Biological Applications. Cell Biochemistry and Biophysics 2003; 39(2) 133-144.

[95] Gibson, M.I., Slowing the Growth of Ice with Synthetic Macromolecules: Beyond Antifreeze(Glyco) Proteins. Polymer Chemistry 2010; 1(8) 1141-1152.

[96] Lin, Y., Duman, J.G., and DeVries, A.L., Studies on the Structure and Activity of Low Molecular Weight Glycoproteins from an Antarctic Fish. Biochemical and Biophysical Research Communications 1972; 46(1) 87-92.

[97] Hew, C.L., Slaughter, D., Fletcher, G.L., and Joshi, S.B., Antifreeze Glycoproteins in the Plasma of Newfoundland Atlantic Cod (*Gadus morhua*). Canadian Journal of Zoology 1981; 59(11) 2186-2192.

[98] Burcham, T.S., Osuga, D.T., Rao, B.N., Bush, C.A., and Feeney, R.E., Purification and Primary Sequences of the Major Arginine-Containing Antifreeze Glycopeptides from the Fish *Eleginus gracilis*. The Journal of Biological Chemistry 1986; 261(14) 6384-6389.

[99] Geoghegan, K.F., Osuga, D.T., Ahmed, A.I., Yeh, Y., and Feeney, R.E., Antifreeze Glycoproteins from Polar Fish: Structural Requirements for Function of Glycopeptide 8. The Journal of Biological Chemistry 1980; 255(2) 663-667.

[100] O'Grady, S.M., Schrag, J.D., Raymond, J.A., and Devries, A.L., Comparison of Antifreeze Glycopeptides from Arctic and Antarctic Fishes. Journal of Experimental Zoology 1982; 224(2) 177-185.

[101] Fletcher, G.L., Hew, C.L., and Joshi, S.B., Isolation and Characterization of Antifreeze Glycoproteins from the Frostfish, *Microgadus tomcod.* Canadian Journal of Zoology 1982; 60(3) 348-355.

[102] Franks, F. and Morris, E.R., Blood Glycoprotein from Antarctic Fish Possible Conformational Origin of Antifreeze Activity. Biochimica et Biophysica Acta (BBA) - General Subjects 1978; 540(2) 346-356.

[103] Bush, C.A., Feeney, R.E., Osuga, D.T., Ralapati, S., and Yeh, Y.I.N., Antifreeze Glycoprotein. Conformational Model Based on Vacuum Ultraviolet Circular Dichroism Data. International Journal of Peptide and Protein Research 1981; 17(1) 125-129.

[104] Ahmed, A.I., Feeney, R.E., Osuga, D.T., and Yeh, Y., Antifreeze Glycoproteins from an Antarctic Fish. Quasi-Elastic Light Scattering Studies of the Hydrodynamic Conformations of Antifreeze Glycoproteins. The Journal of Biological Chemistry 1975; 250(9) 3344-3347.

[105] Berman, E., Allerhand, A., and DeVries, A.L., Natural Abundance Carbon 13 Nuclear Magnetic Resonance Spectroscopy of Antifreeze Glycoproteins. The Journal of Biological Chemistry 1980; 255(10) 4407-4410.

[106] Lane, A.N., Hays, L.M., Crowe, L.M., Crowe, J.H., and Feeney, R.E., Conformational and Dynamic Properties of a 14 Residue Antifreeze Glycopeptide from Antarctic Cod. Protein Science 1998; 7(7) 1555-1563.

[107] Tsvetkova, N.M., Phillips, B.L., Krishnan, V.V., Feeney, R.E., Fink, W.H., Crowe, J.H., Risbud, S.H., Tablin, F., and Yeh, Y., Dynamics of Antifreeze Glycoproteins in the Presence of Ice. Biophysical Journal 2002; 82(1) 464-473.

[108] Bush, C.A. and Feeney, R.E., Conformation of the Glycotripeptide Repeating Unit of Antifreeze Glycoprotein of Polar Fish as Determined from the Fully Assigned Proton N.M.R. Spectrum. International Journal of Peptide and Protein Research 1986; 28(4) 386-397.

[109] Bush, C.A., Ralapati, S., Matson, G.M., Yamasaki, R.B., Osuga, D.T., Yeh, Y., and Feeney, R.E., Conformation of the antifreeze glycoprotein of polar fish. Archives of Biochemistry and Biophysics 1984; 232(2) 624-631.

[110] Rao, B.N.N. and Bush, C.A., Comparison by ^{1}H-NMR Spectroscopy of the Conformation of the 2600 Dalton Antifreeze Glycopeptide of Polar Cod with That of the High Molecular Weight Antifreeze Glycoprotein. Biopolymers 1987; 26(8) 1227-1244.

[111] Tyshenko, M.G., Doucet, D., Davies, P.L., and Walker, V.K., The Antifreeze Potential of the Spruce Budworm Thermal Hysteresis Protein. Nature Biotechnology 1997; 15(9) 887-890.

[112] Hew, C.L., Kao, M.H., So, Y.-P., and Lim, K.-P., Presence of Cystine-Containing Antifreeze Proteins in the Spruce Bud Worm, *Choristoneura fumiferana.* Canadian Journal of Zoology 1983; 61(10) 2324-2328.

[113] Graham, L.A., Liou, Y.-C., Walker, V.K., and Davies, P.L., Hyperactive Antifreeze Protein from Beetles. Nature 1997; 388(6644) 727-728.

[114] Schneppenheim, R. and Theede, H., Isolation and Characterization of Freezing-Point Depressing Peptides from Larvae of *Tenebrio molitor*. Comparative Biochemistry and Physiology Part B: Comparative Biochemistry 1980; 67(4) 561-568.

[115] Duman, J.G., Li, N., Verleye, D., Goetz, F.W., Wu, D.W., Andorfer, C.A., Benjamin, T., and Parmelee, D.C., Molecular Characterization and Sequencing of Antifreeze Proteins from Larvae of the Beetle *Dendroides canadensis*. Journal of Comparative Physiology B 1998; 168(3) 225-232.

[116] Graham, L.A. and Davies, P.L., Glycine-Rich Antifreeze Proteins from Snow Fleas. Science 2005; 310(5747) 461.

[117] Smallwood, M., Worrall, D., Byass, L., Elias, L., Ashford, D., Doucet, C.J., Holt, C., Telford, J., Lillford, P., and Bowles, D.J., Isolation and Characterization of a Novel Antifreeze Protein from Carrot (*Daucus carota*). Biochemical Journal 1999; 340(2) 385-391.

[118] Huang, T., Nicodemus, J., Zarka, D.G., Thomashow, M.F., Wisniewski, M., and Duman, J.G., Expression of an Insect (*Dendroides canadensis*) Antifreeze Protein in *Arabidopsis thaliana*; Results in a Decrease in Plant Freezing Temperature. Plant Molecular Biology 2002; 50(3) 333-344.

[119] Sidebottom, C., Buckley, S., Pudney, P., Twigg, S., Jarman, C., Holt, C., Telford, J., McArthur, A., Worrall, D., Hubbard, R., and Lillford, P., Heat-Stable Antifreeze Protein from Grass. Nature 2000; 406(6793) 256.

[120] Pudney, P.D.A., Buckley, S.L., Sidebottom, C.M., Twigg, S.N., Sevilla, M.-P., Holt, C.B., Roper, D., Telford, J.H., McArthur, A.J., and Lillford, P.J., The Physico-Chemical Characterization of a Boiling Stable Antifreeze Protein from a Perennial Grass (*Lolium perenne*). Archives of Biochemistry and Biophysics 2003; 410(2) 238-245.

[121] Kumble, K.D., Demmer, J., Fish, S., Hall, C., Corrales, S., DeAth, A., Elton, C., Prestidge, R., Luxmanan, S., Marshall, C.J., and Wharton, D.A., Characterization of a Family of Ice-Active Proteins from the Ryegrass, *Lolium perenne*. Cryobiology 2008; 57(3) 263-268.

[122] John, U.P., Polotnianka, R.M., Sivakumaran, K.A., Chew, O., Mackin, L., Kuiper, M.J., Talbot, J.P., Nugent, G.D., Mautord, J., Schrauf, G.E., and Spangenberg, G.C., Ice Recrystallization Inhibition Proteins (IRIPs) and Freeze Tolerance in the Cryophilic Antarctic Hair Grass *Deschampsia antarctica* E. Desv. Plant, Cell & Environment 2009; 32(4) 336-348.

[123] Hon, W.C., Griffith, M., Mlynarz, A., Kwok, Y.C., and Yang, D., Antifreeze Proteins in Winter Rye are Similar to Pathogenesis-Related Proteins. Plant Physiology 1995; 109(3) 879-889.

[124] Tremblay, K., Ouellet, F., Fournier, J., Danyluk, J., and Sarhan, F., Molecular Characterization and Origin of Novel Bipartite Cold-Regulated Ice Recrystallization Inhibition Proteins from Cereals. Plant and Cell Physiology 2005; 46(6) 884-891.

[125] Duman, J.G. and Olsen, T.M., Thermal Hysteresis Protein Activity in Bacteria, Fungi, and Phylogenetically Diverse Plants. Cryobiology 1993; 30(3) 322-328.

[126] Hoshino, T., Kiriaki, M., and Nakajima, T., Novel Thermal Hysteresis Proteins from Low Temperature Basidiomycete, *Coprinus psychromorbidus*. Cryoletters 2003; 24(3) 135-142.

[127] Gilbert, J.A., Davies, P.L., and Laybourn-Parry, J., A Hyperactive, Ca^{2+}-Dependent Antifreeze Protein in an Antarctic Bacterium. FEMS Microbiology Letters 2005; 245(1) 67-72.

[128] Sun, X., Griffith, M., Pasternak, J.J., and Glick, B.R., Low Temperature Growth, Freezing Survival, and Production of Antifreeze Protein by the Plant Growth Promoting Rhizobacterium *Pseudomonas putida* GR12-2. Canadian Journal of Microbiology 1995; 41(9) 776-784.

[129] Yamashita, Y., Nakamura, N., Omiya, K., Nishikawa, J., Kawahara, H., and Obata, H., Identification of an Antifreeze Lipoprotein from *Moraxella* sp. of Antarctic Origin. Bioscience, Biotechnology, and Biochemistry 2002; 66(2) 239-247.

[130] Kawahara, H., Nakano, Y., Omiya, K., Muryoi, N., Nishikawa, J., and Obata, H., Production of Two Types of Ice Crystal-Controlling Proteins in Antarctic Bacterium. Journal of Bioscience and Bioengineering 2004; 98(3) 220-223.

[131] Graether, S.P., Kuiper, M.J., Gagné, S.M., Walker, V.K., Jia, Z., Sykes, B.D., and Davies, P.L., β-Helix Structure and Ice-Binding Properties of a Hyperactive Antifreeze Protein from an Insect. Nature 2000; 406(6793) 325-328.

[132] Liou, Y.-C., Tocilj, A., Davies, P.L., and Jia, Z., Mimicry of Ice Structure by Surface Hydroxyls and Water of a β-Helix Antifreeze Protein. Nature 2000; 406(6793) 322-324.

[133] Raymond, J.A. and DeVries, A.L., Adsorption Inhibition as a Mechanism of Freezing Resistance in Polar Fishes. Proceedings of the National Academy of Sciences of the United State of America 1977; 74(6) 2589-2593.

[134] Knight, C.A. and DeVries, A.L., Effects of a Polymeric, Nonequilibrium "Antifreeze" Upon Ice Growth from Water. Journal of Crystal Growth 1994; 143(3-4) 301-310.

[135] Wilson, P.W., Explaining Thermal Hysteresis by the Kelvin Effect. Cryoletters 1993; 14 31-36.

[136] Knight, C.A., Cheng, C.C., and DeVries, A.L., Adsorption of α-Helical Antifreeze Peptides on Specific Ice Crystal Surface Planes. Biophysical Journal 1991; 59(2) 409-418.

[137] Knight, C.A., Driggers, E., and DeVries, A.L., Adsorption to Ice of Fish Antifreeze Glycopeptides 7 and 8. Biophysical Journal 1993; 64(1) 252-259.

[138] Chakrabartty, A. and Hew, C.L., The Effect of Enhanced α-Helicity on the Activity of a Winter Flounder Antifreeze Polypeptide. European Journal of Biochemistry 1991; 202(3) 1057-1063.

[139] Knight, C.A., Hallett, J., and DeVries, A.L., Solute Effects on Ice Recrystallization: An Assessment Technique. Cryobiology 1988; 25(1) 55-60.

[140] Tomczak, M.M., Marshall, C.B., Gilbert, J.A., and Davies, P.L., A Facile Method for Determining Ice Recrystallization Inhibition by Antifreeze Proteins. Biochemical and Biophysical Research Communications 2003; 311(4) 1041-1046.

[141] Yu, S.O., Brown, A., Middleton, A.J., Tomczak, M.M., Walker, V.K., and Davies, P.L., Ice Restructuring Inhibition Activities in Antifreeze Proteins with Distinct Differences in Thermal Hysteresis. Cryobiology 2010; 61(3) 327-334.

[142] Yagci, Y.E., Antonietti, M., and Börner, H.G., Synthesis of Poly(Tartar Amides) as Bio-Inspired Antifreeze Additives. Macromolecular Rapid Communications 2006; 27(19) 1660-1664.

[143] Baruch, E. and Mastai, Y., Antifreeze Properties of Polyglycidol Block Copolymers. Macromolecular Rapid Communications 2007; 28(23) 2256-2261.

[144] Mastai, Y., Rudloff, J., Cölfen, H., and Antonietti, M., Control Over the Structure of Ice and Water by Block Copolymer Additives. ChemPhysChem 2002; 3(1) 119-123.

[145] Inada, T. and Lu, S.-S., Inhibition of Recrystallization of Ice Grains by Adsorption of Poly(Vinyl Alcohol) onto Ice Surfaces. Crystal Growth and Design 2003; 3(5) 747-752.

[146] Jackman, J., Noestheden, M., Moffat, D., Pezacki, J.P., Findlay, S., and Ben, R.N., Assessing Antifreeze Activity of AFGP 8 Using Domain Recognition Software. Biochemical and Biophysical Research Communications 2007; 354(2) 340-344.

[147] Wilson, P.W., Beaglehole, D., and DeVries, A.L., Antifreeze Glycopeptide Adsorption on Single Crystal Ice Surfaces Using Ellipsometry. Biophysical Journal 1993; 64(6) 1878-1884.

[148] Raymond, J.A., Wilson, P., and DeVries, A.L., Inhibition of Growth of Nonbasal Planes in Ice by Fish Antifreezes. Proceedings of the National Academy of Sciences of the United States of America 1989; 86(3) 881-885.

[149] Knight, C.A., DeVries, A.L., and Oolman, L.D., Fish Antifreeze Protein and the Freezing and Recrystallization of Ice. Nature 1984; 308(5956) 295-296.

[150] DeVries, A.L., The Role of Antifreeze Glycopeptides and Peptides in the Freezing Avoidance of Antarctic Fishes. Comparative Biochemistry and Physiology Part B: Comparative Biochemistry 1988; 90(3) 611-621.

[151] Scotter, A.J., Marshall, C.B., Graham, L.A., Gilbert, J.A., Garnham, C.P., and Davies, P.L., The Basis for Hyperactivity of Antifreeze Proteins. Cryobiology 2006; 53(2) 229-239.

[152] Yeh, Y. and Feeney, R.E., Antifreeze Proteins: Structures and Mechanisms of Function. Chemical Reviews 1996; 96(2) 601-618.

[153] Hall, D.G. and Lips, A., Phenomenology and Mechanism of Antifreeze Peptide Activity. Langmuir 1999; 15(6) 1905-1912.

[154] Wen, D. and Laursen, R.A., A Model for Binding of an Antifreeze Polypeptide to Ice. Biophysical Journal 1992; 63(6) 1659-1662.

[155] Osuga, D.T., Feeney, R.E., Yeh, Y., and Hew, C.-L., Co-Functional Activities of Two Different Antifreeze Proteins: The Antifreeze Glycoprotein from Polar Fish and the Nonglycoprotein from a Newfoundland Fish. Comparative Biochemistry and Physiology Part B: Comparative Biochemistry 1980; 65(2) 403-406.

[156] Osuga, D.T., Ward, F.C., Yeh, Y., and Feeney, R.E., Cooperative Functioning between Antifreeze Glycoproteins. The Journal of Biological Chemistry 1978; 253(19) 6669-72.

[157] Inada, T. and Lu, S.-S., Thermal Hysteresis Caused by Non-Equilibrium Antifreeze Activity of Poly(Vinyl Alcohol). Chemical Physics Letters 2004; 394(4-6) 361-365.

[158] Wierzbicki, A., Taylor, M.S., Knight, C.A., Madura, J.D., Harrington, J.P., and Sikes, C.S., Analysis of Shorthorn Sculpin Antifreeze Protein Stereospecific Binding to (2 - 1 0) Faces of Ice. Biophysical Journal 1996; 71(1) 8-18.

[159] Chao, H., Houston, M.E., Hodges, R.S., Kay, C.M., Sykes, B.D., Loewen, M.C., Davies, P.L., and Sönnichsen, F.D., A Diminished Role for Hydrogen Bonds in Antifreeze Protein Binding to Ice. Biochemistry 1997; 36(48) 14652-14660.

[160] Haymet, A.D.J., Ward, L.G., Harding, M.M., and Knight, C.A., Valine Substituted Winter Flounder `Antifreeze': Preservation of Ice Growth Hysteresis. FEBS Letters 1998; 430(3) 301-306.

[161] Zhang, W. and Laursen, R.A., Structure-Function Relationships in a Type I Antifreeze Polypeptide. The Journal of Biological Chemistry 1998; 273(52) 34806-34812.

[162] Baardsnes, J., Kondejewski, L.H., Hodges, R.S., Chao, H., Kay, C., and Davies, P.L., New Ice-Binding Face for Type I Antifreeze Protein. FEBS Letters 1999; 463(1) 87-91.

[163] Marshall, C.B., Daley, M.E., Graham, L.A., Sykes, B.D., and Davies, P.L., Identification of the Ice-Binding Face of Antifreeze Protein from *Tenebrio molitor*. FEBS Letters 2002; 529(2) 261-267.

[164] Davies, P.L., Baardsnes, J., Kuiper, M.J., and Walker, V.K., Structure and Function of Antifreeze Proteins. Philosophical Transactions of The Royal Society B 2002; 357(1423) 927-935.

[165] Doxey, A.C., Yaish, M.W., Griffith, M., and McConkey, B.J., Ordered Surface Carbons Distinguish Antifreeze Proteins and Their Ice-Binding Regions. Nature Biotechnology 2006; 24(7) 852-855.

[166] Tachibana, Y., Fletcher, G.L., Fujitani, N., Tsuda, S., Monde, K., and Nishimura, S.-I., Antifreeze Glycoproteins: Elucidation of the Structural Motifs That are Essential for Antifreeze Activity. Angewandte Chemie International Edition 2004; 43(7) 856-862.

[167] Wilkinson, B.L., Stone, R.S., Capicciotti, C.J., Thaysen-Andersen, M., Matthews, J.M., Packer, N.H., Ben, R.N., and Payne, R.J., Total Synthesis of Homogeneous Antifreeze Glycopeptides and Glycoproteins. Angewandte Chemie International Edition 2012; 51(15) 3606-3610.

[168] Carpenter, J.F. and Hansen, T.N., Antifreeze Protein Modulates Cell Survival During Cryopreservation: Mediation Through Influence on Ice Crystal Growth. Proceedings of the National Academy of Sciences of the United States of America 1992; 89(19) 8953-8957.

[169] Chao, H., Davies, P.L., and Carpenter, J.F., Effects of Antifreeze Proteins on Red Blood Cell Survival During Cryopreservation. The Journal of Experimental Biology 1996; 199(9) 2071-2076.

[170] Rubinsky, B., Arav, A., and DeVries, A.L., The Cryoprotective Effect of Antifreeze Glycopeptides from Antarctic Fishes. Cryobiology 1992; 29(1) 69-79.

[171] Wen, D. and Laursen, R.A., Structure-Function Relationships in an Antifreeze Polypeptide. The Role of Neutral, Polar Amino Acids. The Journal of Biological Chemistry 1992; 267(20) 14102-14108.

[172] Eniade, A., Purushotham, M., Ben, R.N., Wang, J.B., and Horwath, K., A Serendipitous Discovery of Antifreeze Protein-Specific Activity in *C*-Linked Antifreeze Glycoprotein Analogs. Cell Biochemistry and Biophysics 2003; 38(2) 115-124.

[173] Ben, R.N., Eniade, A.A., and Hauer, L., Synthesis of a *C*-linked Antifreeze Glycoprotein (AFGP) Mimic: Probes for Investigating the Mechanism of Action. Organic Letters 1999; 1(11) 1759-1762.

[174] Galema, S.A., Engberts, J.B.F.N., Hoeiland, H., and Foerland, G.M., Informative Thermodynamic Properties of the Effect of Stereochemistry on Carbohydrate Hydration. Journal of Physical Chemistry 1993; 97(26) 6885-6889.

[175] Galema, S.A. and Hoeiland, H., Stereochemical Aspects of Hydration of Carbohydrates in Aqueous Solutions. 3. Density and Ultrasound Measurements. Journal of Physical Chemistry 1991; 95(13) 5321-5326.

[176] Galema, S.A., Howard, E., Engberts, J.B.F.N., and Grigera, J.R., The Effect of Stereochemistry Upon Carbohydrate Hydration. A Molecular Dynamics Simulation of β-D-Galactopyranose and (α,β)-D- Talopyranose. Carbohydrate Research 1994; 265(2) 215-225.

[177] Tam, R.Y., Ferreira, S.S., Czechura, P., Chaytor, J.L., and Ben, R.N., Hydration Index - A Better Parameter for Explaining Small Molecule Hydration in Inhibition of Ice Recrystallization. Journal of the American Chemical Society 2008; 130(51) 17494-17501.

[178] Tam, R.Y., Rowley, C.N., Petrov, I., Zhang, T., Afagh, N.A., Woo, T.K., and Ben, R.N., Solution Conformation of *C*-Linked Antifreeze Glycoprotein Analogues and Modulation of Ice Recrystallization. Journal of the American Chemical Society 2009; 131(43) 15745-15753.

[179] Heggemann, C., Budke, C., Schomburg, B., Majer, Z., Wißbrock, M., Koop, T., and Sewald, N., Antifreeze Glycopeptide Analogues: Microwave-Enhanced Synthesis and Functional Studies. Amino Acids 2010; 38(1) 213-222.

[180] Miller, N., Williams, G.M., and Brimble, M.A., Synthesis of Fish Antifreeze Neoglycopeptides Using Microwave-Assisted "Click Chemistry". Organic Letters 2009; 11(11) 2409-2412.

[181] Norgren, A.S., Budke, C., Majer, Z., Heggemann, C., Koop, T., and Sewald, N., On-Resin Click-Glycoconjugation of Peptoids. Synthesis 2009; 2009(3) 488-494.

[182] Capicciotti, C.J., Trant, J.F., Leclère, M., and Ben, R.N., Synthesis of *C*-Linked Triazole-Containing AFGP Analogues and Their Ability to Inhibit Ice Recrystallization. Bioconjugate Chemistry 2011; 22(4) 605-616.

[183] Budke, C. and Koop, T., Ice Recrystallization Inhibition and Molecular Recognition of Ice Faces by Poly(Vinyl Alcohol). ChemPhysChem 2006; 7(12) 2601-2606.

[184] Horng, J.-C. and Raines, R.T., Stereoelectronic Effects on Polyproline Conformation. Protein Science 2006; 15(1) 74-83.

[185] Stokes, R.H. and Robinson, R.A., Interactions in Aqueous Nonelectrolyte Solutions. I. Solute-Solvent Equilibria. The Journal of Physical Chemistry 1966; 70(7) 2126-2131.

[186] Suggett, A., Ablett, S., and Lillford, P.J., Molecular Motion and Interactions in Aqueous Carbohydrate Solutions. II. Nuclear-Magnetic-Relaxation Studies. Journal of Solution Chemistry 1976; 5(1) 17-31.

[187] Tait, M.J., Suggett, A., Franks, F., Ablett, S., and Quickenden, P.A., Hydration of Monosaccharides: A Study by Dielectric and Nuclear Magnetic Relaxation. Journal of Solution Chemistry 1972; 1(2) 131-151.

[188] Uedaira, H. and Uedaira, H., Sugar-Water Interaction from Diffusion Measurements. Journal of Solution Chemistry 1985; 14(1) 27-34.

[189] Kabayama, M.A., Patterson, D., and Piche, L., The Thermodynamics of Mutarotation of Some Sugars: I. Measurement of the Heat of Mutarotation by Microcalorimetry. Canadian Journal of Chemistry 1958; 36(3) 557-562.

[190] Walkinshaw, M.D., Variation in the Hydrophilicity of Hexapyranose Sugars Explains Features of the Anomeric Effect. Journal of the Chemical Society, Perkin Transactions 2 1987(12) 1903-1906.

[191] Miyajima, K., Machida, K., and Nakagaki, M., Hydrophobic Indexes for Various Monosaccharides. Bulletin of the Chemical Society of Japan 1985; 58(9) 2595-2599.

[192] Franks, F., Solute-Water Interactions: Do Polyhydroxy Compounds Alter the Properties of Water? Cryobiology 1983; 20(3) 335-345.

[193] Suggett, A., Molecular Motion and Interactions in Aqueous Carbohydrate Solutions. III. A Combined Nuclear Magnetic and Dielectric-Relaxation Strategy. Journal of Solution Chemistry 1976; 5(1) 33-46.

[194] Danford, M.D. and Levy, H.A., The Structure of Water at Room Temperature. Journal of the American Chemical Society 1962; 84(20) 3965-3966.

[195] Warner, D.T., Some Possible Relationships of Carbohydrates and Other Biological Components with the Water Structure at 37°. Nature 1962; 196(4859) 1055-1058.

[196] Chaytor, J.L., Tokarew, J.M., Wu, L.K., Leclère, M., Tam, R.Y., Capicciotti, C.J., Guolla, L., von Moos, E., Findlay, C.S., Allan, D.S., and Ben, R.N., Inhibiting Ice Recrystallization and Optimization of Cell Viability after Cryopreservation. Glycobiology 2012; 22(1) 123-133.

[197] Balcerzak, A.K., Ferreira, S.S., Trant, J.F., and Ben, R.N., Structurally Diverse Disaccharide Analogs of Antifreeze Glycoproteins and Their Ability to Inhibit Ice Recrystallization. Bioorganic & Medicinal Chemistry Letters 2012; 22(4) 1719-1721.

[198] Chaytor, J.L. and Ben, R.N., Assessing the Ability of a Short Fluorinated Antifreeze Glycopeptide and a Fluorinated Carbohydrate Derivative to Inhibit Ice Recrystallization. Bioorganic & Medicinal Chemistry Letters 2010; 20(17) 5251-5254.

[199] Jurisicova, A., Varmuza, S., and Casper, R.F., Involvement of Programmed Cell Death in Preimplantation Embryo Demise. Human Reproduction Update 1995; 1(6) 558-566.

[200] Fowler, A. and Toner, M., Cryo-Injury and Biopreservation. Annals of the New York Academy of Sciences 2005; 1066(1) 119-135.

[201] Hobbs, P.V. Ice Physics (1st ed.). Oxford: Oxford University Press, USA; 1975

[202] Mazur, P. Life in the Frozen State - Principles of Cryobiology (Boca Baton, FL: CRC Press; 2004.

[203] Song, Y.C., Khirabadi, B.S., Lightfoot, F., Brockbank, K.G.M., and Taylor, M.J., Vitreous Cryopreservation Maintains the Function of Vascular Grafts. Nature Biotechnology 2000; 18(3) 296-299.

[204] Karlsson, J.O., Cravalho, E.G., Borel Rinkes, I.H., Tompkins, R.G.Y., M. L., and Toner, M., Nucleation and Growth of Ice Crystals Inside Cultured Hepatocytes During

Freezing in the Presence of Dimethyl Sulfoxide. Biophysical Journal 1993; 65(6) 2524-2536.

[205] Toner, M., Caravalho, E.G., and Karel, M., Thermodynamcis and Kinetics of Intracellular Ice Formation During Freezing of Biological Cells. Journal of Applied Physics 1990; 67(3) 1582-1593.

[206] Mazur, P., Equilibrium, Quasi-Equilibrium and Non-Equilibrium Freezing of Mammalian Embryos. Cell Biophysics 1990; 17(1) 53-92.

[207] Fowler, A. and Toner, M., Prevention of Hemolysis in Rapidly Frozen Erythrocytes by Using a Laser Pulse. Annals of the New York Academy of Sciences 1998; 858 245-252.

[208] Farrant, J., Molyneux, P., Hasted, J.B., Peares, P., and Echlin, P., Water Transport and Cell Survival in Cryobiological Procedures (and Discssion). Philosophical Transactions of The Royal Society B 1977; 278(959) 191-205.

[209] Acker, J.P. and McGann, L.E., Innocuous Intracellular Ice Improves Survival of Frozen Cells. Cell Transplantation 2002; 11(6) 563-571.

[210] Ramløv, H.E.A., Wharton, D.A., and Wilson, P.W., Recrystallization in a Freezing Tolerant Antarctic Nematode, *Pnagrolaimus davidi* and an Alpine Weta, *Hemideina maori* (Orthopetra; Stenopelmatidae). Cryobiology 1996; 33(6) 607-613.

[211] Rubinsky, B., Arav, A., and Fletcher, G.L., Hypothermic Protection — A Fundamental Property of "Antifreeze" Proteins. Biochemical and Biophysical Research Communications 1991; 180(2) 566-571.

[212] Rubinsky, B., Arav, A., Mattioli, M., and DeVries, A.L., The Effect of Antifreeze Glycopeptides on Membrane Potential Changes at Hypothermic Temperatures. Biochemical and Biophysical Research Communications 1990; 173(3) 1369-1374.

[213] Hays, L.M., Feeney, R.E., Crowe, L.M., Crowe, J.H., and Oliver, A.E., Antifreeze Glycoproteins Inhibit Leakage from Liposomes During Thermotropic Phase Transitions. Proceedings of the National Academy of Sciences of the United States of America 1996; 93(13) 6835-6840.

[214] Oliver, A.E., Tablin, F., Crowe, J.H., Tsvetkova, N.M., Fisk, E.L., Walker, N.J., Hays, L.M., and Crowe, L.M., Antifreeze Glycoproteins can Protect Human Platelets from Cold-Induced Activation and Prevent Lateral-Phase Separation in Membranes Below the Phase-Transition Temperature. Cryobiology 1997; 35(4) 333-334.

[215] Tablin, F., Oliver, A.E., Walker, N.J., Crowe, L.M., and Crowe, J.H., Membrane Phase Transition of Intact Human Platelets: Correlation with Cold-Induced Activation. Journal of Cellular Physiology 1996; 168(2) 305-313.

[216] Hincha, D.K., DeVries, A.L., and Schmitt, J.M., Cryotoxicity of Antifreeze Proteins and Glycoproteins to Spinach Thylakoid Membranes — Comparison with Cryotoxic

Sugar Acids. Biochimica et Biophysica Acta (BBA) - Biomembranes 1993; 1146(2) 258-264.

[217] Payne, S.R., Oliver, J.E., and Upreti, G.C., Effect of Antifreeze Proteins on the Motility of Ram Spermatozoa. Cryobiology 1994; 31(2) 180-184.

[218] Younis, A.I., Rooks, B., Khan, S., and Gould, K.G., The Effects of Antifreeze Peptide III (AFP) and Insulin Transferrin Selenium (ITS) on Cryopreservation of Chimpanzee (*Pan troglodytes*) Spermatozoa. Journal of Andrology 1998; 19(2) 207-214.

[219] Rubinsky, B., Arav, A., and DeVries, A.L., Cryopreservation of Oocytes Using Directional Cooling and Antifreeze Glycoproteins. Cryoletters 1991; 12 93-106.

[220] Rubinsky, B., Arav, A., and DeVries, A.L., The Cryoprotective Effect of Antifreeze Glycopeptides from Antarctic Fishes. Cryobiology. 1992; 29(1) 69-79.

[221] O'Neil, L., Paynter, S.J., Fuller, B.J., Shaw, R.W., and DeVries, A.L., Vitrification of Mature Mouse Oocytes in a 6 M Me_2SO Solution Supplemented with Antifreeze Glycoproteins: The Effect of Temperature. Cryobiology 1998; 37(1) 59-66.

[222] Arav, A., Ramsbottom, G., Baguisi, A., Rubinsky, B., Roche, J.F., and Boland, M.P., Vitrification of Bovine and Ovine Embryos with the MDS Technique and Antifreeze Proteins. Cryobiology 1993; 30(6) 621-622.

[223] Chen, L.R., Huang, W.Y., Luoh, Y.S., and Wu, M.C., Cryopreservation of Porcine Oocytes Before and After Polar Body Formation by Antifreeze Protein Type III. Journal of Taiwan Liverstock Research 1995; 28 169-179.

[224] Palasz, A.T., Gustafasson, H., Rodriguez-Martinez, H., Gusta, L., Larsson, B., and Mapletoft, R.J., Successful Vitrification of IVF Bovine Blastocytes with Antifreeze Plant Proteins. Cryobiology 1995; 32(6) 572-572.

[225] Naidenko, T., Cryopreservation of *Crassostrea gigas* Oocytes, Embryos and Larvae Using Antioxidant Echinochrome A and Antifreeze Protein AFP-I. Cryoletters 1997; 18 375-382.

[226] Lagneaux, D., Huhtinen, M., Koskinen, E., and Palmer, E., Effect of Anti-Freeze Protein (AFP) on the Cooling and Freezing of Equine Embryos as Measured by DAPI-Staining. Equine Veterinary Journal 1997; 29(S25) 85-87.

[227] Pegg, D.E., Antifreeze Proteins. Cryobiology 1992; 29(6) 774.

[228] Rubinsky, B. and DeVries, A.L., Effect of Ice Crystal Habit on the Viability of Glycerol-Protected Red Blood Cells. Cryobiology 1989; 26(6) 580.

[229] Hansen, T.N., Smith, K.M., and Brockbank, K.G.M., Type I Antifreeze Protein Attenuates Cell Recoveries Following Cryopreservation. Transplantation Proceedings 1993; 25(6) 3182-4.

[230] Wang, T., Zhu, Q., Yang, X., Layne, J.R., and DeVries, A.L., Antifreeze Glycoproteins from Antarctic Notothenioid Fishes Fail to Protect the Rat Cardiac Explant during Hypothermic and Freezing Preservation. Cryobiology 1994; 31(2) 185-192.

[231] Mugano, J.A., Wang, T., Layne, J.R., DeVries, A.L., and Lee, R.E., Antifreeze Glycoproteins Promote Lethal Intracellular Freezing of Rat Cardiomyocytes at High Subzero Temperatures. Cryobiology 1995(6) 556-557.

[232] Larese, A., Acker, J., Muldrew, K., Yang, H.Y., and McGann, L., Antifreeze Proteins Induce Intracellular Nucleation. Cryoletters 1996; 17 172-182.

[233] Wang, J.H., Bian, H.W., Huang, C.N., and Ge, J.G., Studies on the Application of Antifreeze Proteins in Cryopreservation of Rice Embryogenic Suspension Cells. Acta Biologiae Experimentalis Sinica 1999; 32(3) 271-276.

[234] Carpenter, J.F. and Hansen, T.N., Antifreeze Protein Modulates Cell Survival During Cryopreservation: Mediation Through Influence on Ice Crystal Growth. Proceedings of the National Academy of Sciences of the United States of America 1992; 89(19) 8953-7.

[235] Chao, H., Davies, P.L., and Carpenter, J.F., Effects of Antifreeze Proteins on Red Blood Cell Survival During Cryopreservation. Journal of Experimental Biology 1996; 199(9) 2071-6.

[236] Leclère, M., Kwok, B.K., Wu, L.K., Allan, D.S., and Ben, R.N., *C*-Linked Antifreeze Glycoprotein (C-AFGP) Analogues as Novel Cryoprotectants. Bioconjugate Chemistry 2011; 22(9) 1804-1810.